Communications and Control Engineering

Springer
London
Berlin
Heidelberg
New York
Barcelona
Budapest
Hong Kong
Milan
Paris
Santa Clara
Singapore
Tokyo

Published titles include:

Sampled-Data Control Systems
J. Ackermann

Interactive System Identification
T. Bohlin

The Riccatti Equation
S. Bittanti, A.J. Laub and J.C. Willems (Eds)

Analysis and Design of Stream Ciphers
R.A. Rueppel

Sliding Modes in Control Optimization
V.I. Utkin

Fundamentals of Robotics
M. Vukobratović

Parametrizations in Control, Estimation and Filtering Problems: Accuracy Aspects
M. Gevers and G. Li

Parallel Algorithms for Optimal Control of Large Scale Linear Systems
Zoran Gajić and Xuemin Shen

Loop Transfer Recovery: Analysis and Design
A. Saberi, B.M. Chen and P. Sannuti

Markov Chains and Stochastic Stability
S.P. Meyn and R.L. Tweedie

Robust Control: Systems with Uncertain Physical Parameters
J. Ackermann in co-operation with A. Bartlett, D. Kaesbauer, W. Sienel and R. Steinhauser

Optimization and Dynamical Systems
U. Helmke and J.B. Moore

Optimal Sampled-Data Control Systems
Tongwen Chen and Bruce Francis

Nonlinear Control Systems (3rd edition)
Alberto Isidori

Theory of Robot Control
C. Canudas de Wit, B. Siciliano and G. Bastin (Eds)

Fundamental Limitations in Filtering and Control
María M. Seron, Julio Braslavsky and Graham C. Goodwin

R. Sepulchre, M. Janković and P.V. Kokotović

Constructive Nonlinear Control

With 41 Figures

Springer

Dr R. Sepulchre
CESAME, Université Catholique de Louvain,
4-6 Avenue Georges Lemaitre, B-1348 Louvain-la-Neuve, Belgium

Dr M. Janković
Ford Motor Company, Scientific Research Laboratories,
P.O. Box 2053, MD 2036 SRL, Dearborn, MI 48121-2053, USA

Professor P.V. Kokotović
Department of Electrical and Computer Engineering, University of California,
Santa Barbara, CA 93106-9560, USA

ISBN-13:978-1-4471-1245-7 Springer-Verlag Berlin Heidelberg New York

British Library Cataloguing in Publication Data
Sepulchre, Rodolphe
Constructive nonlinear control. - (Communications and control engineering series)
1.Nonlinear control theory 2.Adaptive control systems
I.Title II.Janković, Mrdjan III.Kokotović, Petar V.
ISBN-13:978-1-4471-1245-7 e-ISBN-13:978-1-4471-0967-9
DOI: 10.1007/978-1-4471-0967-9

Library of Congress Cataloging-in-Publication Data
A catalog record for this book is available from the Library of Congress

Softcover reprint of the hardcover 1st edition 1997

Typesetting: Camera ready by authors

69/3830-543210 Printed on acid-free paper

Preface

In this book several streams of nonlinear control theory are merged and directed towards a constructive solution of the feedback stabilization problem. Analytic, geometric and asymptotic concepts are assembled as design tools for a wide variety of nonlinear phenomena and structures. Differential-geometric concepts reveal important structural properties of nonlinear systems, but allow no margin for modeling errors. To overcome this deficiency, we combine them with analytic concepts of passivity, optimality and Lyapunov stability. In this way geometry serves as a guide for construction of design procedures, while analysis provides robustness tools which geometry lacks.

Our main tool is passivity. As a common thread, it connects all the chapters of the book. Passivity properties are induced by *feedback passivation* designs. Until recently, these designs were restricted to weakly minimum phase systems with relative degree one. Our recursive designs remove these restrictions. They are applicable to wider classes of nonlinear systems characterized by feedback, feedforward, and interlaced structures.

After the introductory chapter, the presentation is organized in two major parts. The basic nonlinear system concepts - passivity, optimality, and stability margins - are presented in Chapters 2 and 3 in a novel way as design tools. Most of the new results appear in Chapters 4, 5, and 6. For cascade systems, and then, recursively, for larger classes of nonlinear systems, we construct design procedures which result in feedback systems with optimality properties and stability margins.

The book differs from other books on nonlinear control. It is more design-oriented than the differential-geometric texts by Isidori [43] and Nijmeijer and Van der Schaft [84]. It complements the books by Krstić, Kanellakopoulos and Kokotović [61] and Freeman and Kokotović [26], by broadening the class of systems and design tools. The book is written for an audience of graduate students, control engineers, and applied mathematicians interested in control theory. It is self-contained and accessible with a basic knowledge of control theory as in Anderson and Moore [1], and nonlinear systems as in Khalil [56].

For clarity, most of the concepts are introduced through and explained by examples. Design applications are illustrated on several physical models of practical interest.

The book can be used for a first level graduate course on nonlinear control, or as a collateral reading for a broader control theory course. Chapters 2, 3, and 4 are suitable for a first course on nonlinear control, while Chapters 5 and 6 can be incorporated in a more advanced course on nonlinear feedback design.

* * *

The book is a result of the postdoctoral research by the first two authors with the third author at the Center for Control Engineering and Computation, University of California, Santa Barbara. In the cooperative atmosphere of the Center, we have been inspired by, and received help from, many of our colleagues. The strongest influence on the content of the book came from Randy Freeman and his ideas on inverse optimality. We are also thankful to Dirk Aeyels, Mohammed Dahleh, Miroslav Krstić, Zigang Pan, Laurent Praly and Andrew Teel who helped us with criticism and advice on specific sections of the book. Gang Tao generously helped us with the final preparation of the manuscript. Equally generous were our graduate students Dan Fontaine with expert execution of figures, Srinivasa Salapaka and Michael Larsen with simulations, and Kenan Ezal with proofreading.

Our families contributed to this project by their support and endurance. Ivana, Edith, Simon and Filip often saw their fathers absent or absent-minded. Our wives, Natalie, Seka, and Anna unwaveringly carried the heaviest burden. We thank them for their infinite stability margins.

* * *

The support for research that led to this book came from several sources. Ford Motor Company supported us financially and encouraged one of its researchers (MJ) to continue this project. Support was also received form BAEF and FNRS, Belgium (RS). The main support for this research program (PK) are the grants NSF ECS-9203491 and AFOSR F49620-95-1-0409.

Rodolphe Sepulchre
Mrdjan Janković
Petar Kokotović

Santa Barbara, California, August 1996

Contents

Chapter 1

Introduction

Control theory has been extremely successful in dealing with linear time-invariant models of dynamic systems. A blend of state space and frequency domain methods has reached a level at which feedback control design is systematic, not only with disturbance-free models, but also in the presence of disturbances and modeling errors. There is an abundance of design methodologies for linear models: root locus, Bode plots, LQR-optimal control, eigenstructure assignment, H-infinity, μ-synthesis, linear matrix inequalities, etc. Each of these methods can be used to achieve stabilization, tracking, disturbance attenuation and similar design objectives.

The situation is radically different for nonlinear models. Although several nonlinear methodologies are beginning to emerge, none of them taken alone is sufficient for a satisfactory feedback design. A question can be raised whether a single design methodology can encompass all nonlinear models of practical interest, and whether the goal of developing such a methodology should even be pursued. The large diversity of nonlinear phenomena suggests that, with a single design approach most of the results would end up being unnecessarily conservative. To deal with diverse nonlinear phenomena we need a comparable diversity of design *tools and procedures*. Their construction is the main topic of this book.

Once the "tools and procedures" attitude is adopted, an immediate task is to determine the areas of applicability of the available tools, and critically evaluate their advantages and limitations. With an arsenal of tools one is encouraged to construct design procedures which exploit structural properties to avoid conservativeness. Geometric and analytic concepts reveal these properties and are the key ingredients of every design procedure in this book.

Analysis is suitable for the study of stability and robustness, but it often disregards structure. On the other hand, geometric methods are helpful in determining structural properties, such as relative degree and zero dynamics, but, taken alone, do not guarantee *stability margins*, which are among the prerequisites for *robustness*. In the procedures developed in this book, the geometry makes the analysis constructive, while the analysis makes the geometry more robust.

Chapters 2 and 3 present the main geometric and analytic tools needed for the design procedures in Chapters 4, 5, and 6. Design procedures in Chapter 4 are constructed for several types of cascades, and also serve as building blocks in the construction of recursive procedures in Chapters 5 and 6.

The main recursive procedures are *backstepping* and *forwarding*. While backstepping is known from [61], forwarding is a procedure recently developed by the authors [46, 95]. This is its first appearance in a book. An important feature of this procedure is that it endows the systems with certain optimality properties and desirable stability margins.

In this chapter we give a brief preview of the main topics discussed in this book.

1.1 Passivity, Optimality, and Stability

1.1.1 From absolute stability to passivity

Modern theory of feedback systems was formed some 50-60 years ago from two separate traditions. The Nyquist-Bode *frequency domain methods*, developed for the needs of feedback amplifiers, became a tool for servomechanism design during the Second World War. In this tradition, feedback control was an outgrowth of linear network theory and was readily applicable only to linear time-invariant models.

The second tradition is more classical and goes back to Poincaré and Lyapunov. This tradition, subsequently named the *state-space approach*, employs the tools of nonlinear mechanics, and addresses both linear and nonlinear models. The main design task is to achieve stability in the sense of Lyapunov of feedback loops which contain significant nonlinearities, especially in the actuators. A seminal development in this direction was the *absolute stability* problem of Lurie [70].

In its simplest form, the absolute stability problem deals with a feedback loop consisting of a linear block in the forward path and a nonlinearity in the feedback path, Figure 1.1. The nonlinearity is specified only to the extent that it belongs to a "sector", or, in the multivariable case, to a "cone". In other words, the admissible nonlinearities are linearly bounded. One of the absolute

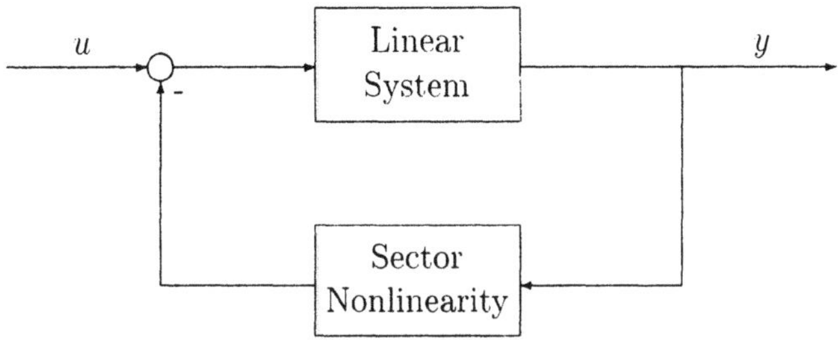

Figure 1.1: The absolute stability problem.

stability results is a Lyapunov function construction for this class of systems. The stability property is "absolute" in the sense that it is preserved for any nonlinearity in the sector. Hence, a "sector stability margin" is guaranteed.

During a period of several years, the frequency domain methods and the absolute stability analysis coexisted as two separate disciplines. Breakthroughs by Popov in the late 1950's and early 1960's dramatically changed the landscape of control theory. While Popov's stability criterion [87] was of major importance, even more important was his introduction of the concept of *passivity* as one of the fundamental feedback properties [88].

Until the work of Popov, passivity was a network theory concept dealing with rational transfer functions which can be realized with passive resistances, capacitances and inductances. Such transfer functions are restricted to have *relative degree* (excess of the number of poles over the number of zeros) not larger than one. They are called *positive real* because their real parts are positive for all frequencies, that is, their phase lags are always less than 90 degrees. A key feedback stability result from the 1960's, which linked passivity with the existence of a quadratic Lyapunov function for a linear system, is the celebrated Kalman-Yakubovich-Popov (KYP) lemma also called *Positive Real Lemma*. It has spawned many significant extensions to nonlinear systems and adaptive control.

1.1.2 Passivity as a phase characteristic

The most important passivity result, and also one of the fundamental laws of feedback, states that *a negative feedback loop consisting of two passive systems is passive*. This is illustrated in Figure 1.2. Under an additional detectability condition *this feedback loop is also stable.*

To appreciate the content of this brief statement, assume first that the two

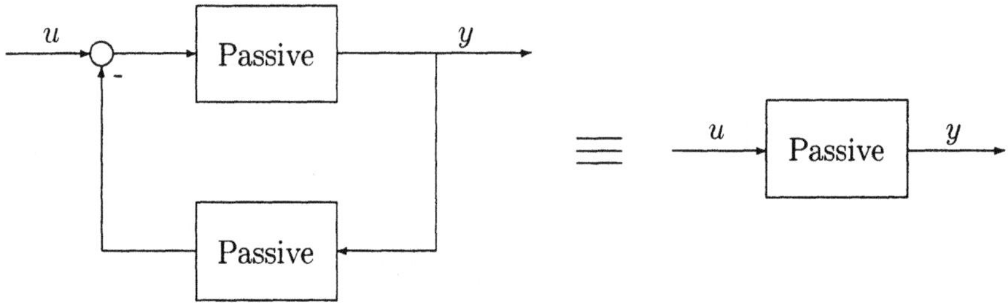

Figure 1.2: The fundamental passivity result.

passive blocks in the feedback connection of Figure 1.2 are linear. Then their transfer functions are positive real, that is, with the phase lag not larger than 90 degrees. Hence, the phase lag over the entire feedback loop is not larger than 180 degrees. By the Nyquist-Bode criterion, such a linear feedback loop is stable for all feedback gains, that is, it possesses an "infinite gain margin".

When the two blocks in the feedback loop are nonlinear, the concept of passivity can be seen to extend the Nyquist-Bode 180 degree phase lag criterion to nonlinear systems. For nonlinear systems, passivity can be therefore interpreted as a "phase" property, a complement of the gain property characterized by various small gain theorems such as those presented in [18].

In the early 1970's, Willems [120] systematized passivity (and dissipativity) concepts by introducing the notions of *storage function* $S(x)$ and *supply rate* $w(u, y)$, where x is the system state, u is the input, and y is the output. A system is passive if it has a positive semidefinite storage function $S(x)$ and a bilinear supply rate $w(u, y) = u^T y$, satisfying the inequality

$$S(x(T)) - S(x(0)) \leq \int_0^T w(u(t), y(t))\, dt \tag{1.1.1}$$

for all u and $T \geq 0$. Passivity, therefore, is the property that the increase in storage S is not larger than the integral amount supplied. Restated in the derivative form

$$\dot{S}(x) \leq w(u, y) \tag{1.1.2}$$

passivity is the property that the rate of increase of storage is not higher than the supply rate. In other words, any storage increase in a passive system is due solely to external sources. The relationship between passivity and Lyapunov stability can be established by employing the storage $S(x)$ as a Lyapunov function. We will make a constructive use of this relationship.

1.1.3 Optimal control and stability margins

Another major development in the 1950's and 1960's was the birth of optimal control twins: Dynamic Programming and Maximum Principle. An optimality result crucial for feedback control was the solution of the optimal linear-quadratic regulator (LQR) problem by Kalman [50] for linear systems $\dot{x} = Ax + Bu$. The well known optimal control law has the form $u = -B^T P x$, where x is the state, u is the control and P is the symmetric positive definite solution of a matrix algebraic Riccati equation. The matrix P determines the optimal value $x^T P x$ of the cost functional, which, at the same time, is a Lyapunov function establishing the asymptotic stability of the optimal feedback system.

A remarkable connection between optimality and passivity, established by Kalman [52], is that a linear system can be optimal only if it has a passivity property with respect to the output $y = B^T P x$. Furthermore, optimal linear systems have infinite gain margin and phase margin of 60 degrees.

These *optimality, passivity, and stability margin* properties have been extended to nonlinear systems which are affine in control:

$$\dot{x} = f(x) + g(x)u \tag{1.1.3}$$

A feedback control law $u = k(x)$ which minimizes the cost functional

$$J = \int_0^\infty (l(x) + u^2)dt \tag{1.1.4}$$

where $l(x)$ is positive semidefinite and u is a scalar, is obtained by minimizing the Hamiltonian function

$$\mathcal{H}(x, u) = l(x) + u^2 + \frac{\partial V}{\partial x}(f(x) + g(x)u) \tag{1.1.5}$$

If a differentiable optimal value function $V(x)$ exists, then the optimal control law is in the "$L_g V$-form":

$$u = k(x) = -\frac{1}{2} L_g V(x) = -\frac{1}{2}\frac{\partial V}{\partial x} g(x) \tag{1.1.6}$$

The optimal value function $V(x)$ also serves as a Lyapunov function which, along with a detectability property, guarantees the asymptotic stability of the optimal feedback system. The connection with passivity was established by Moylan [80] by showing that, as in the linear case, the optimal system has an infinite gain margin thanks to its passivity property with respect to the output $y = L_g V$.

In Chapters 2 and 3 we study in detail the design tools of passivity and optimality, and their ability to provide desirable stability margins. A particular case of interest is when $V(x)$ is a Lyapunov function for $\dot{x} = f(x)$, which is stable but not asymptotically stable. In this case, the control law $u = -L_gV$ adds additional "damping". This *damping control* is again in the "L_gV-form". It is often referred to as "Jurdjevic-Quinn feedback" [49] and will frequently appear in this book.

What this book does not include are methods applicable only to linearly bounded nonlinearities. Such methods, including various small gain theorems [18], H-infinity designs with bounded uncertainties [21], and linear matrix inequality algorithms [7] are still too restrictive for the nonlinear systems considered in this book. Progress has been made in formulating nonlinear small gain theorems by Mareels and Hill [71], Jiang, Teel and Praly [48], among others, and in using them for design [111]. Underlying to these efforts, and to several results of this book, is the concept of *input-to-state stability (ISS)* of Sontag [103] and its relationship to dissipativity. The absolute stability tradition has also continued with a promising development by Megretski and Rantzer [76], where the static linear constraints are replaced by *integral quadratic constraints.*

1.2 Feedback Passivation

1.2.1 Limitations of feedback linearization

Exciting events in nonlinear control theory of the 1980's marked a rapid development of differential-geometric methods which led to the discovery of several structural properties of nonlinear systems. The interest in geometric methods was sparked in the late 70's by "feedback linearization," in which a nonlinear system is completely or partially transformed into a linear system by a state diffeomorphism and a feedback transformation.

However, feedback linearization may result not only in wasteful controls, but also in nonrobust systems. Feedback linearizing control laws often destroy inherently stabilizing nonlinearities and replace them with destabilizing terms. Such feedback systems are without any stability margins, because even the smallest modeling errors may cause a loss of stability.

A complete or partial feedback linearization is performed in two steps. First, a change of coordinates (diffeomorphism) is found in which the system appears "the least nonlinear." This step is harmless. In the second step, a control is designed to cancel all the nonlinearities and render the system linear. This step can be harmful because it often replaces a stabilizing nonlinearity by its wasteful and dangerous negative.

Fortunately, the harmful second step of feedback linearization is avoidable. For example, a control law minimizing a cost functional like (1.1.4) does not cancel useful nonlinearities. On the contrary, it employs them, especially for large values of x which are penalized more. This motivated Freeman and Kokotović [25] to introduce an "inverse optimal" design in which they replace feedback linearization by robust backstepping and achieve a form of worst-case optimality. Because of backstepping, this design is restricted to a lower-triangular structure with respect to nonlinearities which grow faster than linear. A similar idea of employing optimality to avoid wasteful cancellations is pursued in this book, but in a different setting and for a larger class of systems, including the systems that cannot be linearized by feedback.

1.2.2 Feedback passivation and forwarding

Lyapunov designs in this book achieve stability margins by exploiting the connections of stability, optimality and passivity. Geometric tools are used to characterize the system structure and to construct Lyapunov functions.

Most of the design procedures in this book are based on *feedback passivation.* For the partially linear cascade, including the Byrnes-Isidori normal form [13], the problem of achieving passivity by feedback was first posed and solved by Kokotović and Sussmann [59]. A general solution to the feedback passivation problem was given by Byrnes, Isidori and Willems [15] and is further refined in this book.

Because of the pursuit of feedback passivation, the geometric properties of primary interest are the relative degree of the system and the stability of its zero dynamics. The concepts of relative degree and zero dynamics, along with other geometric tools are reviewed in Appendix A. A comprehensive treatment of these concepts can be found in the books by Isidori [43], Nijmeijer and van der Schaft [84], and Marino and Tomei [73].

Achieving passivity with feedback is an appealing concept. However, in the construction of feedback passivation designs which guarantee stability margins, there are two major challenges. The first challenge is to avoid nonrobust cancellations. In this book this is achieved by rendering the passivating control optimal with respect to a cost functional (1.1.4). It is intuitive that highly penalized control effort will not be wasted to cancel useful nonlinearities, as confirmed by the stability margins of optimal systems in Chapter 3.

The second challenge of feedback passivation is to make it constructive. This is difficult because, to establish passivity, which is an input-output concept, we must select an output y and construct a positive semidefinite storage function $S(x)$ for the supply rate $u^T y$. In the state feedback stabilization the search for an output is a part of the design procedure. This search is guided by the structural properties: in a passive system the relative degree must not be larger than one and the zero dynamics must not be unstable ("nonminimum phase"). Like in the linear case, the nonlinear relative degree and the nonlinear zero-dynamics subsystem are invariant under feedback. If the zero-dynamics subsystem is unstable, the entire system cannot be made passive by feedback. For feedback passivation one must search for an output with respect to which the system will not only have relative degree zero or one, but also be "weakly minimum phase" (a concept introduced in [92] to include some cases in which the zero-dynamics subsystem is only stable, rather than asymptotically stable).

Once an output has been selected, a positive semidefinite storage function $S(x)$ must be found for the supply rate $u^T y$. For our purpose this storage function serves as a Lyapunov function. It is also required to be the optimal value of a cost functional which penalizes the control effort.

One of the perennial criticisms of Lyapunov stability theory is that it is not constructive. Design procedures developed in this book remove this defficiency for classes of systems with special structures. *Backstepping* solves the stabilization problem for systems having a lower-triangular structure, while *forwarding* does the same for systems with an upper-triangular structure. This methodology, developed by the authors [46, 95], evolved from an earlier nested saturation design by Teel [109] and recent results by Mazenc and Praly [75].

1.3 Cascade Designs

1.3.1 Passivation with composite Lyapunov functions

The design procedures in this book are first developed for cascade systems. The cascade is "partially linear" if one of the two subsystems is linear, that is

$$\begin{aligned} \dot{z} &= f(z) + \psi(z,\xi), \quad \psi(z,0) = 0 \\ \dot{\xi} &= A\xi + Bu \end{aligned} \tag{1.3.1}$$

where (A, B) is a stabilizable pair. Even when the subsystem $\dot{z} = f(z)$ is GAS, it is the interconnection term $\psi(z,\xi)$ which determines whether the entire cascade is stabilizable or not.

Applying the result that a feedback connection of two passive systems is passive, the cascade (1.3.1) can be rendered passive if it can be represented as a feedback interconnection of two passive systems. To this end, an output $y_1 = h_1(\xi) = C\xi$ is obtained for the ξ-subsystem by a factorization of the interconnection term:

$$\psi(z,\xi) = \tilde{\psi}(z,\xi)h_1(\xi) \tag{1.3.2}$$

The output y_1 of the ξ-subsystem is the input of the z-subsystem. We let $W(z)$ be the z-subsystem Lyapunov function such that $L_f W(z) \leq 0$. Then for the input $h_1(\xi)$, the z-subsystem is passive with respect to the output $y_2 = L_{\tilde{\psi}} W$ and $W(z)$ is its storage function. It is now sufficient that the ξ-subsystem with the output $y_1 = h_1(\xi) = C\xi$ can be made passive by a feedback transformation $u = Kx + Gv$. Then a composite Lyapunov function for the whole cascade is $V(z,\xi) = W(z) + \xi^T P\xi$, where $P > 0$ satisfies the Positive Real Lemma for the $(A+BK, BG, B^T P)$. Such a matrix P exists if the linear subsystem (A, B, C) is *feedback passive.* Because the relative degree and the zero dynamics are invariant under feedback, a structural restriction on (A,B,C) is to be relative degree one and weakly minimum phase.

A similar construction of a composite Lyapunov function

$$V(z,\xi) = W(z) + U(\xi) \tag{1.3.3}$$

is possible when both subsystems in the cascade are nonlinear

$$\begin{aligned} \dot{z} &= f(z) + \tilde{\psi}(z,\xi)h_1(\xi) \\ \dot{\xi} &= a(\xi, u) \end{aligned} \tag{1.3.4}$$

and when the assumption on $\dot{z} = f(z)$ is relaxed to be only GS (globally stable), with a Lyapunov function $W(z)$ such that $L_f W(z) \leq 0$. Again, the z-subsystem is passive with the input-output pair $u_2 = h_1(\xi)$ and $y_2 = L_{\tilde{\psi}} W$. The entire cascade is rendered passive if the ξ-subsystem with output $y_1 = h_1(\xi)$ is made passive by feedback. As in the linear case, the relative degree and zero-dynamics restrictions must be satisfied and a storage function $U(\xi)$ must be found.

In Chapter 4 several versions of such passivation designs are employed to stabilize translational oscillations of a platform using a rotating actuator.

1.3.2 A structural obstacle: peaking

One of the novelties of this book is the treatment in Chapter 4 of an often overlooked obstacle to global and semiglobal stabilization – *the peaking phenomenon.* In its simplest form this phenomenon occurs in the linear system

$\dot{\xi} = A\xi + Bu$ when the gain K in the state feedback $u = K\xi$ is chosen to place the eigenvalues of $A + BK$ to the left of $Re\{s\} = -a < 0$. For a fast convergence of ξ to zero, the value of a must be large, that is, the gain K must be high.

Each state component ξ_i is bounded by $\gamma_i e^{-at}$ where γ_i depends not only on the initial condition $\xi(0)$, but also on the rate of decay a, that is $\gamma_i = \tilde{\gamma}_i a^{\pi_i}$. The *peaking states* are those ξ_i's for which the *peaking exponent* π_i is one or larger, while for the nonpeaking states this exponent is zero. In a partially linear cascade (1.3.1), an undesirable effect of peaking in the linear subsystem is that it limits the size of the achievable stability region, as we now illustrate. In the cascade

$$\begin{aligned} \dot{z} &= -z + yz^2 \\ \dot{\xi}_1 &= \xi_2 \\ \dot{\xi}_2 &= u, \qquad y = c_1\xi_1 + c_2\xi_2 \end{aligned} \tag{1.3.5}$$

the z-equation can be solved explicitly:

$$z(t) = e^{-t}z(0)[1 - z(0)\int_0^t e^{-\tau}y(\tau)\, d\tau]^{-1}$$

Clearly, to avoid the escape of $z(t)$ to infinity in finite time, it is necessary that the following bound be satisfied

$$z(0)\int_0^\infty e^{-t}y(t)\, dt \le 1 \tag{1.3.6}$$

With partial-state feedback $u = k_1\xi_1 + k_2\xi_2$ the decay of $y(t)$ is exponential, $|y(t)| \le \gamma e^{-at}$, and the bound (1.3.6) is satisfied if

$$\frac{z(0)\gamma}{a+1} \le 1 \tag{1.3.7}$$

If $y(t)$ is not peaking, that is if γ does not grow with a, then $z(0)$ can be allowed to be as large as desired by making a sufficiently large. Thus, when y is a *nonpeaking output* of the linear subsystem, that is, when y can be forced to decay arbitrarily fast without peaking, then the entire cascade can be semiglobally stabilized.

Even when y is a nonpeaking output, not every feedback law will achieve fast decay of y without peaking, as we illustrate with the "high-gain" design

$$u = -a^2\xi_1 - 2a\xi_2 \tag{1.3.8}$$

for ξ-subsystem in (1.3.5). This high-gain control law places the eigenvalues at $\lambda_1 = \lambda_2 = -a$. A simple calculation shows that in this case ξ_1 is a nonpeaking state, while ξ_2 is peaking with $\pi_2 = 1$. Thus, $y = \xi_1$ satisfies (1.3.7) and the semiglobal stability is achieved. On the other hand, when $y = \xi_2$ the bound (1.3.6) for $(\xi_1(0), \xi_2(0)) = (1, 0)$ is

$$\frac{z(0)a^2}{a^2+1} \leq 1$$

and semiglobal stability cannot be achieved: no increase of a will allow $z(0)$ to be larger than one.

To see that $y = \xi_2$ is in fact a nonpeaking output we now use the "two time-scale" design

$$u = -\xi - (a + \frac{1}{a})\xi_2 \qquad (1.3.9)$$

which, for large a, renders $\lambda_2 = -a$ "fast", and $\lambda_1 = -\frac{1}{a}$ "slow." A simple calculation shows that, with feedback (1.3.9), the output $y = \xi_2$ still has the fast decay rate a, but is nonpeaking, that is, it satisfies the bound (1.3.7) which guarantees semiglobal stability.

We have thus demonstrated that with either $y = \xi_1$ (or $y = \xi_2$) semiglobal stabilization of the cascade (1.3.5) is possible with partial-state feedback design (1.3.8) (or (1.3.9)), each rendering the decay of y arbitrarily fast without peaking.

Can global stabilization also be achieved? The answer is affirmative, but for this we must use full-state feedback $u(\xi_1, \xi_2, z)$. For $y = \xi_2$ we can design such a feedback law using passivation discussed in the preceding section, while for $y = \xi_1$, we can use a backstepping design, to be discussed later. These two full-state feedback designs satisfy the bound (1.3.6) for all $z(0)$ by forcing $y(t)$ to depend on $z(t)$ and to contribute to the stabilization process via the interconnection term yz^2.

In the discussion thus far we have mentioned the control laws which avoid output peaking for $y = \xi_1$ and $y = \xi_2$ in (1.3.5). However, it can be shown that output peaking cannot be avoided if $y = \xi_1 - \xi_2$. In this case, neither global nor semiglobal stabilization of the cascade (1.3.5) is possible. With $y = \xi_1 - \xi_2$ the double integrator is "strictly" nonminimum phase and all such systems are *peaking systems.*

For the cascade (1.3.1), with $\dot{z} = f(z)$ being GAS, the peaking phenomenon and the structure of the interconnection term $\psi(z, \xi)$ determine whether global or semiglobal stabilization is possible. If the interconnection term $\psi(z, \xi)$ contains peaking states multiplied with functions of z which grow faster than linear, global stabilization may be impossible. To determine whether this is the case, the interconnection is factored as $\tilde{\psi}(z, \xi_0)C\xi$, where $C\xi$ is treated as the output of the linear subsystem and ξ_0 denotes the nonpeaking states.

Now the problem is to stabilize the linear subsystem while preventing the peaking in the output $C\xi$. The class of output *nonpeaking* linear systems is characterized in Chapter 4 where it is shown that strictly nonminimum phase linear systems are peaking systems. Our new analysis encompasses both fast and slow peaking.

We reiterate that peaking is an obstacle not only to global stabilization, but also to more practical semiglobal stabilization which is defined as the possibility to guarantee any prespecified bounded stability region. Our analysis of peaking in Chapter 4 applies and extends earlier results by Mita [79], Francis and Glover [20], and the more recent results by Sussmann and Kokotović [105], and Lin and Saberi [67].

1.4 Lyapunov Constructions

1.4.1 Construction of the cross-term

The most important part of our design procedures is the construction of a Lyapunov function for an uncontrolled subsystem. In Chapter 5 this task is addressed with a structure-specific approach and a novel Lyapunov construction is presented for the cascade

$$(\Sigma_0)\left\{\begin{array}{rcl} \dot{z} &=& f(z)+\psi(z,\xi) \\ \dot{\xi} &=& a(\xi) \end{array}\right. \tag{1.4.1}$$

where $\dot{z} = f(z)$ is globally stable and $\dot{\xi} = a(\xi)$ is globally asymptotically stable and locally exponentially stable. Such constructions have not appeared in the literature until the recent work by Mazenc and Praly [75] and the authors [46]. Chapter 5 presents a comprehensive treatment of several exact and approximate Lyapunov constructions.

The main difficulty in constructing a Lyapunov function for (Σ_0) is due to the fact that $\dot{z} = f(z)$ is only globally stable, rather than globally asymptotically stable, so that simple composite Lyapunov functions such as $W(z)+U(\xi)$ in (1.3.3) are not suitable.

Our main construction is aimed at finding the cross-term $\Psi(z,\xi)$ for a more general Lyapunov function

$$V_0(z,\xi) = W(z) + \Psi(z,\xi) + U(\xi)$$

where $W(z)$ and $U(\xi)$ are the Lyapunov functions of the subsystems. The cross-term $\Psi(z,\xi)$ is needed to achieve nonpositivity of

$$\dot{V}_0 = L_f W + L_\psi W + \dot{\Psi} + L_a U$$

Because $L_\psi W$ is indefinite, $\dot{\Psi}$ is constructed to eliminate it, that is $\dot{\Psi} = -L_\psi W$. In Chapter 5 we prove the existence and continuity of $\Psi(z,\xi)$ under the conditions

$$\|\frac{\partial W}{\partial z}\| \, \|z\| \le cW(z), \quad \text{as } \|z\| \to \infty \tag{1.4.2}$$
$$\|\psi(z,\xi)\| \le \gamma_1(\|\xi\|)\|z\| + \gamma_2(\|\xi\|) \tag{1.4.3}$$

The first condition restricts the growth of W to be polynomial. The second condition restricts the growth of the interconnection term $\psi(z,\xi)$ to be linear in $\|z\|$. These conditions are structural and cannot be removed without additional restrictions on $f(z)$ and $\psi(z,\xi)$. An expression for $\Psi(z,\xi)$, which for special classes of cascades can be obtained explicitly, is the line integral

$$\Psi(z,\xi) = \int_0^\infty L_\psi W(\tilde{z}(s;(z,\xi)), \tilde{\xi}(s,\xi))ds \tag{1.4.4}$$

along the solution of (Σ_0) which starts at (z,ξ). In general, this integral is either precomputed, or implemented with on-line numerical integrations. Approximate evaluations of $\Psi(z,\xi)$ from a PDE can also be employed.

1.4.2 A benchmark example

As an illustration of the explicit construction of the cross-term $\Psi(z,\xi)$ and its use in a passivation design we consider the system

$$\begin{aligned} \dot{x}_1 &= x_2 + \theta x_3^2 \\ \dot{x}_2 &= x_3 \\ \dot{x}_3 &= \tilde{u} \end{aligned} \tag{1.4.5}$$

We first let $\theta = 1$ and later allow θ to be an unknown constant parameter. This system cannot be completely linearized by a change of coordinates and feedback. For $y = x_2 + x_3$ it has the relative degree one and can be written as

$$\begin{aligned} \dot{x}_1 &= x_2 + x_2^2 + (2x_2 + y)y \\ \dot{x}_2 &= -x_2 + y \\ \dot{y} &= -y + u \end{aligned} \tag{1.4.6}$$

where we have set $\tilde{u} = -2y + x_2 + u$. To proceed with a passivation design we observe that the zero-dynamics subsystem

$$\begin{aligned} \dot{x}_1 &= x_2 + x_2^2 \\ \dot{x}_2 &= -x_2 \end{aligned}$$

is stable, but not asymptotically stable. For this subsystem we need a Lyapunov function and, to construct it, we consider x_1 as z, x_2 as ξ and view the zero-dynamics subsystem as the cascade system (Σ_0). For $W = x_1^2$ the line-integral (1.4.4) yields the explicit expression

$$\Psi(x_1, x_2) = (x_1 + x_2 + \frac{x_2^2}{2})^2 - x_1^2$$

which, along with $U(x_2) = x_2^2$, results in the Lyapunov function

$$V_0(x_1, x_2) = (x_1 + x_2 + \frac{x_2^2}{2})^2 + x_2^2$$

Returning to the normal form (1.4.6) we get the cascade (1.3.1), in the notation (z_1, z_2, ξ) instead of (x_1, x_2, y). The interconnection term $\psi^T = [2x_2+y, 1]^T y$ is already factored because $y = \xi$ and the ξ-subsystem is passive with the storage function $S(y) = y^2$. Applying the passivation design from Section 1.3.1, where $V_0(x_1, x_2)$ plays the role of $W(z)$ and $\tilde{\psi}^T = [2x_2+y, 1]^T$, the resulting feedback control is

$$u = -\frac{\partial V_0}{\partial x_1}(2x_2 + y) - \frac{\partial V_0}{\partial x_2}$$

Using $V = V_0(x_1, x_2) + y^2$ as a Lyapunov function it can be verified that the designed feedback system is globally asymptotically stable. It is instructive to observe that this design exploits two nested cascade structures: first, the zero-dynamics subsystem is itself a cascade; and second, it is also the nonlinear part of the overall cascade (1.4.6).

An alternative approach, leading to recursive forwarding designs in Chapter 6, is to view the same system (1.4.5) as the cascade of the double integrator $\dot{x}_2 = x_3$, $\dot{x}_3 = \tilde{u}$ with the x_1-subsystem. The double integrator part is first made globally exponentially stable by feedback, say $u = -x_2 - 2x_3 + v$. It is easy to verify that with this feedback the whole system is globally stable. To proceed with the design, a Lyapunov function $V(x)$ is to be constructed for the whole system such that, with respect to the passivating output $y = \frac{\partial V}{\partial x_3}$, the system satisfies a detectability condition. The global asymptotic stability of the whole system can then be achieved with the damping control $v = -\frac{\partial V}{\partial x_3}$. Again, the key step is the construction of the cross-term Ψ for the Lyapunov function $V(x)$. In this case the cross-term is

$$\Psi(x_1, x_2, x_3) = \frac{1}{2}(x_1 + 2x_2 + x_3 + \frac{1}{2}(x_2^2 + x_3^2)^2)^2 - \frac{1}{2}x_1^2$$

and results in

$$V(x) = \frac{1}{2}(x_1 + 2x_2 + x_3 + \frac{1}{2}(x_2^2 + x_3^2)^2)^2 + \frac{1}{2}x_2^2 + \frac{1}{2}x_3^2$$

which is the desired Lyapunov function for (1.4.5) with $u = -x_2 - 2x_3$.

1.4.3 Adaptive control

While adaptive control is not a major topic of this book, the Lyapunov construction in Chapter 5 is extended to nonlinear systems with unknown constant parameters, such as the system (1.4.5) with unknown θ. Without a known bound on θ, the global stabilization problem for this benchmark system has not been solved before. Its solution can now be obtained by constructing the same control law as if θ were known. Then the unknown parameter is replaced by its estimate, and the Lyapunov function is augmented by a term penalizing the parameter estimation error. Finally, a parameter update law is designed to make the time-derivative of the augmented Lyapunov function negative. This step, in general, requires that the estimates be overparameterized. Thus, for the above example, instead of one, estimates of two parameters are needed. This adaptive design is presented in Chapter 5.

1.5 Recursive Designs

1.5.1 Obstacles to passivation

With all its advantages, feedback passivation has not yet become a widely used design methodology. Many passivation attempts have been frustrated by the requirements that the system must have a relative degree one and be weakly minimum phase. As the dimension of the system increases, searching for an output which satisfies these requirements rapidly becomes an unwieldy task. Even for a highly structured system such as

$$\begin{aligned} \dot{z} &= f(z) + \tilde{\psi}(z, \xi_i)\xi_i, \quad i \in \{1, \ldots, n\} \\ \dot{\xi}_1 &= \xi_2 \\ \dot{\xi}_2 &= \xi_3 \\ &\vdots \\ \dot{\xi}_n &= u, \end{aligned} \tag{1.5.1}$$

with globally asymptotically stable $\dot{z} = f(z)$, feedback passivation is difficult because each candidate output $y = \xi_i$ fails to satisfy at least one of the two passivity requirements. Thus, if $y = \xi_1$, the system is minimum phase, but it has a relative degree n. On the other hand, if $y = \xi_n$, the relative degree is one, but the system is not weakly minimum phase because the zero-dynamics subsystem contains an unstable chain of integrators. For all other choices $y = \xi_i$, neither the relative degree one, nor the weak minimum phase requirement are satisfied.

The recursive step-by-step constructions in Chapter 6 circumvent the structural obstacles to passivation. At each step, only a subsystem is considered, for which the feedback passivation is feasible. Each of the two recursive procedures, backstepping and forwarding, removes one of the obstacles to feedback passivation.

1.5.2 Removing the relative degree obstacle

Backstepping removes the relative degree one restriction. This is illustrated with the cascade (1.5.1) with $i = 1$, that is with $y = \xi_1$. With this output, the relative degree one requirement is not satisfied for the entire system. To avoid this difficulty, the backstepping procedure first isolates the subsystem

$$\begin{aligned} \dot{z} &= f(z) + \tilde{\psi}(z, \xi_1)\xi_1, \\ \dot{\xi}_1 &= u_1, \\ y_1 &= \xi_1 \end{aligned} \tag{1.5.2}$$

With u_1 as the input, this system has relative degree one and is weakly minimum phase. Therefore, we can construct a Lyapunov function $V_1(z, \xi_1)$ and a stabilizing feedback $u_1 = \alpha_1(z, \xi_1)$. In the second step, this subsystem is augmented by the ξ_2-integrator:

$$\begin{aligned} \dot{z} &= f(z) + \tilde{\psi}(z, \xi_1)\xi_1, \\ \dot{\xi}_1 &= \xi_2 \\ \dot{\xi}_2 &= u_2, \\ y_2 &= \xi_2 - \alpha_1(z, \xi_1) \end{aligned} \tag{1.5.3}$$

and the stabilizing feedback $\alpha_1(z, \xi_1)$ from the preceding step is used to define the new passivating output y_2. With this output and the input u_2 the augmented subsystem has relative degree one because

$$\dot{y}_2 = u_2 - \frac{\partial \alpha_1}{\partial z}(f(z) + \tilde{\psi}(z, \xi_1)\xi_1) - \frac{\partial \alpha_1}{\partial \xi_1}\xi_2 \tag{1.5.4}$$

By construction, the augmented subsystem is also minimum phase, because its zero-dynamics subsystem is (1.5.2) with stabilizing feedback $u_1 = \alpha_1(z, \xi_1)$. Moreover, $V_1(z, \xi_1)$ is a Lyapunov function for the zero-dynamics subsystem. By augmenting V_1 with y_2^2 we obtain the composite Lyapunov function

$$V_2(z, \xi) = V_1(z, \xi_1) + y_2^2 = V_1(z, \xi_1) + (\xi_2 - \alpha_1(z, \xi_1))^2$$

which now serves for the construction of the new feedback $u_2 = \alpha_2(z, \xi_1, \xi_2)$.

For the case $n = 2$, the relative degree obstacle to feedback passivation has thus been overcome in two steps. The procedure is pursued until the output has a relative degree one with respect to the true input u.

In this way, *backstepping* extends feedback passivation design to a system with any relative degree by recursively constructing an output which eventually satisfies the passivity requirements. At each step, the constructed output is such that the entire system is minimum phase. However, the relative degree one requirement is satisfied only at the last step of the procedure.

Backstepping has already become a popular design procedure, particularly successful in solving global stabilization and tracking problems for nonlinear systems with unknown parameters. This adaptive control development of backstepping is presented in the recent book by Krstic, Kanellakopoulos and Kokotović [61]. Backstepping has also been developed for robust control of nonlinear systems with uncertainties in the recent book by Freeman and Kokotović [26]. Several backstepping designs are also presented in [73].

1.5.3 Removing the minimum phase obstacle

Forwarding is a new recursive procedure which removes the weak minimum phase obstacle to feedback passivation and is applicable to systems not handled by backstepping. For example, backstepping is not applicable to the cascade (1.5.1) with $i = n$, because with $y = \xi_n$ the zero-dynamics subsystem contains an unstable chain of integrators. The forwarding procedure circumvents this obstacle step-by-step. It starts with the cascade

$$\begin{aligned} \dot{z} &= f(z) + \tilde{\psi}(z, \xi_n)\xi_n, \\ \dot{\xi}_n &= u_n, \\ y_n &= \xi_n \end{aligned} \tag{1.5.5}$$

which ignores the unstable part of the zero dynamics. This subsystem satisfies both passivation requirements, so that a Lyapunov function $V_n(z, \xi_n)$ and a stabilizing feedback $u_n = \alpha_n(z, \xi_n)$ are easy to construct. The true control input is denoted by u_n to indicate that the first step of forwarding starts with the ξ_n-equation. The second step moves "forward" from the input, that is it includes the ξ_{n-1}-equation:

$$\begin{aligned} \dot{\xi}_{n-1} &= \xi_n \\ \dot{z} &= f(z) + \tilde{\psi}(z, \xi_n)\xi_n \\ \dot{\xi}_n &= u_n(z, \xi_n) \end{aligned} \tag{1.5.6}$$

This new subsystem has the structure of (1.4.1): it is the cascade of a stable system $\dot{\xi}_{n-1} = 0$ with the globally asymptotically stable system (z, ξ_n), the interconnection term being just the state ξ_n. The construction with a cross-term is used to obtain a Lyapunov function $V_{n-1}(z, \xi_n, \xi_{n-1})$ which is nonincreasing along the solutions of (1.5.6). This means that the system

$$\begin{aligned} \dot{\xi}_{n-1} &= \xi_n \\ \dot{z} &= f(z) + \tilde{\psi}(z, \xi_n)\xi_n, \\ \dot{\xi}_n &= u_n(z, \xi_n) + u_{n-1}, \\ y_{n-1} &= L_g V_{n-1} \end{aligned} \tag{1.5.7}$$

with the input-output pair (u_{n-1}, y_{n-1}) is passive, and the damping control $u_{n-1} = -y_{n-1}$ can be used to achieve *global asymptotic* stability.

By recursively adding a new state equation to an already stabilized subsystem, a Lyapunov function $V_1(z, \xi_n, \ldots, \xi_1)$ is constructed and the entire cascade is rendered feedback passive with respect to the output $y = L_g V_1$. This output is the last one in a sequence of outputs constructed at each step. With respect to each of these outputs, the entire system has relative degree one, but the weak minimum phase requirement is satisfied only at the last step. At each intermediate step, the zero dynamics of the entire system are unstable.

This description shows that with *forwarding* the weak minimum phase requirement of feedback passivation is relaxed by allowing instability of the zero dynamics, characterized by repeated eigenvalues on the imaginary axis. Because of the peaking obstacle, this weak nonminimum phase requirement cannot be further relaxed without imposing some other restrictions.

1.5.4 System structures

For convenience, backstepping and forwarding have been introduced using a system consisting of a nonlinear z-subsystem and a ξ-integrator chain. However, these procedures are applicable to larger classes of systems.

Backstepping is applicable to the systems in the following *feedback (lower-triangular) form*:

$$\begin{aligned} \dot{z} &= f(z) + \psi(z, \xi_1)\xi_1 \\ \dot{\xi}_1 &= a_1(\xi_1, \xi_2) \\ \dot{\xi}_2 &= a_2(\xi_1, \xi_2, \xi_3) \\ &\vdots \\ \dot{\xi}_n &= a_n(\xi_1, \xi_2, \ldots, \xi_n, u) \end{aligned} \tag{1.5.8}$$

which, for the input-output pair (u, ξ_1), has relative degree n.

Likewise, forwarding is not restricted to systems in which the unstable part of the zero-dynamics subsystem is a chain of integrators. Forwarding only requires that the added dynamics satisfy the assumptions for the construction of the cross-term Ψ. Therefore, the systems which can be stabilized by forwarding have the following *feedforward (upper-triangular) form:*

$$\begin{array}{rcl}\dot{\xi}_1 &=& f_1(\xi_1) + \psi_1(\xi_1, \xi_2, \ldots, \xi_n, z, u) \\ \dot{\xi}_2 &=& f_2(\xi_2) + \psi_2(\xi_2, \ldots, \xi_n, z, u) \\ &\vdots& \\ \dot{\xi}_{n-1} &=& f_{n-1}(\xi_{n-1}) + \psi_{n-1}(\xi_{n-1}, \xi_n, z, u) \\ \dot{z} &=& f(z) + \psi(\xi_n, z)\xi_n \\ \dot{\xi}_n &=& u \end{array} \tag{1.5.9}$$

where $\xi_i^T = [\xi_{i1}, \ldots, \xi_{iq}]$, the subsystems $\dot{\xi}_i = f_i(\xi_i)$ are stable, and the interconnections terms ψ_i satisfy a growth condition in ξ_i.

It is important to stress that, without further restrictions on the z-subsystem, the triangular forms (1.5.8) and (1.5.9) are necessary, as illustrated by the following example:

$$\begin{array}{rcl}\dot{x}_0 &=& (-1 + x_1)x_0^3 \\ \dot{x}_1 &=& x_2 + x_3^2 \\ \dot{x}_2 &=& x_3 \\ \dot{x}_3 &=& u \end{array} \tag{1.5.10}$$

Because the (x_1, x_2, x_3)-subsystem is not lower-triangular, backstepping is not applicable. The entire system is upper-triangular, but the growth condition imposed by forwarding is violated by the interconnection term $x_0^3 x_1$. In fact, it can be shown that (1.5.10) is not globally stabilizable.

Broader classes of systems can be designed by interlacing steps of backstepping and forwarding. Such *interlaced systems* are characterized by structural conditions which only restrict the system interconnections, that is, the states which enter the different nonlinearities. We show in Chapter 6 that, when a nonlinear system lacks this structural property, additional conditions, like restrictions on the growth of the nonlinearities, must be imposed to guarantee global stabilizability.

Backstepping and forwarding designs can be executed to guarantee that a cost functional including a quadratic cost on the control is minimized. Stability margins are therefore guaranteed for the designed systems.

1.5.5 Approximate asymptotic designs

The design procedures discussed thus far guarantee global stability properties with desirable stability margins. However, their complexity increases with the dimension of the system, and, for higher-order systems, certain simplified designs are of interest. They require a careful trade-off analysis because the price paid for such simplifications may be a significant reduction in performance and robustness.

Simplifications of backstepping and forwarding, presented in Chapter 6, are two distinct slow-fast designs. They are both *asymptotic* in the sense that in the limit, as a design parameter ϵ tends to zero, the separation of time scales is complete. They are also *geometric*, because the time-scale properties are induced by a particular structure of invariant manifolds.

Asymptotic approximations to *backstepping* employ high-gain feedback to create invariant manifolds. The convergence to the manifold is fast, while the behavior in the manifold is slower. The relationship of such asymptotic designs with backstepping is illustrated on the cascade

$$\begin{aligned} \dot{z} &= f(z) + \tilde{\psi}(z, \xi_1)\xi_1, \\ \dot{\xi}_1 &= \xi_2 \\ \dot{\xi}_2 &= u_2, \\ y_2 &= \xi_2 - \alpha_1(z, \xi_1) \end{aligned} \tag{1.5.11}$$

where y_2 is the error between ξ_2 and the "control law " $\alpha_1(z, \xi_1)$ designed to stabilize the (z, ξ_1)-subsystem using ξ_2 as the "virtual control". In backstepping the actual control law is designed to render the cascade (1.5.11) passive from the input u_2 to the output y_2. Such a control law is of considerable complexity because it implements the analytical expressions of the time-derivatives $\dot{z}$ and $\dot{\xi}_1$, available from the first two equations of (1.5.11). A major simplification is to disregard these derivatives and to use the high-gain feedback

$$u_2 = -ky_2 := -\frac{1}{\epsilon} y_2$$

where ϵ is sufficiently small. The resulting feedback system is

$$\begin{aligned} \dot{z} &= f(z) + \tilde{\psi}(z, \xi_1)\xi_1, \\ \dot{\xi}_1 &= \alpha_1(z, \xi_1) + y_2 \\ \epsilon\dot{y}_2 &= -y_2 - \epsilon\left(\frac{\partial \alpha_1}{\partial z}\dot{z} + \frac{\partial \alpha_1}{\partial \xi_1}\dot{\xi}_1\right) \end{aligned} \tag{1.5.12}$$

This system is in a standard singular perturbation form and, therefore, it has a slow invariant manifold in an ϵ-neighborhood of the plane $y_2 \equiv 0$. In this manifold the behavior of the whole system (1.5.12) is approximately described by the reduced (z, ξ_1)-subsystem. An estimate of the stability region, which is no longer global, is made by using the level sets of a composite Lyapunov function.

The key feature of this design is that the existence of the slow manifold is enforced by feedback with high-gain $\frac{1}{\epsilon}$. In recursive designs, several nested manifolds are enforced by increasing gains leading to multiple time scales. The high-gain nature of these designs is their major drawback: it may lead to instability due to the loss of robustness to high-frequency unmodeled dynamics as discussed in Chapter 3.

The simplification of *forwarding* employs *low-gain* and *saturated feedback* to allow a design based on the Jacobian linearization of the system. This is the *saturation design* of Teel [109], which was the first constructive result in the stabilization of systems in the upper-triangular form (1.5.9). Its relation to forwarding is illustrated on the benchmark system

$$\begin{aligned}\dot{x}_1 &= x_2 + x_3^2\\ \dot{x}_2 &= x_3\\ \dot{x}_3 &= -x_2 - 2x_3 + v\end{aligned} \tag{1.5.13}$$

One step of forwarding yields the stabilizing feedback

$$v = -(x_1 + 2x_2 + x_3 + \frac{1}{2}(x_1 + x_2 + 2x_3)^2)(1 + 2x_3) \tag{1.5.14}$$

obtained from the cross-term

$$\Psi(x_1, x_2, x_3) = \int_0^\infty \tilde{x}_1(s)(\tilde{x}_2(s) + \tilde{x}_3^2(s))\, ds$$

If we replace the control law (1.5.14) by its linear approximation saturated at a level ϵ, we obtain the simpler control law

$$v = -\sigma_\epsilon(x_1 + 2x_2 + x_3) \tag{1.5.15}$$

where σ_ϵ denotes the saturation

$$\begin{aligned}\sigma_\epsilon(s) &= s, & &\text{for } |s| \le \epsilon\\ &= \text{sign}(s)\,\epsilon, & &\text{for } |s| \ge \epsilon\end{aligned} \tag{1.5.16}$$

A justification for the approximation (1.5.15) comes from the exponential stability of the linear subsystem $\dot{x}_2 = x_3$, $\dot{x}_3 = -2x_3 - x_2$. The ϵ-saturated control law (1.5.15) lets all the solutions of (1.5.13) approach an ϵ-neighborhood of the x_1-axis, that is the manifold $x_2 = x_3 = 0$. Along this manifold, the nonlinear term x_3^2 can be neglected because it is of higher-order and the behavior of the entire system in this region is described by

$$\dot{\zeta} = -\sigma_\epsilon(\zeta) + \mathcal{O}(\epsilon^2), \quad \zeta = x_1 + 2x_2 + x_3 \tag{1.5.17}$$

The convergence of ζ is slow, but ζ eventually enters an ϵ-neighborhood of the origin. In this neighborhood, the control law (1.5.15) no longer saturates and the local exponential stability of the system ensures the convergence of the solutions to zero.

The key feature of the saturation design is the existence of a manifold (for the uncontrolled system $v = 0$) to which all the solutions converge and along which the design of a stabilizing feedback is simplified. With a low-gain saturated feedback, the approach to the manifold is preserved, and, at the same time, the simplified control law achieves a slow stabilization along the manifold. In recursive designs, this convergence towards several nested invariant manifolds is preserved when the saturation levels are decreased, which leads to multiple time scales.

For more general systems in the upper-triangular form (1.5.9), the stabilization achieved with the saturation design is no longer global, but the stability region can be rendered as large as desired with smaller ϵ. The fact that, for a desired stability region, ϵ may have to be very small, shows potential drawbacks of this design.

The first drawback is that, while approaching the slow manifold, the system operates essentially "open-loop" because the ϵ-saturated feedback is negligible as long as x_2 and x_3 are large. During this transient, the state x_1 remains bounded but may undergo a very large overshoot. The control law will have a stabilizing effect on x_1 only after the solution has come sufficiently close to the slow manifold. Even then the convergence is slow because the control law is ϵ-saturated.

The second drawback is that an additive disturbance larger than ϵ will destroy the convergence properties of the equation (1.5.17). Both of these drawbacks suggest that the saturation design should not be pursued if the saturation level ϵ is required to be too small.

Even with their drawbacks, the simplified high-gain and saturation designs presented in Chapter 6 are of practical interest because they reveal structural limitations and provide conservative estimates of achievable performance.

Backstepping and forwarding are not conservative because they employ the knowledge of system nonlinearities and avoid high gains for small signals and low gains for large signals. With guaranteed stability margins they guard against static and dynamic uncertainties. Progressive simplifications of backstepping and forwarding offer a continuum of design procedures which the designer can use for his specific needs.

1.6 Book Style and Notation

1.6.1 Style

Throughout this book we have made an effort to avoid a dry "definition-theorem" style. While definitions are used as the precise form of expression, they are often simplified. Some assumptions obvious from the context, such as differentiability, are explicitly stated only when they are critical.

Examples are used to clarify new concepts prior or after their definitions. They also precede and follow propositions and theorems, not only as illustrations, but often as refinements and extensions of the presented results.

The "example-result-example" style is in the spirit of the book's main goal to enrich the repertoire of nonlinear design tools and procedures. Rather than insisting on a single methodology, the book assembles and employs structure-specific design tools from both analysis and geometry. When a design procedure is constructed, it is presented as one of several possible constructions, pliable enough to be "deformed" to fit the needs of an actual problem.

The main sources of specific results are quoted in the text. Comments on history and additional references appear at the end of each chapter.

1.6.2 Notation and acronyms

A function $f : \mathbb{R}^n \to \mathbb{R}^q$ is C^k if its partial derivatives exist and are continuous up to order k, $1 \leq k < \infty$. A C^0 function is continuous. A C^∞ function is *smooth*, that is, it has continuous partial derivatives of any order. The same notation is used for vector fields in $\mathbb{R}^n$. All the results are presented under the differentiability assumptions which lead to the shortest and clearest proofs.

This book does not require the formalism of differential geometry and employs Lie derivatives only for notational convenience. If $f : \mathbb{R}^n \to \mathbb{R}^n$ is a vector field and $h : \mathbb{R}^n \to \mathbb{R}$ is a scalar function, the notation $L_f h$ is used for $\frac{\partial f}{\partial x} h(x)$. It is recursively extended to

$$L_f^k h(x) = L_f(L_f^{k-1} h(x)) = \frac{\partial f}{\partial x} L_f^{k-1} h(x)$$

A C^0 function $\gamma : \mathbb{R}^+ \to \mathbb{R}^+$ is said to belong to class $\mathcal{K}$, in short $\gamma \in \mathcal{K}$, if it is strictly increasing and $\gamma(0) = 0$. It is said to belong to class $\mathcal{K}_\infty$ if, in addition, $\gamma(r) \to \infty$ as $r \to \infty$.

Table 1.1: List of acronyms.

GS	global stability	CLF	control Lyapunov function
GAS	global asymptotic stability	ZSD	zero-state detectability
LES	local exponential stability	ZSO	zero-state observability
OFP	output feedback passivity	SISO	single-input single-output
IFP	input feedforward passivity	MIMO	multi-input multi-output

A C^0 function $\beta : \mathbb{R}^+ \times \mathbb{R}^+ \to \mathbb{R}^+$ is said to belong to class $\mathcal{KL}$ if for each fixed s the function $\beta(\cdot, s)$ belongs to class $\mathcal{K}$, and for each fixed r, the function $\beta(r, \cdot)$ is decreasing and $\beta(r, s) \to 0$ as $s \to \infty$.

For the reader's convenience, Table 1.1 contains a list of acronyms used throughout the book.

Chapter 2

Passivity Concepts as Design Tools

Only a few system theory concepts can match *passivity* in its physical and intuitive appeal. This explains the longevity of the passivity concept from the time of its first appearance some 60 years ago, to its current use as a tool for nonlinear feedback design. The pioneering results of Lurie and Popov, summarized in the monographs by Aizerman and Gantmacher [3], and Popov [88], were extended by Yakubovich [121], Kalman [51], Zames [123], Willems [120], and Hill and Moylan [37], among others. The first three sections of this chapter are based on these references from which we extract, and at times reformulate, the most important concepts and system properties to be used in the rest of the book.

We begin by defining and illustrating the concepts of storage function, supply rate, dissipativity and passivity in Section 2.1. The most useful aspect of these concepts, discussed in Section 2.2, is that they reveal the properties of parallel and feedback interconnections in which excess of passivity in one subsystem can compensate for the shortage in the other.

After these preparatory sections, we proceed to establish, in Section 2.3, the relationship between different forms of passivity and stability. Particularly important are the conditions for stability of feedback interconnections. In Section 2.4, we present a characterization of systems which can be rendered passive by feedback. The concept of *feedback passive* systems has evolved from recent work of Kokotović and Sussmann [59], and Byrnes, Isidori, and Willems [15]. It is one of the main tools for our cascade and passivation designs.

2.1 Dissipativity and Passivity

2.1.1 Classes of systems

Although the passivity concepts apply to wider classes of systems, we restrict our attention to dynamical systems modeled by ordinary differential equations with an input vector u and an output vector y:

$$(H) \quad \left\{ \begin{array}{ll} \dot{x} = f(x, u), & x \in I\!\!R^n \\ y = h(x, u), & u, y \in I\!\!R^m \end{array} \right. \tag{2.1.1}$$

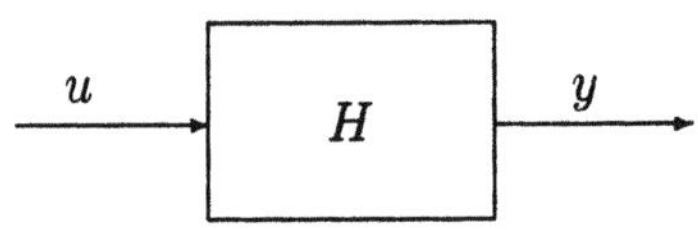

Figure 2.1: Input-output representation of (2.1.1).

We will be concerned with the case when the state $x(t)$, as a function of time, is uniquely determined by its initial value $x(0)$ and the input function $u(t)$. We assume that $u : I\!\!R^+ \to I\!\!R^m$ belongs to an input set U of functions which are bounded on all bounded subintervals of $I\!\!R^+$. In feedback designs u becomes a function of x, so the assumption $u \in U$ cannot be a priori verified. The satisfaction of this assumption for initial conditions in the region of interest will have to be a posteriori guaranteed by the design.

Another restriction in this chapter is that the system (2.1.1) is "square," that is, its input and output have the same dimension m. Finally, an assumption made for convenience is that the system (2.1.1) has an equilibrium at the origin, that is, $f(0,0) = 0$, and $h(0,0) = 0$.

We will find it helpful to visualize the system (2.1.1) as the input-output block diagram in Figure 2.1. In such block diagrams the dependence on the initial state $x(0)$ will not be explicitly stressed, but must not be overlooked.

The system description (2.1.1) includes as special cases the following three classes of systems:

- Nonlinear systems affine in the input:

$$\begin{array}{l} \dot{x} = f(x) + g(x)u \\ y = h(x) + j(x)u \end{array} \tag{2.1.2}$$

- Static nonlinearity:

$$y = \varphi(u) \tag{2.1.3}$$

- Linear systems:

$$\begin{aligned} \dot{x} &= Ax + Bu \\ y &= Cx + Du \end{aligned} \tag{2.1.4}$$

For static nonlinearity $y = \varphi(u)$, the state space is void. In the case of linear systems, we will let the system H be represented by its transfer function $H(s) := D + C(sI - A)^{-1}B$ where $s = \sigma + j\omega$ is the complex variable.

2.1.2 Basic concepts

For an easy understanding of the concepts of dissipativity and passivity it is convenient to imagine that H is a physical system with the property that its energy can be increased only through the supply from an external source. From an abundance of real-life examples let us think of baking a potato in a microwave oven. As long as the potato is not allowed to burn, its energy can increase only as supplied by the oven. A similar observation can be made about an RLC-circuit connected to an external battery. The definitions given below are abstract generalizations of such physical properties.

Definition 2.1 *(Dissipativity)*
Assume that associated with the system H is a function $w : I\!R^m \times I\!R^m \to I\!R$, called the *supply rate*, which is locally integrable for every $u \in U$, that is, it satisfies $\int_{t_0}^{t_1} |w(u(t), y(t))|\, dt < \infty$ for all $t_0 \leq t_1$. Let X be a connected subset of $I\!R^n$ containing the origin. We say that the system H is *dissipative* in X with the supply rate $w(u, y)$ if there exists a function $S(x)$, $S(0) = 0$, such that for all $x \in X$

$$S(x) \geq 0 \text{ and } S(x(T)) - S(x(0)) \leq \int_0^T w(u(t), y(t))\, dt \tag{2.1.5}$$

for all $u \in U$ and all $T \geq 0$ such that $x(t) \in X$ for all $t \in [0, T]$. The function $S(x)$ is then called a *storage function.* □

Definition 2.2 *(Passivity)*
System H is said to be *passive* if it is dissipative with supply rate $w(u, y) = u^T y$. □

We see that passivity is dissipativity with bilinear supply rate. In our circuit example, the storage function S is the energy, w is the input power, and $\int_0^T w(u(t), y(t))\, dt$ is the energy supplied to the system from the external sources. The system is dissipative if the increase in its energy during the interval $(0, T)$ is not bigger than the energy supplied to the system during that interval.

If the storage function $S(x)$ is differentiable, we can write (2.1.5) as

$$\dot{S}(x(t)) \le w(u(t), y(t)) \tag{2.1.6}$$

Again, the interpretation is that the rate of increase of energy is not bigger than the input power.

If H is dissipative, we can associate with it a function $S_a(x)$, called the *available storage*, defined as

$$S_a(x) = \sup_{u, T \ge 0} \left\{ - \int_0^T w(u(t), y(t))\, dt \,\middle|\, x(0) = x \text{ and } \forall t \in [0, T] : \; x(t) \in X \right\} \tag{2.1.7}$$

An interpretation of the available storage $S_a(x)$ is that it is the largest amount of energy which can be extracted from the system given the initial condition $x(0) = x$.

The available storage $S_a(x)$ is itself a storage function and any other storage function must satisfy $S(x) \ge S_a(x)$. This can be seen by rewriting (2.1.5) as

$$S(x(0)) \ge S(x(0)) - S(x(T)) \ge - \int_0^T w(u(t), y(t))\, dt,$$

which yields

$$S(x(0)) \ge \sup_{u, T \ge 0} \left\{ - \int_0^T w(u(t), y(t))\, dt \right\} = S_a(x(0))$$

The properties of $S_a(x)$ are summarized in the following theorem due to Willems [120].

Theorem 2.3 *(Available Storage)*
The system H is dissipative in X with the supply rate $w(u, y)$ if and only if $S_a(x)$ is defined for all $x \in X$. Moreover, $S_a(x)$ is itself a storage function and, if $S(x)$ is another storage function with the same supply rate $w(u, y)$, then $S(x) \ge S_a(x)$. □

For linear passive systems, the available storage function is further characterized in the following theorem by Willems [120] which we quote without proof.

Theorem 2.4 *(Quadratic storage function for linear systems)*
If H is linear and passive, then the available storage function is quadratic $S_a(x) = x^T P x$. The matrix P is the limit $P = \lim_{\epsilon \to 0} P_\epsilon$ of the real symmetric positive semidefinite solution $P_\epsilon \geq 0$ of the Ricatti equation

$$P_\epsilon A + A^T P_\epsilon + (P_\epsilon B - C^T)(D + D^T + \epsilon I)^{-1}(B^T P_\epsilon - C^T) = 0$$

□

The above concepts are now illustrated with several examples.

Example 2.5 *(Integrator as a passive system)*
An integrator is the simplest storage element:

$$\begin{aligned} \dot{x} &= u \\ y &= x \end{aligned}$$

This system is passive with $S(x) = \frac{1}{2}x^2$ as a storage function because $\dot{S} = uy$. Its available storage S_a can be obtained from the following inequalities:

$$\frac{1}{2}x_0^2 = S(x_0) \geq S_a(x_0) = \sup_{u,T} \left\{ -\int_0^T yu \, dt \right\} \geq \int_0^\infty y^2 \, dt = x_0^2 \int_0^\infty e^{-2t} \, dt = \frac{1}{2}x_0^2$$

The second inequality sign is obtained by choosing $u = -y$ and $T = \infty$. Note that, for the choice $u = -y$, the assumption $u \in U$ is a posteriori verified by the fact that with this choice $u(t) = -y(t)$ is a decaying exponential. □

In most of our examples, the domain X of dissipativity will be the entire space $I\!R^n$. However, for nonlinear systems, this is not always the case.

Example 2.6 *(Local passivity)*
The system

$$\begin{aligned} \dot{x} &= (x^3 - kx) + u \\ y &= x \end{aligned}$$

is passive in the interval $X = [-\sqrt{k}, \sqrt{k}] \subset I\!R$ with $S(x) = \frac{1}{2}x^2$ as a storage function because $\dot{S} = x^2(x^2 - k) + uy \leq uy$ for all x in X. However, we can verify that it is not passive in any larger subset of $I\!R^n$: for any constant $\bar{k}$, the input $u = -(\bar{k}^3 - k\bar{k})$ and the initial condition $x = \bar{k}$ yield the constant solution $x(t) \equiv \bar{k}$. If the system is passive, then along this solution, we must have

$$0 = S(x(T)) - S(x(0)) \leq \int_0^T u(t)y(t) \, dt = -\bar{k}^2(\bar{k}^2 - k)T$$

This is violated for $\bar{k} \notin [-\sqrt{k}, \sqrt{k}]$, and hence, the system is not passive outside X. □

Example 2.7 *(RLC circuit)*
In the absence of a good model of a potato as a dynamical system, our next example is a circuit consisting of an inductor L in parallel with a series connection of a resistor R and a capacitor C. External voltage v applied to the inductor is the input, and the total current i is the output. Considering inductor current i_L and capacitor voltage v_C as the state variables, the circuit equations written in the form (2.1.1) are:

$$\begin{aligned} \dot{i}_L &= \frac{1}{L}v \\ \dot{v}_C &= \frac{1}{RC}(v - v_C) \\ i &= i_L + \frac{1}{R}(v - v_C) \end{aligned} \tag{2.1.8}$$

The energy stored in the inductor is $\frac{1}{2}Li_L^2$ and the energy stored in the capacitor is $\frac{1}{2}Cv_C^2$. Therefore, the total energy in the circuit is

$$E = \frac{1}{2}Li_L^2 + \frac{1}{2}Cv_C^2$$

and its rate of change is

$$\dot{E} = vi - \frac{1}{R}(v - v_C)^2 \leq vi$$

Thus the system (2.1.8) is dissipative, and the bilinear form of the supply rate $w(v, i) = vi$ means that it is passive. Physically, the supply rate vi is the power supplied by the voltage source. It is of interest to observe that the system obtained by considering i as the input and v as the output is also passive with respect to the same supply rate.

□

Example 2.8 *(Mass-spring-damper system)*
A system made of passive elements may not be passive for some input-output pairs, as illustrated by a mass-spring-damper system, with an external force acting on the mass considered as the input u. The state equations for the mass position x and velocity v are

$$\begin{aligned} \dot{x} &= v \\ \dot{v} &= -\tfrac{k}{m}x - \tfrac{b}{m}v + \tfrac{1}{m}u \end{aligned}$$

where $k > 0$ is the spring constant, $m > 0$ is the mass, and $b > 0$ is the viscous friction coefficient. The energy is

$$E = \frac{1}{2}mv^2 + \frac{1}{2}kx^2$$

and its rate of change is

$$\dot{E} = uv - bv^2 \leq uv \tag{2.1.9}$$

Thus, when the velocity is considered as the output, the mass-spring-damper system is passive. Its storage function is the energy E and the supply rate is the input power uv. However, the same system is *not passive* if the position x is taken to be the output $y = x$, so that the transfer function is

$$H(s) = \frac{1}{ms^2 + bs + k}$$

The output $y(t)$ for the input $u(t) = \sin(\omega t)$ with $x(0) = 0$, $v(0) = 0$, is $y(t) = A(\omega)\sin(\omega t + \phi(\omega))$ where $A(\omega) > 0$ is the magnitude and $\phi(\omega)$ the phase of $H(j\omega)$. Passivity of the system would imply

$$S(x(\frac{2\pi}{\omega})) - S(0) \leq \int_0^{\frac{2\pi}{\omega}} A(\omega)\sin(\omega t + \phi(\omega))\sin(\omega t)\, dt$$

for some storage function $S(x)$. Because $S(0) = 0$ and $S(x(T)) > 0$, this would require that

$$0 \leq \frac{2\pi}{\omega} A(\omega)\cos(\phi(\omega)) \tag{2.1.10}$$

However, for ω sufficiently large, $\phi(\omega)$ drops below -90° so that $\cos(\phi(\omega)) < 0$. This contradicts (2.1.10), which shows that the mass-spring-damper system with the mass position as the output and the force acting on the mass as the input, cannot be passive. As we shall see, the same conclusion is immediate from the fact that the relative degree of $H(s)$ is larger than one.

□

2.2 Interconnections of Passive Systems

2.2.1 Parallel and feedback interconnections

Our design methods will exploit the structure of systems formed as interconnections of subsystems with certain passivity properties. The two basic structures, feedback and parallel, are presented in Figure 2.2.

Assuming that both H_1 and H_2 are in the form (2.1.1), we first must make sure that the interconnection is also in the form (2.1.1) for which well-posedness can be deduced from the standard results on the existence of solutions of ordinary differential equations. This is obviously true for the parallel interconnection, which constitutes the new system

$$\begin{aligned} \dot{x}_1 &= f_1(x_1, u) \\ \dot{x}_2 &= f_2(x_2, u) \\ y &= h_1(x_1, u) + h_2(x_2, u) \end{aligned}$$

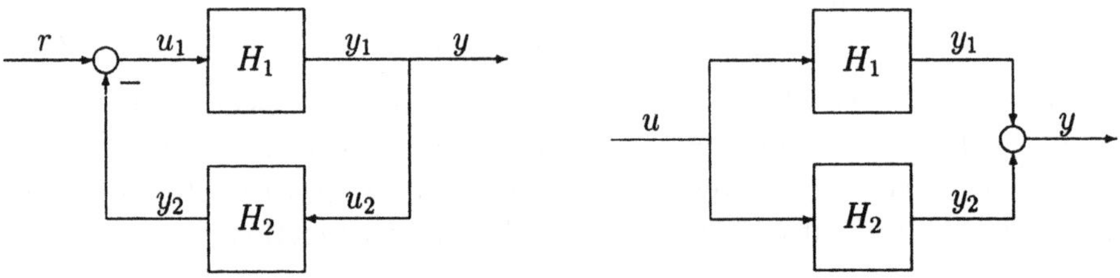

Figure 2.2: Feedback and parallel interconnections.

However, the feedback interconnection may not be in the form (2.1.1), and may fail to have a well-defined solution even locally if h_1 depends on u_1 and h_2 depends on u_2. A static feedback loop created by the two throughputs may obliterate the dynamics of H_1 and H_2 so that their differential equations cannot be satisfied, except, possibly, for some special initial conditions.

Example 2.9 *(Ill-posedness of feedback interconnections)*
It is easy to see that with $d_1 = 0$, $d_2 \neq 0$ the feedback interconnection of

$$H_1 : \begin{array}{rcl} \dot{x}_1 & = & -x_1 + d_1 u_1 \\ y_1 & = & x_1 + d_1 u_1 \end{array} \qquad H_2 : \; y_2 = d_2 u_2$$

represents a system of the form (2.1.1); hence, it is well-posed.

However, if $d_1 = -1$, $d_2 = 1$, the feedback interconnection is ill-posed because of the static loop which imposes the constraint $x_1(t) \equiv r(t)$ and violates the state equation of H_1. This can be readily seen from the fact that the interconnection conditions

$$u_1(t) = -y_2(t) + r(t), \qquad u_2(t) = y_1(t)$$

along with the output functions

$$y_1(t) = x_1(t) - u_1(t), \qquad u_2(t) = y_2(t)$$

result in $y_1 = y_2$. Hence, $x_1(t) \equiv r(t)$, which leaves no room for the dynamics of $\dot{x}_1 = -x_1 + u_1$, except in the special case when $x_1(0) = r(0)$.

□

To avoid ill-posedness of the feedback interconnection, it is sufficient to require that at least one throughput be zero. Thus, when $\frac{\partial h_1}{\partial u_1} \equiv 0$, that is when $y_1 = h_1(x_1)$, the feedback interconnection defines a new system

$$\begin{aligned} \dot{x}_1 &= f_1(x_1, -h_2(x_2, h_1(x_1)) + r) =: \tilde{f}(x_1, x_2, r) \\ \dot{x}_2 &= f_2(x_2, h(x_1)) \\ y &= h_1(x_1) \end{aligned}$$

which is in the form (2.1.1), and hence, well-posed. Unless explicitly stated otherwise, all feedback interconnections in this book will satisfy

$$\text{either } \frac{\partial h_1}{\partial u_1} \equiv 0, \quad \text{or } \frac{\partial h_2}{\partial u_2} \equiv 0. \tag{2.2.1}$$

We now present interconnection passivity properties which will be frequently used in this book.

Theorem 2.10 *(Interconnections of passive systems)*
Suppose that H_1 and H_2 are passive. Then the two systems, one obtained by the parallel interconnection, and the other obtained by the feedback interconnection, are both passive.

Proof: By passivity of H_1 and H_2, there exist $S_1(x_1)$ and $S_2(x_2)$ such that $S_i(x_i(T)) - S_i(x_i(0)) \leq \int_0^T u_i^T y_i \, dt$, $i = 1, 2$. Define $x := (x_1, x_2)$ and $S(x) = S_1(x_1) + S_2(x_2)$ and note that $S(x)$ is positive semidefinite.

For the parallel interconnection the output is $y = y_1 + y_2$, so that

$$S(x(T)) - S(x(0)) \leq \int_0^T (u^T y_1 + u^T y_2) \, dt = \int_0^T u^T y \, dt$$

This proves that the parallel interconnection is passive.

For the feedback interconnection we have

$$S(x(T)) - S(x(0)) \leq \int_0^T (u_1^T y_1 + u_2^T y_2) \, dt$$

Substituting $u_2 = y_1$ and $u_1 = r - y_2$ we obtain

$$S(x(T)) - S(x(0)) \leq \int_0^T r^T y_1 \, dt$$

which proves that the feedback interconnection is passive.

□

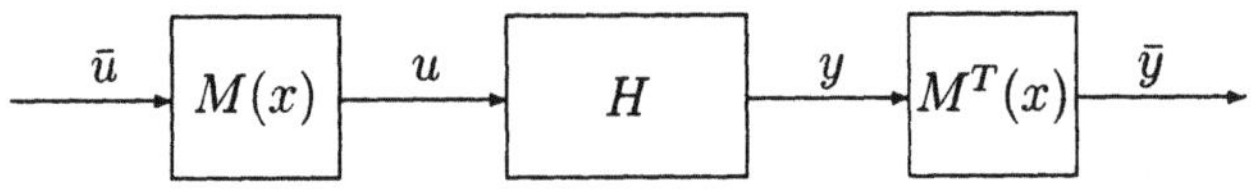

Figure 2.3: Pre- and post-multiplication by a state-dependent matrix.

A transformation of the input and output, which often appears in interconnections, is depicted in Figure 2.3. For a matrix $M(x)$ depending on the state of the system, the new input and output satisfy $u = M(x)\bar{u}$ and $\bar{y} = M^T(x)y$. It is not difficult to see that, if H is passive with $S(x)$, then the transformed system is also passive with the same storage function:

$$S(x(T)) - S(x(0)) \leq \int_0^T u^T y \, dt = \int_0^T \bar{u}^T M^T(x) y \, dt = \int_0^T \bar{u}^T \bar{y} \, dt$$

The passivity property of H remains the same even if the matrix M is a function of the state of the other system in the interconnection. We will encounter such a situation in Chapter 4.

Proposition 2.11 *(Interconnections with pre- and post-multiplication)*
Let M be a matrix which depends on the states of the systems H_1 and H_2. Then the parallel and feedback interconnections of H_1 and H_2 remain passive if either one or both of the systems H_1 and H_2 are pre-multiplied by $M(x_1, x_2)$ and post-multiplied by $M^T(x_1, x_2)$.

□

2.2.2 Excess and shortage of passivity

What can happen when one of the systems in the interconnection is not passive? Can an "excess of passivity" of the other system assure that the interconnection is passive? To answer these questions let us select a system which clearly is not passive. The simplest system of this kind is the constant negative gain $y = -ku$, where $k > 0$. This system is static, its state space is void, and the only possible storage function is $S = 0$. With $yu = -ku^2$ as the supply rate, the integral in (2.1.5) is negative, which violates the definition of passivity. An analogous multivariable system is the matrix gain $-kI$ where I is the $m \times m$ identity and $k > 0$ is a scalar.

Let H be passive and consider its parallel interconnection with $-kI$. For this interconnection to be passive, its supply rate $u^T\bar{y}$ must satisfy (2.1.5). Since $\bar{y} = y - ku$ we have

$$u^Ty = u^T\bar{y} + ku^Tu$$

It follows that the parallel interconnection of H with $-kI$ is passive if H is dissipative with respect to the supply rate $w(u,y) = u^Ty - \nu u^Tu$, with $\nu \geq k$. This is verified by rewriting the dissipation inequality for H as

$$S(x(T)) - S(x(0)) \leq \int_0^T u^T\bar{y}\,dt - (\nu - k)\int_0^T u^Tu\,dt$$

Thus, if $\nu \geq k$ then the interconnection is passive with $S(x)$ as the storage function.

The analogous situation arises in the feedback interconnection of H with $-kI$. The input to the system H is $u = r + ky$. The interconnected system is passive if H is dissipative with respect to the supply rate

$$w(u,y) = u^Ty - \rho y^Ty \tag{2.2.2}$$

with $\rho \geq k$, because

$$S(x(T)) - S(x(0)) \leq \int_0^T y^Tr\,dt - (\rho - k)\int_0^T y^Ty\,dt \leq \int_0^T y^Tr\,dt$$

In each of the two cases a particular "excess of passivity" of H has compensated for the lack of passivity of $-kI$ and guaranteed the passivity of the interconnection. The opposite situation arises when the system H is not passive, but has a certain dissipativity property; for example, if the constant ρ in the supply rate (2.2.2) is negative. The feedback interconnection of H with the matrix gain kI may still be passive if $k + \rho > 0$ because then

$$S(x(T)) - S(x(0)) \leq \int_0^T y^Tr\,dt - (\rho + k)\int_0^T y^Ty\,dt \leq \int_0^T y^Tr\,dt$$

In this case ρ being negative indicates a "shortage of passivity" which can be compensated by output feedback $u = -kI + r$. Similarly, a "shortage of passivity" of H, which is dissipative with the supply rate $w(u,y) = u^Ty - \nu u^Tu$, $\nu < 0$, can be compensated by feeding forward the input: $\bar{y} = y + ku$, $k + \nu > 0$.

The possibility of achieving passivity of interconnections which combines systems with "excess" and "shortage" of passivity motivates us to introduce the following definition.

Definition 2.12 *(Excess/Shortage of Passivity)*
System H is said to be

- Output Feedback Passive (OFP) if it is dissipative with respect to $w(u,y) = u^T y - \rho\, y^T y$ for some $\rho \in I\!R$.
- Input Feedforward Passive (IFP) if it is dissipative with respect to $w(u,y) = u^T y - \nu\, u^T u$ for some $\nu \in I\!R$. □

We quantify the excess and shortage properties with the notation IFP(ν) and OFP(ρ). According to our convention, positive sign of ρ and ν means that the system has an excess of passivity. In this case, the concepts of IFP and OFP coincide with Input Strict Passivity and Output Strict Passivity introduced by Hill and Moylan [38]. Conversely, negative sign of ρ and ν means that the system has a shortage of passivity.

Another common concept in passivity theory is *strict passivity* defined in [18] by requiring that

$$\int_0^T u^T y\, dt \geq \nu \int_0^T u^T u\, dt + \beta$$

for some $\nu > 0$ and $\beta \in \mathbb{R}$. This concept coincides with IFP with positive ν.

Example 2.13 *(Excess of passivity - feedforward)*
Consider a system represented by the transfer function $H(s) = \frac{s+1}{s}$. Its minimal realization in Figure 2.4 consists of an integrator in parallel with a positive unity gain. This system is IFP(1) because, when connected in parallel with

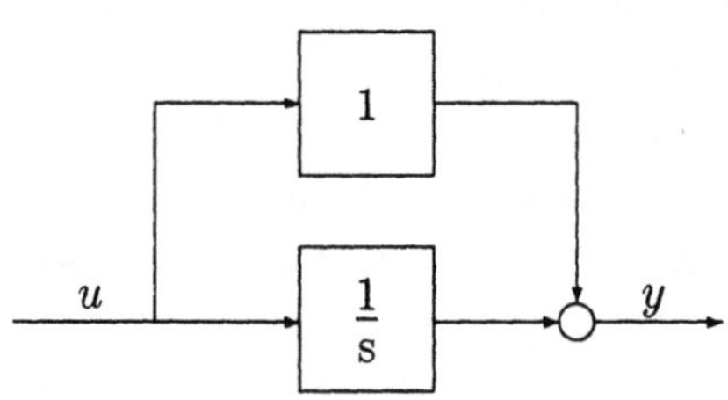

Figure 2.4: An illustration that $\frac{s+1}{s}$ is IFP(1).

a negative unity gain, it becomes an integrator which is passive. The "excess of passivity" is provided by the feedforward path with positive gain. To show this analytically we use

$$\begin{aligned} \dot{x} &= u \\ y &= x + u \end{aligned}$$

and the storage function $S(x) = \frac{1}{2}x^2$. Then $\dot{S} = xu = uy - u^2$ proves the IFP(1) property.

□

Example 2.14 *(Excess of passivity - feedback)*
The system

$$\dot{x} = -x + u$$
$$y = \arctan(x)$$

with the storage function $S(x) = \int_0^x \arctan(z)dz$ is OFP(1) because it is dissipative with the supply rate $uy - y^2$. This is clear from $\dot{S} = \arctan x(-x+u) \leq -y^2 + yu$. Let us interpret this conclusion with the help of the block diagram in Figure 2.5.

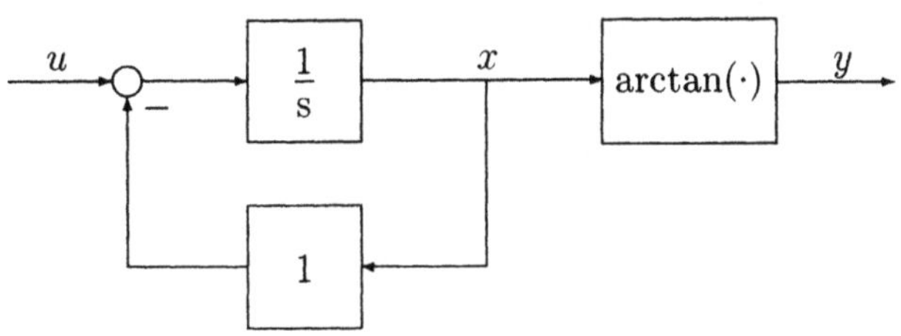

Figure 2.5: A system which is OFP(1) because $|y| \leq |x|$.

The excess of passivity in this case is provided by the negative unity gain feedback around the integrator. A positive unity gain feedback from y does not destroy passivity because $|y| \leq |x|$. □

Example 2.15 *(Sector nonlinearity)*
Consider a static nonlinearity $y = \varphi(u)$, where $\varphi(\cdot)$ in Figure 2.6 belongs to a sector $[\alpha, \beta]$:

$$\alpha u^2 \leq u\varphi(u) \leq \beta u^2, \qquad 0 \leq \alpha \leq \beta$$

If the inequalities are strict, we use the notation (α, β).

The state space of this system is void and the only choice for the storage function is $S \equiv 0$. By manipulating the bounds on φ we obtain

$$uy - \alpha u^2 \geq 0 \quad \text{and} \quad uy - \tfrac{1}{\beta}y^2 \geq 0$$

Thus, the sector nonlinearity $y = \varphi(u)$ is IFP(α) as well as OFP($\frac{1}{\beta}$). □

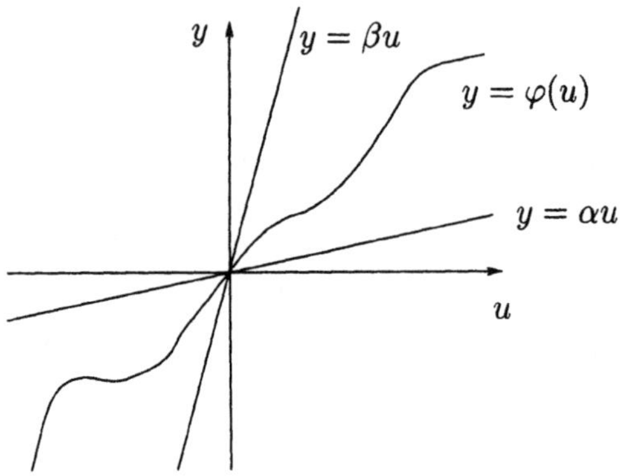

Figure 2.6: Sector nonlinearity $\varphi(\cdot)$.

Example 2.16 *(Shortage of passivity)*
The system

$$\dot{x} = x + u$$
$$y = x$$

is OFP(-1) with the storage function $S(x) = \frac{1}{2}x^2$ because $\dot{S} = y^2 + uy$. Clearly, $k = 1$ is exactly the amount of output feedback required to compensate for the "shortage of passivity," that is to make the system passive. □

The following scaling property of OFP and IFP systems will be useful in later chapters.

Proposition 2.17 *(IFP/OFP Scaling)*
For the systems H and αH, where α is a constant, the following statements are true:

(i) If H is OFP(ρ) then αH is OFP($\frac{1}{\alpha}\rho$).

(ii) If H is IFP(ν) then αH is IFP($\alpha\nu$).

Proof: The output y_α of the system αH is just $y_\alpha = \alpha y$ where y is the output of H. Define a storage function for αH by $S_\alpha = \alpha S$. Then (i) follows from

$$\begin{aligned} S_\alpha(x(T)) - S_\alpha(x(0)) &= \alpha(S(x(T)) - S(x(0))) \leq \alpha \int_0^T (u^T y - \rho\, y^T y)\, dt \\ &= \int_0^T (u^T y_\alpha - \frac{1}{\alpha}\rho\, y_\alpha^T y_\alpha)\, dt \end{aligned}$$

The proof of (ii) is similar.

□

An excess/shortage of passivity in Definition 2.12 is quantified by *linear* feedback or feedforward terms, ρy or νu. For nonlinear systems such properties may hold only locally, that is in some neighborhood of $x = 0$. For global properties of nonlinear systems a possible extension of the excess/shortage definitions would be to replace ρy and νu by

$$\begin{aligned} \rho(y) &= [\rho_1(y_1), \ldots, \rho_m(y_m)]^T, \\ \nu(u) &= [\nu_1(u_1), \ldots, \nu_m(u_m)]^T, \end{aligned} \tag{2.2.3}$$

where $\rho_i(y_i)$, $\nu_i(u_i)$ are in the sector $(0, +\infty)$ or $(-\infty, 0)$, $i = 1, \ldots, m$. Instead of extended definitions, we will use $\rho(y)$ and $\nu(u)$ as needed in specific problems like in the following example.

Example 2.18 *(Nonlinear excess/shortage of passivity)*
For the system

$$\begin{aligned} \dot{x} &= x^3 + u \\ y &= x \end{aligned}$$

a linear feedback $u = -\rho y + \bar{u}$ cannot achieve passivity outside the set $[-\sqrt{\rho}, \sqrt{\rho}]$. It was indeed shown in Example 2.6 that the system

$$\begin{aligned} \dot{x} &= x^3 - \rho x + \bar{u} \\ y &= x \end{aligned}$$

is passive only in the interval $[-\sqrt{\rho}, \sqrt{\rho}]$. However, the nonlinear output feedback $\rho(y) = -ky^3$ achieves passivity for all $k \geq 1$, because the system

$$\begin{aligned} \dot{x} &= (1-k)x^3 + \bar{u}, \quad k \geq 1 \\ y &= x \end{aligned}$$

has a storage function $S(x) = \frac{1}{2}x^2$ which satisfies $\dot{S} \leq \bar{u}y$. □

We conclude our discussion of passivity concepts with an illustration of their usefulness in feedback stabilization. As will be shown in the next section, passivity implies stability, and one way to stabilize a plant is to achieve passivity of the feedback interconnection of the plant-controller feedback loop. In Figure 2.7 the controller is H_1 and the plant is H_2. If the plant is unstable and, therefore, not passive, but known to be OFP($-\rho$) with $\rho > 0$, this shortage of passivity can be compensated for by a negative ρ-feedback around H_2 which makes this feedback subsystem passive. To preserve the overall feedback interconnection unchanged, a feedforward $-\rho I$ is connected in parallel

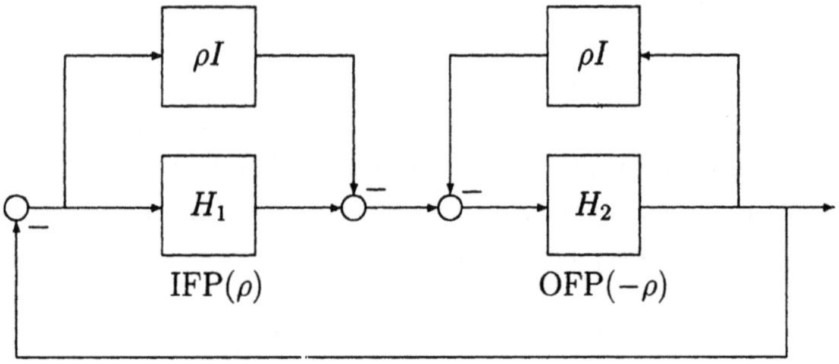

Figure 2.7: Feedback interconnection of the controller H_1 and the plant H_2.

with the controller H_1. If the controller H_1 is IFP(ρ), that is if it has the excess of passivity ρ, then its parallel connection with $-\rho I$ is passive. Thus, a shortage of passivity (and lack of stability) of the plant H_2 has been compensated for by the excess of passivity of the controller H_1. The net effect is the same as in a feedback interconnection of two passive systems.

2.3 Lyapunov Stability and Passivity

2.3.1 Stability and convergence theorems

Lyapunov stability and *input-output stability* are widely used in control theory. This book mostly employs Lyapunov stability, which we now briefly review. To begin with, we remind the reader that Lyapunov stability and asymptotic stability are properties not of a dynamical system as a whole, but rather of its individual solutions. Consider the time-invariant system

$$\dot{x} = f(x) \tag{2.3.1}$$

where $x \in I\!R^n$ and $f : I\!R^n \to I\!R^n$ is locally Lipschitz continuous. The solution of (2.3.1) which starts from x_0 at time $t_0 \in I\!R$ is denoted as $x(t; x_0, t_0)$, so that $x(t_0; x_0, t_0) = x_0$. Because the solutions of (2.3.1) are invariant under a translation of t_0, that is, $x(t + T; x_0, t_0 + T) = x(t; x_0, t_0)$, the stability properties of $x(t; x_0, t_0)$ are *uniform*, that is they do not depend on t_0. Without loss of generality, we assume $t_0 = 0$ and write $x(t; x_0)$ instead of $x(t; x_0, 0)$.

Lyapunov stability is a continuity property of $x(t; x_0, t_0)$ with respect to x_0. If the initial state x_0 is perturbed to $\tilde{x}_0$, then, for stability, the perturbed solution $x(t;\tilde{x}_0)$ is required to stay close to $x(t;x_0)$ for all $t \geq 0$. In addition, for asymptotic stability, the error $x(t;\tilde{x}_0) - x(t;x_0)$ is required to vanish as $t \to \infty$. So, the solution $x(t;x_0)$ of (2.3.1) is

- *bounded*, if there exists a constant $K(x_0)$ such that

$$\|x(t;x_0)\| \leq K(x_0),\ \forall t \geq 0;$$

- *stable*, if for each $\epsilon > 0$ there exists $\delta(\epsilon) > 0$ such that

$$\|\tilde{x}_0 - x_0\| < \delta \Rightarrow \|x(t;\tilde{x}_0) - x(t;x_0)\| < \epsilon,\ \forall t \geq 0;$$

- *attractive*, if there exists an $r(x_0) > 0$ such that

$$\|\tilde{x}_0 - x_0\| < r(x_0) \Rightarrow \lim_{t\to\infty} \|x(t;\tilde{x}_0) - x(t;x_0)\| = 0;$$

- *asymptotically stable*, if it is stable and attractive;
- *unstable*, if it is not stable.

Some solutions of a given system may be stable and some unstable. In particular, (2.3.1) may have stable and unstable *equilibria*, that is, constant solutions $x(t;x_e) \equiv x_e$ satisfying $f(x_e) = 0$. The above definitions of stability properties of an equilibrium x_e involve only initial states *close* to x_e, that is they are *local*. If an equilibrium is attractive, then it has a *region of attraction* - a set Ω of initial states x_0 such that $x(t;x_0) \to x_e$ as $t \to \infty$ for all $x_0 \in \Omega$. Our attention will be focused on *global* stability properties (GS and GAS):

- x_e is GS – *globally stable* – if it is stable and if all the solutions of (2.3.1) are bounded.
- x_e is GAS – *globally asymptotically stable* – if it is asymptotically stable and its region of attraction is $\mathbb{R}^n$.

In certain situations we will need *exponential* stability for which we stress its local character:

- x_e is *locally exponentially stable (LES)*, if there exist positive constants α, γ and r such that

$$\|x_0 - x_e\| < r \Rightarrow \|x(t;x_0) - x_e\| \leq \gamma \exp(-\alpha t)\|x_0 - x_e\|$$

Any equilibrium under investigation can be translated to the origin by redefining the state x as $z = x - x_e$. For simplicity, we will assume that the translation has been performed, that is, $f(0) = 0$, and thus the equilibrium under investigation is $x_e = 0$. When, for brevity, we say that "the system (2.3.1) is GS or GAS", we mean that its equilibrium $x_e = 0$ is GS or GAS. While global asymptotic stability of $x_e = 0$ prevents the existence of other equilibria, the reader should keep in mind that it is not so with global stability. When we say that the *system* (2.23) is globally stable, we refer to global stability of $x_e = 0$.

Example 2.19 *(Global stability - several equilibria)*
The scalar system

$$\dot{x} = -x(x-1)(x-2)$$

has three equilibria: $x_e = 0, +1, +2$. The equilibria $x_e = 0$ and $x_e = 2$ are asymptotically stable, while $x_e = +1$ is unstable. Both $x_e = 0$ and $x_e = 2$ are globally stable. □

The direct method of Lyapunov aims at determining the stability properties of $x(t; x_0)$ from the properties of $f(x)$ and its relationship with a positive definite function $V(x)$. Global results are obtained if this function is *radially unbounded*: $V(x) \to \infty$ as $\|x\| \to \infty$. From among many classical stability tools we will mostly use those due to Barbashin, Krasovsky, LaSalle, and Yoshizawa [6, 63, 122], which, specialized for our needs, are now formulated as two theorems and one corollary:

Theorem 2.20 *(Stability)*
Let $x = 0$ be an equilibrium of (2.3.1) and suppose f is locally Lipschitz continuous. Let $V : \mathbb{R}^n \to \mathbb{R}^+$ be a C^1 positive definite and radially unbounded function $V(x)$ such that

$$\dot{V} = \frac{\partial V}{\partial x}(x) f(x) \leq 0, \qquad \forall x \in \mathbb{R}^n$$

Then $x = 0$ is globally stable (GS) and all solutions of (2.3.1) converge to the set E where $\dot{V}(x) \equiv 0$. If $\dot{V}$ is negative definite, then $x = 0$ is globally asymptotically stable (GAS). □

For a sharper characterization of convergence properties we employ the concept of *invariant sets*. A set M is called an invariant set of (2.3.1) if any solution $x(t)$ that belongs to M at some time t_1 belongs to M for all future and past time:

$$x(t_1) \in M \Rightarrow x(t) \in M, \quad \forall t \in \mathbb{R}$$

A set P is *positively invariant* if this is true for all future time only:

$$x(t_1) \in P \Rightarrow x(t) \in P, \ \ \forall t \geq t_1$$

An important result describing convergence to an invariant set is LaSalle's Invariance Principle.

Theorem 2.21 *(Invariance Principle: convergence)*
Let Ω be a positively invariant set of (2.3.1). Suppose that every solution starting in Ω converges to a set $E \subset \Omega$ and let M be the largest invariant set contained in E. Then, every bounded solution starting in Ω converges to M as $t \to \infty$. □

An application of the Invariance Principle is the following asymptotic stability condition.

Corollary 2.22 *(Asymptotic stability)*
Under the assumptions of Theorem 2.20, let $E = \{x \in I\!R^n \mid \dot{V}(x) = 0\}$. If no solution other than $x(t) \equiv 0$ can stay for all t in E, then the equilibrium $x = 0$ is globally asymptotically stable (GAS). □

Theorem 2.21 has been the most dependable work horse in the analysis of nonlinear time-invariant systems. While the main stability theorem (Theorem 2.20) establishes that the solutions are bounded and converge to the set E where $\dot{V} \equiv 0$, Theorem 2.21 sharpens this result by establishing the convergence to a subset of E. Thanks to its invariance, this subset can be found by examining only those solutions which, having started in E, remain in E for all t.

In control systems, such invariance and convergence results are made possible by system's observability properties. Typically, the convergence of the system output y to zero is established first, and then the next task is to investigate whether some (or all) of the states converge to zero. For this task we need to examine only the solutions satisfying $y(t) \equiv 0$. If it is known beforehand that $y(t) \equiv 0$ implies $x(t) \equiv 0$, then the asymptotic stability of $x = 0$ is established, as in Corollary 2.22. An example will help us to visualize the situation.

Example 2.23 *(Invariant set and observability)*
Consider the system

$$\begin{aligned} \dot{x}_1 &= x_2 \\ \dot{x}_2 &= -x_1 - \phi(x_2) \end{aligned} \tag{2.3.2}$$

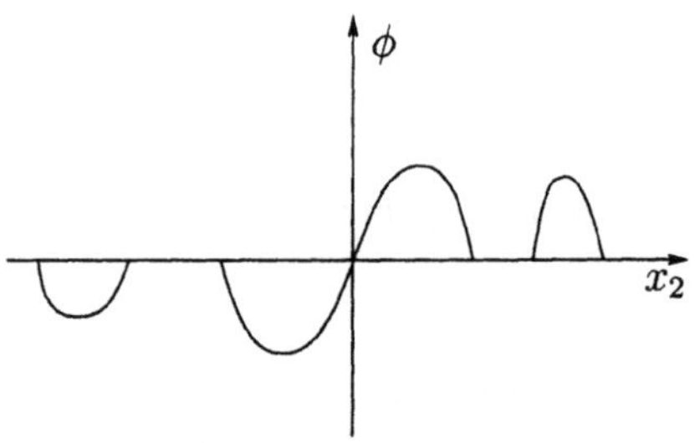

Figure 2.8: The function $\phi(\cdot)$.

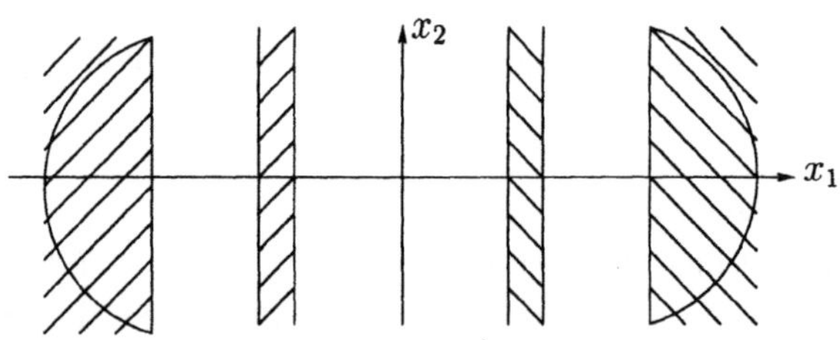

Figure 2.9: The shaded strips defined by $\phi(x_2) = 0$.

where $\phi(s)$ is shown in Figure 2.8. Using the simplest Lyapunov function $V = \frac{1}{2}x_1^2 + \frac{1}{2}x_2^2$, we obtain

$$\dot{V} = -x_2\phi(x_2) \leq 0 \tag{2.3.3}$$

By the main stability theorem, the solutions are bounded and converge to the set E in in Figure 2.9 which consists of the axis $x_2 = 0$ and the vertical shaded strips.

Let us treat $\sqrt{x_2\phi(x_2)}$ as the output y so that $\dot{V} = -y^2$. Corollary 2.22 instructs us to investigate only the solutions for which $y(t) = 0$ for all t. It is not hard to see that this excludes all the shaded strips in Figure 2.9 because on them the system behaves like a harmonic oscillator and its solution leaves every strip in finite time. In other words, none of these strips contains an invariant set. We are left with $x_2(t) \equiv 0$, which forces $x_1(t) \equiv 0$, and proves asymptotic stability of $(x_1, x_2) = (0, 0)$.

The observability interpretation for the system (2.3.2) with the output $y = \sqrt{x_2\phi(x_2)}$ is that $y(t) \equiv 0$ implies $x(t) \equiv 0$. This is the "zero state observability" property defined in the next section.

□

2.3.2 Stability with semidefinite Lyapunov functions

We now discuss how to prove stability with a Lyapunov function which is positive *semidefinite*, rather than positive *definite*. For this we need the notion of *conditional stability*. The stability properties of a solution $x(t; x_0)$, $x_0 \in Z \subset \mathbb{R}^n$, are said to be conditional to Z if the perturbed initial condition $\tilde{x}_0$ is also restricted to Z. So, the solution $x(t; x_0)$ of (2.3.1) is

- stable *conditionally to* Z, if $x_0 \in Z$ and for each $\epsilon > 0$ there exists $\delta(\epsilon) > 0$ such that

$$\|\tilde{x}_0 - x_0\| < \delta \text{ and } \tilde{x}_0 \in Z \Rightarrow \|x(t; \tilde{x}_0) - x(t; x_0)\| < \epsilon, \ \forall t \geq 0;$$

- attractive *conditionally to* Z, if $x_0 \in Z$ and there exists an $r(x_0)$ such that

$$\|\tilde{x}_0 - x_0\| < r(x_0) \ \text{ and } \ \tilde{x}_0 \in Z \Rightarrow \lim_{t \to \infty} \|x(t; \tilde{x}_0) - x(t; x_0)\| = 0$$

- asymptotically stable *conditionally to* Z, if it is both stable and attractive conditionally to Z.

- globally asymptotically stable *conditionally to* Z, if it is asymptotically stable conditionally to Z and $r(x_0) = +\infty$.

Although weaker than stability, conditional stability may help us to prove stability as in the following theorem due to Iggidr, Kalitine, and Outbib [42].

Theorem 2.24 *(Stability with positive semidefinite V)*
Let $x = 0$ be an equilibrium of $\dot{x} = f(x)$ and let $V(x)$ be a C^1 positive *semidefinite* function such that $\dot{V} \leq 0$. Let Z be the largest positively invariant set contained in $\{x \mid V(x) = 0\}$. If $x = 0$ is asymptotically stable conditionally to Z, then $x = 0$ is stable.

Proof: The proof is by contradiction. Suppose that $x = 0$ is unstable. Then, for $\epsilon > 0$ small enough, there exist a sequence $(x_i)_{i \geq 1} \to 0$ in $\mathbb{R}^n$ and a sequence $(t_i)_{i \geq 1}$ in $\mathbb{R}^+$ such that

$$\forall t \in [0, t_i) : \|x(t; x_i)\| < \epsilon, \ \ \|x(t_i; x_i)\| = \epsilon \tag{2.3.4}$$

The new sequence $z_i = x(t_i; x_i)$ belongs to a compact set, so a subsequence converges to z with $\|z\| = \epsilon$. Because $x = 0$ is an equilibrium and f is locally Lipschitz continuous, continuity of the solutions implies that $t_i \to \infty$ as $i \to \infty$. We now establish two properties of the solution starting at z and evolving backward in time, that is for all $\tau \leq 0$:

(i) $\|x(\tau; z)\| \leq \epsilon$;

(ii) $V(x(\tau; z)) = 0$.

We prove (i) by contradiction. Let $\tau_1 < 0$ such that $\|x(\tau_1; z)\| > \epsilon$ and pick a constant $\nu > 0$ small enough such that $\|x(\tau_1; z)\| > \epsilon + \nu$. By continuity of the solutions, there exists a constant $\delta = \delta(\nu) > 0$ such that

$$\|z - \tilde{z}\| < \delta \Rightarrow \|x(\tau_1; z) - x(\tau_1; \tilde{z})\| < \nu$$

For i sufficiently large, we have $\|z_i - z\| < \delta$ and $t_i > t_i + \tau_1 > 0$. But this implies $\|x(\tau_1; z_i)\| = \|x(t_i + \tau_1; x_i)\| > \epsilon$ which contradicts (2.3.4).

To prove (ii), we use the continuity of $V(x)$ which, because $V(0) = 0$, implies that $V(x_i) \to 0$ as $i \to \infty$. Because V is nonincreasing along the solutions, we also have that

$$\forall t \geq 0 : \lim_{i \to \infty} V(x(t; x_i)) = 0$$

Now, if we pick any $\tau < 0$, then there exists i such that $t_i + \tau > 0$. Therefore,

$$V(x(\tau; z)) = \lim_{i \to \infty} V(x(\tau; z_i)) = \lim_{i \to \infty} V(x(t_i + \tau; x_i)) = 0$$

It remains to prove that (i) and (ii) cannot hold if the equilibrium $x = 0$ is asymptotically stable conditionally to Z. Because $\epsilon > 0$ can be chosen arbitrary small, we can assume without loss of generality that for any initial condition $x_0 \in Z$ with $\|x_0\| \leq \epsilon$ the solution converges to zero. So, there exists a constant $T = T(\epsilon) > 0$, independent of x_0, such that $\|x(T; x_0)\| \leq \frac{\epsilon}{2}$. Because of (ii), one possible choice for x_0 is $x(-T; z)$. But then $\frac{\epsilon}{2} \geq \|x(T; x_0)\| = \|x(T - T; z)\| = \|z\| = \varepsilon$ which is a contradiction.

□

We discuss two typical situations in which *global* stability is established with semidefinite Lyapunov functions.

Example 2.25 *(Global invariant manifold)*
The conditions of Theorem 2.24 are satisfied by the system

$$\begin{aligned} \dot{x}_1 &= -x_1^3 + x_1 x_2 \\ \dot{x}_2 &= -x_2 \end{aligned} \qquad (2.3.5)$$

with the Lyapunov function $V(x) = x_2^2$. The equilibrium $(x_1, x_2) = (0,0)$ is globally asymptotically stable conditionally to the set $x_2 = 0$, which is a global invariant manifold of (2.3.5). The system reduced to this manifold is $\dot{x}_1 = -x_1^3$. This manifold is also the largest positively invariant set of (2.3.5) contained in $V(x) = 0$. By Theorem 2.24, the equilibrium $(x_1, x_2) = (0,0)$ is stable because $\dot{V} = -2x_2^2 \leq 0$.

To prove global asymptotic stability of the origin, we first show that all the off-manifold solutions are bounded. With $x_2(t) = e^{-t}x_2(0)$ the solutions of

$$\dot{x}_1 = -x_1^3 + x_1 x_2 \tag{2.3.6}$$

are bounded. This follows from

$$\frac{d}{dt}x_1^2 = -2x_1^4 + 2x_1^2 e^{-t}x_2(0) \leq -x_1^4 + e^{-2t}x_2^2(0)$$

as $x_1^2(t)$ must decrease if $|x_1(t)| > \sqrt{e^{-t}x_2(0)}$. Thanks to this "bounded input – bounded state" property of (2.3.6), the equilibrium $(x_1, x_2) = (0,0)$ is globally stable. By Theorem 2.21 it is also GAS because in the set where $\dot{V} = -2x_2^2 = 0$, $x_1 \to 0$.

□

In our next example global boundedness is established with a radially unbounded Lyapunov function which is only positive semidefinite.

Example 2.26 *(Semidefinite, radially unbounded Lyapunov function)*
Defining $\varphi(x_1) = 0$ for $|x_1| \leq 1$, $x_1 - 1$ for $x_1 > 1$, $x_1 + 1$ for $x_1 < -1$, we analyze stability of the system

$$\begin{aligned} \dot{x}_1 &= -x_1^3 - x_1^2\varphi(x_1) + x_1 x_2 \\ \dot{x}_2 &= -x_1\varphi(x_1) - x_2 \end{aligned} \tag{2.3.7}$$

For this purpose we use the Lyapunov function

$$V = \frac{1}{2}\varphi^2(x_1) + \frac{1}{2}x_2^2$$

which is radially unbounded and positive semidefinite because $V = 0$ in $Z = \{|x_1| \leq 1, x_2 = 0\}$. It is easy to see that Z is a positively invariant set of (2.3.7) which in this set reduces to $\dot{x}_1 = -x_1^3$. Therefore, $(x_1, x_2) = (0,0)$ is asymptotically stable conditionally to Z. To satisfy Theorem 2.24, we verify that $\dot{V} \leq 0$. Noting that $\frac{d}{dx_1}(\varphi^2(x_1)) = 2\varphi(x_1)$ we get

$$\begin{aligned} \dot{V} &= \varphi(x_1)(-x_1^3 - x_1^2\varphi(x_1) + x_1 x_2) - x_1 x_2 \varphi(x_1) - x_2^2 \\ &= -x_1^3\varphi(x_1) - x_1^2\varphi^2(x_1) - x_2^2 \leq 0 \end{aligned}$$

Hence, the equilibrium $(x_1, x_2) = (0,0)$ is stable. The boundedness of the solutions of (2.3.7) follows from the fact that V is radially unbounded. This proves global stability. To establish asymptotic stability we note that the solutions converge to the set where $\dot{V} = 0$. This set is again Z. By Theorem 2.21, the equilibrium $(x_1, x_2) = (0,0)$ is GAS because all the solutions in Z converge to $(0,0)$.

□

The above examples clearly indicate the three steps in the proof of GAS. Local stability is first established either with a positive definite or a positive semidefinite Lyapunov function, as in Theorems 2.20 and 2.24. In the second step the global boundedness is guaranteed via the convergence to a global invariant manifold and a bounded-input bounded-state property, or with a radially unbounded Lyapunov function. Finally, asymptotic sability is established with $\dot{V} < 0$ as in Theorem 2.20, or with $\dot{V} \leq 0$ and the Invariance Principle (Theorem 2.21).

2.3.3 Stability of passive systems

The definitions of dissipativity and passivity do not require that the storage function S be positive definite. They are also satisfied if S is only positive *semidefinite.* As a consequence, in the presence of an unobservable unstable part of the system, they allow $x = 0$ to be unstable. For instance, the unstable system $\dot{x}_1 = x_1$, $\dot{x}_2 = u$, $y = x_2$ is passive with the storage function $S = \frac{1}{2}x_2^2$.

For dissipativity to imply Lyapunov stability, we must exclude such situations. In linear systems this is achieved with a detectability assumption, which requires that the unobservable part of the system be asymptotically stable. We now define an analogous concept for nonlinear systems.

Definition 2.27 *(Zero-state detectability and observability)*
Consider the system H with zero input, that is $\dot{x} = f(x,0)$, $y = h(x,0)$, and let $Z \subset I\!R^n$ be its largest positively invariant set contained in $\{x \in I\!R^n \mid y = h(x,0) = 0\}$. We say that H is zero-state detectable (ZSD) if $x = 0$ is asymptotically stable conditionally to Z. If $Z = \{0\}$, we say that H is zero-state observable (ZSO). □

Whenever we use the ZSD property to establish a global result, we assume that $x = 0$ is GAS conditionally to Z. One of the benefits from this detectability property is that passivity and stability are connected even when the storage function $S(x)$ is only positive semidefinite. The main benefit, however, is that asymptotic stability is achieved with the simplest feedback $u = -y$. To avoid the well-posedness issue in (iii), we assume that the throughput is absent: $y = h(x)$.

Theorem 2.28 *(Passivity and stability)*
Let the system H be passive with a C^1 storage function S and $h(x,u)$ be C^1 in u for all x. Then the following properties hold:

(i) If S is positive definite, then the equilibrium $x = 0$ of H with $u = 0$ is stable.

(ii) If H is ZSD, then the equilibrium $x = 0$ of H with $u = 0$ is stable.

(iii) When there is no throughput, $y = h(x)$, then the feedback $u = -y$ achieves asymptotic stability of $x = 0$ if and only if H is ZSD.

When the storage function S is radially unbounded, these properties are global.

Proof: (i) If H is passive, then with $u = 0$, the storage function $S(x)$ satisfies $\dot{S}(x) \leq 0$. If S is positive definite, the equilibrium $x = 0$ of $\dot{x} = f(x,0)$ is stable by Theorem 2.20.

(ii) To prove stability of $x = 0$ when S is only positive semidefinite, we first show that

$$S(x) = 0 \Rightarrow h(x,0) = 0 \tag{2.3.8}$$

Because $S(x) \geq 0$ for all x, $\dot{S}(x) \leq u^T y = u^T h(x,u)$ must be nonnegative for all u whenever $S(x) = 0$. Because $h(x,u)$ is C^1 in u, we let $y = h(x,u) = h(x,0) + \eta(x,u)u$. We obtain that, for all $x \in \{x|\ S(x) = 0\}$ and all u,

$$0 \leq \dot{S}(x) \leq u^T h(x,0) + u^T \eta^T(x,u)u \tag{2.3.9}$$

The only possibility for (2.3.9) to be satisfied for all u is that $h(x,0) = 0$ whenever $S(x) = 0$.

As a consequence, the largest positively invariant set Z of $\dot{x} = f(x,0)$ contained in $\{x|\ S(x) = 0\}$ is also contained in $\{x|\ h(x,0) = 0\}$. By the ZSD assumption, $x = 0$ is asymptotically stable conditionally to Z. Therefore, the assumptions of Theorem 2.24 are satisfied, which proves stability of $x = 0$.

(iii) Because h is independent of u, the feedback loop with $u = -y$ is well posed. For $u = -y$, the time-derivative of S satisfies

$$\dot{S}(x) \leq -y^T y \leq 0$$

The stability part is established as in the proof of (ii). By Theorem 2.21, the bounded solutions of $\dot{x} = f(x, -y)$ converge to the largest invariant set of $\dot{x} = f(x, 0)$ contained in $E = \{x|\ h(x) = 0\}$. If H is ZSD, this set is $x = 0$, which proves asymptotic stability.

Conversely, if the equilibrium $x = 0$ of $\dot{x} = f(x, -y)$ is asymptotically stable, then it is asymptotically stable conditionally to any subset Z. In particular, this is the case when Z is the largest positively invariant set contained in $E = \{x|y = h(x) = 0\}$ which proves that H is ZSD.

Finally, if $S(x)$ is radially unbounded and $\dot{S}(x) \leq 0$, all solutions are bounded, so the stability properties are global. □

Example 2.29 *(Local stabilization with $u = -y$)*
The system

$$
\begin{aligned}
\dot{x}_1 &= f_1(x_1, x_2, u) \\
\dot{x}_2 &= u \\
y &= x_2
\end{aligned}
$$

is passive with the positive semidefinite storage function $S(x_1, x_2) = \frac{1}{2}x_2^2$ since $\dot{S} = uy$. It is ZSD if and only if the equilibrium $x_1 = 0$ of $\dot{x}_1 = f_1(x_1, 0, 0)$ is asymptotically stable. By Theorem 2.28, this is a necessary and sufficient condition for local stabilization of the equilibrium $(x_1, x_2) = (0, 0)$ using the feedback $u = -y$.

□

In our stability studies, we will usually deduce stability from the positive definiteness of the storage function and then use the ZSD property to establish *asymptotic* stability. Occasionally, we will use parts of Theorem 2.28 which allow the storage function to be positive semidefinite.

2.3.4 Stability of feedback interconnections

Theorem 2.28 will now be extended to the stability properties of feedback interconnections.

Theorem 2.30 *(Feedback interconnection of dissipative systems)*
Assume that the systems H_1 and H_2 are dissipative with the supply rates

$$w_i(u_i, y_i) = u_i^T y_i - \rho_i^T(y_i) y_i - \nu_i^T(u_i) u_i, \quad i = 1, 2 \tag{2.3.10}$$

where $\nu_i(.)$ and $\rho_i(.)$ are the nonlinear functions defined in (2.2.3). Furthermore assume that they are ZSD and that their respective storage functions $S_1(x_1)$ and $S_2(x_2)$ are C^1. Then the equilibrium $(x_1, x_2) = (0, 0)$ of the feedback interconnection in Figure 2.2 with $r \equiv 0$, is

(i) stable, if $\nu_1^T(v)v + \rho_2^T(v)v \geq 0$ and $\nu_2^T(v)v + \rho_1^T(v)v \geq 0$ for all $v \in I\!R^m$;

(ii) asymptotically stable, if $\nu_1^T(v)v + \rho_2^T(v)v > 0$ and $\nu_2^T(v)v + \rho_1^T(v)v > 0$ for all $v \in I\!R^m/\{0\}$.

If both $S_1(x_1)$ and $S_2(x_2)$ are radially unbounded, then these properties are global.

Proof: (i) A storage function for the feedback interconnection is $S(x_1, x_2) = S_1(x_1) + S_2(x_2)$. Using the interconnection identities $u_1 = -y_2$, $u_2 = y_1$, the time-derivative of S is

$$\dot{S} \leq -(\nu_2 + \rho_1)^T(y_1)y_1 - (\nu_1 + \rho_2)^T(y_2)y_2 \leq 0$$

If S is positive definite, this proves stability. If S is only semidefinite, we deduce stability from Theorem 2.24. Because $S = 0$ implies $S_1 = S_2 = 0$, the argument in the proof of Theorem 2.28 shows that

$$S(x) = 0 \Rightarrow h_1(x_1, 0) = h_2(x_2, 0) = 0$$

By our standing assumption which assures well-posedness, either h_1 or h_2 or both are independent of the input. Without loss of generality we assume that $h_1(x_1, u_1) = h_1(x_1)$. Hence $S(x) = 0 \Rightarrow y_1 = h_1(x_1) = 0$ and also $S(x) = 0 \Rightarrow y_2 = h_2(x_2, u_2) = h_2(x_2, y_1) = h_2(x_2, 0) = 0$. Using the interconnection identities, we obtain

$$S = 0 \Rightarrow y_1 = y_2 = u_1 = u_2 = 0$$

The largest positively invariant set Z of $\dot{x}_1 = f(x_1, 0)$, $\dot{x}_2 = f_2(x_2, 0)$ in $\{(x_1, x_2) | S(x_1, x_2) = 0\}$ is also included in $\{(x_1, x_2) | y_1 = y_2 = 0\}$. Because H_1 and H_2 are ZSD, the equilibrium $(x_1, x_2) = (0, 0)$ is asymptotically stable conditionally to Z. By Theorem 2.24, this proves stability.

(ii) If $\nu_1^T(v)v + \rho_2^T(v)v > 0$ and $\nu_2^T(v)v + \rho_1^T(v)v > 0$, for all $v \neq 0$, then all bounded solutions converge to the set $\{(x_1, x_2) | y_1 = y_2 = 0\}$. By the Invariance Principle, every bounded solution converges to the largest invariant set in E, that is $(x_1, x_2) = (0, 0)$ since H_1 and H_2 are ZSD. From (i) we know that the solutions are bounded in a neighborhood of $(x_1, x_2) = (0, 0)$ which proves local asymptotic stability.

When $S_1(x_1)$ and $S_2(x_2)$ are radially unbounded, so is $S(x)$, and hence, the stability properties are global.

□

An important special case is when $\rho(y)$ and $\nu(u)$ are linear functions ρy and νu, respectively.

Corollary 2.31 *(Feedback interconnections of OFP and IFP systems)*
If H_1 and H_2 are dissipative with radially unbounded storage functions S_1 and S_2 then the equilibrium $(x_1, x_2) = (0, 0)$ of their feedback interconnection is:

(i) GS, if H_1 and H_2 are passive.

(ii) GAS, if H_1 and H_2 are OFP with $\rho_1, \rho_2 > 0$.

(iii) GAS, if H_1 and H_2 are IFP with $\nu_1, \nu_2 > 0$. □

To further refine Theorem 2.30, we need the following definition, which for a linear system means that the zeros of its transfer function are in the open left-half plane.

Definition 2.32 *(Zero-input detectability)*
The system H is said to be Zero-Input Detectable (ZID) if $y \equiv 0$ implies $u(t) \to 0$ as $t \to \infty$. □

Theorem 2.33 *(Interconnection stability)*
Under the assumptions of Theorem 2.30 the equilibrium $(x_1, x_2) = (0, 0)$ of the feedback interconnection is stable if $\nu_1^T(v)v + \rho_2^T(v)v > 0$ and $\nu_2^T(v)v + \rho_1^T(v)v \geq 0$ for all $v \in I\!R^m$. If, in addition, either H_1 is GAS when $u_1 = 0$, or H_2 is ZID, then $(x_1, x_2) = (0, 0)$ is asymptotically stable. If $S_1(x_1)$ and $S_2(x_2)$ are radially unbounded, these properties are global. The same is true with the interchange of the subscripts 1 and 2.

Proof: Stability is proved as in Theorem 2.30. To examine the convergence of solutions, we deduce from

$$\dot{S} \leq -(\nu_1 + \rho_2)^T(y_2)y_2 \leq 0 \tag{2.3.11}$$

that all bounded solutions converge to

$$E = \{(x_1, x_2) \mid y_2 = u_1 = 0\} \tag{2.3.12}$$

By the invariance theorem (Theorem 2.21), we need to investigate only the solutions which, having started in E, remain in E.

Case 1: H_1 is GAS. If H_1 with $u_1 = 0$ is GAS, then $x_1(t) \to 0$ along each solution which remains in E. Therefore, these solutions converge to

$$E' = \{(x_1, x_2) \mid y_2 = u_1 = x_1 = 0\} \subset E \tag{2.3.13}$$

Applying the Invariance Principle one more time, we examine the convergence of bounded solutions that remain in E'. Along these solutions, $y_2 \equiv u_2 \equiv 0$ because $x_1 \equiv 0$ and $u_1 \equiv 0$ imply $y_1 \equiv u_2 \equiv 0$. By ZSD, this proves that $x_2(t)$ converges to zero.

Case 2: H_2 is ZID. Then, by definition, $u_2(t) \to 0$ along the solutions which remain in E. So, each such solution which is bounded converges to

$$E'' = \{(x_1, x_2) \mid y_2 = u_1 = u_2 = y_1 = 0\} \subset E \tag{2.3.14}$$

Applying the invariance theorem, we only examine bounded solutions that remain in E''. Their convergence to zero follows from the ZSD assumption.

If $S_1(x_1)$ and $S_2(x_2)$ are radially unbounded, all solutions are bounded and the asymptotic stability is global. □

From Theorem 2.33 we now characterize stable feedback interconnections which are of primary importance for the rest of the book.

Theorem 2.34 *(Stability of OFP/IFP feedback interconnections)*
Assume that in the feedback interconnection the system H_1 is GAS and IFP(ν), and the system H_2 is ZSD and OFP(ρ). Then $(x_1, x_2) = (0, 0)$ is asymptotically stable if $\nu + \rho > 0$. If, in addition, their storage functions S_1 and S_2 are radially unbounded, then $(x_1, x_2) = (0, 0)$ is GAS.

□

The above result shows how the shortage of passivity in one system can be compensated for by the excess of passivity in the other system.

Example 2.35 *(OFP/IFP interconnection)*
Let the systems

$$H_1: \begin{array}{rcl} \dot{x}_1 &=& x_2 \\ \dot{x}_2 &=& -x_1 - \phi(x_2) + u_1 \\ y_1 &=& x_2 + u_1 \end{array}, \qquad H_2: \begin{array}{rcl} \dot{x}_3 &=& x_4 \\ \dot{x}_4 &=& -x_3 + \frac{1}{2}x_4 + u_2 \\ y_2 &=& x_4 \end{array}$$

be in the feedback interconnection: $u_1 = -y_2$ and $u_2 = y_1$. With the function ϕ as in Example 2.23, H_1 is GAS with $u_1 \equiv 0$. We also readily verify that H_1 is IFP(1) with $S_1 = \frac{x_1^2}{2} + \frac{x_2^2}{2}$ and H_2 is unstable, ZSD, OFP($-\frac{1}{2}$) with $S_2 = \frac{x_3^2}{2} + \frac{x_4^2}{2}$. By Theorem 2.34 we therefore conclude that the equilibrium $(x_1, x_2, x_3, x_4) = (0, 0, 0, 0)$ of the interconnected system is GAS. □

For asymptotic stability of the interconnection, the conditions "H_1 is GAS" or "H_2 is ZID" in Theorem 2.33 are only sufficient.

Example 2.36 *(Relaxing the GAS and ZID assumptions)*
Consider H_1 and H_2 as in Example 2.35 but with $\phi \equiv 0$. The passivity properties of H_1 and H_2 are unchanged and global stability of the interconnection follows from Theorem 2.34. On the other hand, H_1 is no longer GAS with $u_1 = 0$. Neither is H_2 ZID, because $y_2 \equiv 0$ only implies $x_4 = \dot{x}_3 \equiv 0$ and therefore admits the solution

$$x_3 \equiv u_2 \equiv const \neq 0 \tag{2.3.15}$$

So, Theorem 2.34 cannot be applied. Nevertheless, by the main stability theorem, the solutions converge to the set E where $y_2 = u_1 = 0$. Applying Theorem 2.21, we examine the solutions which remain in E. These solutions verify (2.3.15) and therefore we have $y_1 \equiv const$. In E, $u_1 \equiv 0$ and thus $y_1 \equiv x_2 \equiv const$. Hence $0 \equiv \dot{x}_2 = -x_1 - \phi(x_2)$ from which we conclude that $x_1 \equiv const$. This implies $x_2 \equiv 0$, $x_1 \equiv 0$ and so $0 \equiv y_1 \equiv u_2 \equiv x_3$. The only solution remaining in E for all t is $(x_1, x_2, x_3, x_4) = (0, 0, 0, 0)$ which proves GAS. □

Finally, we illustrate that the feedback interconnection of an OFP(ρ) system with an IFP(ν) system with $\nu + \rho > 0$ need not be GAS.

Example 2.37 *(Lack of asymptotic stability)*
Consider again the situation of Example 2.35 with H_1 replaced by

$$H_1' : \begin{array}{rcl} \dot{x}_1 & = & u_1 \\ y_1 & = & x_1 + u_1 \end{array}$$

Then H_1' is IFP(1) (see Example 2.13) but not GAS with $u_1 = 0$. It can be verified that the feedback interconnection of H_1' and H_2 admits any constant solution of the form $(x_1, x_3, x_4) \equiv (c, c, 0)$. The equilibrium $(x_1, x_2, x_3, x_4) = (0, 0, 0, 0)$ is stable, but not asymptotically stable.

This lack of asymptotic stability for the interconnection of H_1' and H_2 is obvious in the frequency domain. Namely, $H_1'(s)$ has a pole at $s = 0$ while $H_2(s)$ has a zero at $s = 0$, leading to a pole-zero cancellation on the imaginary axis when the feedback loop is closed. □

2.3.5 Absolute stability

A system H is said to be *absolutely stable* if its feedback interconnection with any static nonlinearity in a sector (α, β) is globally asymptotically stable. This property is of interest as a robustness property of feedback systems, and will be used in our study of stability margins in the next chapter.

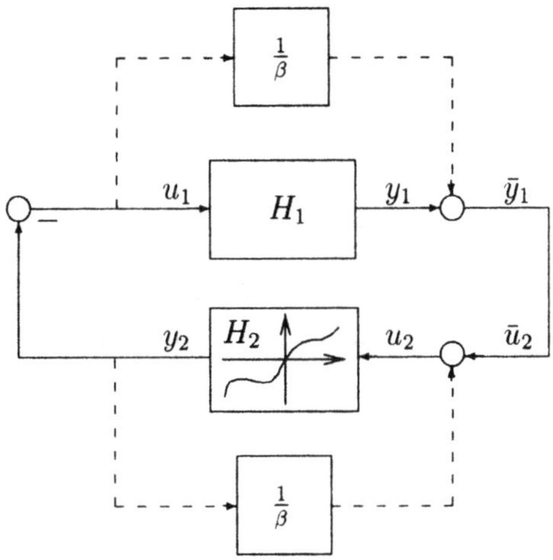

Figure 2.10: Block diagram illustrating absolute stability.

Proposition 2.38 *(Absolute stability)*
Let H_1 be a ZSD system with scalar output $y = h(x)$. Consider its feedback interconnection with a static nonlinearity φ in the sector (α, β), $\beta > 0$. For global asymptotic stability of $x = 0$, it is sufficient that the parallel interconnection of H_1 and $\frac{1}{\beta}$ be OFP($-k$) with a C^1 radially unbounded storage function S, where

$$k = \frac{\alpha\beta}{\beta - \alpha} \tag{2.3.16}$$

Proof: Consider the loop transformation indicated in Figure 2.10 by dotted lines. Denote by $\bar{H}_1$ the parallel interconnection of H_1 and $\frac{1}{\beta}$ and by $\bar{H}_2$ the positive feedback interconnection of the sector nonlinearity block H_2 with $\frac{1}{\beta}$. Then the feedback interconnections of H_1 with H_2, and $\bar{H}_1$ and $\bar{H}_2$, are equivalent. Because $\bar{H}_1$ is OFP($-k$), the storage function S satisfies

$$\dot{S} \leq \bar{y}_1(\bar{u}_1 + k\bar{y}_1) = -\bar{u}_2(\bar{y}_2 - k\bar{u}_2) \tag{2.3.17}$$

Using the linear sector bound and $u_2 = \bar{u}_2 + \frac{1}{\beta}y_2$ we obtain

$$u_2 y_2 \geq \alpha u_2^2 = \alpha u_2(\bar{u}_2 + \frac{1}{\beta}y_2) \Rightarrow u_2(y_2 - \alpha\bar{u}_2 - \frac{\alpha}{\beta}y_2) \geq 0$$

$$u_2(y_2 - \frac{\alpha}{\beta}y_2 - \alpha\bar{u}_2) = \frac{\beta-\alpha}{\alpha}u_2(y_2 - k\bar{u}_2) \Rightarrow u_2(y_2 - k\bar{u}_2) \geq 0 \quad (2.3.18)$$

Because $| y_2 | \leq \beta u_2$ (the inequality being strict for $y_2 \neq 0$), u_2 and $\bar{u}_2$ always have the same sign so that u_2 can be replaced by $\bar{u}_2$ in the inequality (2.3.18), that is

$$\bar{u}_2(y_2 - k\bar{u}_2) \geq 0 \quad (2.3.19)$$

Thus $\bar{H}_2$ is IFP(k), and the excess of passivity of $\bar{H}_2$ compensates for the shortage of passivity of $\bar{H}_1$ to make the interconnection passive. This proves global stability.

To prove asymptotic stability, we note that the inequality (2.3.19) is strict when $y_2\bar{u}_2 \neq 0$. In view of (2.3.17), the solutions converge to the largest invariant set where $\bar{y}_1 = \bar{u}_1 = 0$. In this set the solutions converge to zero because H_1 is ZSD and so is $\bar{H}_1$. This proves that the interconnection is GAS.

□

When H_1 is linear, Proposition 2.38 is known as the *circle criterion* and will be discussed in Chapter 3.

2.3.6 Characterization of affine dissipative systems

Hill and Moylan [37] provided a characterization of input-affine dissipative systems

$$\begin{aligned} \dot{x} &= f(x) + g(x)u \\ y &= h(x) + j(x)u \end{aligned} \quad (2.3.20)$$

which will help us to identify their structural properties.

Theorem 2.39 *(Characterization of IFP and OFP)*
Let S be a C^1 positive semidefinite function. A system H is dissipative with respect to the supply rate

$$w(u,y) = u^T y - \rho\, y^T y - \nu\, u^T u \quad (2.3.21)$$

with the storage function S if and only if there exist functions $q : \mathbb{R}^n \to \mathbb{R}^k$, and $W : \mathbb{R}^n \to \mathbb{R}^{k\times m}$, for some integer k, such that

$$\begin{aligned} L_f S(x) &= -\tfrac{1}{2}q^T(x)q(x) - \rho h^T(x)h(x) \\ L_g S(x) &= h^T(x) - 2\rho h^T(x)j(x) - q^T(x)W(x) \\ W^T(x)W(x) &= -2\nu I + j(x) + j^T(x) - 2\rho j^T(x)j(x) \end{aligned} \quad (2.3.22)$$

Proof: First assume that there exist functions $q(x), W(x)$ which satisfy the conditions (2.3.22). Then, along the solutions of the system (2.3.20),

$$\begin{aligned}
\dot{S} &\leq \dot{S} + \tfrac{1}{2}(q + Wu)^T(q + Wu) \\
&= L_f S + L_g S u + \tfrac{1}{2}(q^T q + 2q^T W u + u^T W^T W u) \\
&= -\rho h^T h + h^T u - 2\rho h^T j u + \tfrac{1}{2}u^T(-2\nu I + j(x) + j^T(x) - 2\rho j^T(x) j(x))u \\
&= u^T y - \rho y^T y - \nu u^T u = w(u, y)
\end{aligned}$$

Thus, the system (2.3.20) is dissipative with the supply rate $w(u, y)$ and $S(x)$ is a storage function.

Conversely, assume that the system (2.3.20) is dissipative with the supply rate $w(u, y)$ and the storage function S, that is

$$\dot{S} \leq w(u, y)$$

Then, by defining $d(x, u) = -\dot{S} + w(u, y)$ we obtain

$$\begin{array}{c}
0 \leq d(x, u) = -\dot{S} + w(u, y) = -L_f S - L_g S u + u^T y - \rho y^T y - \nu u^T u = \\
-L_f S - \rho h^T h - (L_g S + 2\rho h^T j - h^T)u - u^T(\nu I - \tfrac{1}{2}(j + j^T) - \rho j^T j)u
\end{array} \tag{2.3.23}$$

Because $d(x, u)$ is quadratic in u and nonnegative for all u and x, there exist (nonunique) matrix valued functions $q(x)$ and $W(x)$ such that

$$d(x, u) = \frac{1}{2}[q(x) + W(x)u]^T[q(x) + W(x)u] \tag{2.3.24}$$

Then (2.3.22) follows from (2.3.23) and (2.3.24) by equating the terms of the like powers in u. □

For systems without throughput ($j(x) \equiv 0$), the theorem readily extends to the situations in which a nonlinear function $\rho(y)$ is used instead of a linear term ρy, as in Example 2.18.

A structural property of input-affine IFP systems implied by Theorem 2.39 is that they must have relative degree zero.[1]

Corollary 2.40 *(Relative degree zero)*
If the system (2.3.20) is IFP(ν) with $\nu > 0$ and with a C^1 storage function, then the matrix $j(0)$ is nonsingular, that is, the system (2.3.20) has relative degree zero.

Proof: The last equality of (2.3.22) implies that

$$j(x) + j^T(x) \geq W^T(x)W(x) + 2\nu I > 0 \tag{2.3.25}$$

[1] The concept of a *relative degree* for nonlinear systems is presented in Appendix A.

for all x, which implies that $j(x)$ is nonsingular for all x.

□

For a *passive* affine system (2.3.20) without throughput, $j(x) \equiv 0$, the conditions (2.3.22) reduce to

$$L_f S(x) \leq 0 \tag{2.3.26}$$

$$(L_g S)^T(x) = h(x) \tag{2.3.27}$$

If the system is *linear*

$$\begin{aligned} \dot{x} &= Ax + Bu \\ y &= Cx, \end{aligned} \tag{2.3.28}$$

then there exists a quadratic storage function $S(x) = x^T P x$, with $P \geq 0$, and the passivity conditions become algebraic:

$$\begin{aligned} PA + A^T P &\leq 0 \\ B^T P &= C \end{aligned} \tag{2.3.29}$$

The equivalence of these conditions with the frequency-domain characterization of passivity was established by the celebrated Kalman-Yakubovich-Popov Lemma. The KYP Lemma is given here for the case when (A, B, C) is a minimal realization.

Theorem 2.41 (*KYP Lemma*)
If for the linear system (A, B, C) there exists a symmetric positive definite matrix P satisfying (2.3.29), then the transfer function $H(s) = C(sI - A)^{-1}B$ is *positive real*, that is, it satisfies the conditions

(i) $Re(\lambda_i(A)) \leq 0, \quad 1 \leq i \leq n$;

(ii) $H(j\omega) + H^T(-j\omega) \geq 0$ for all $\omega \in I\!R$, $j\omega \neq \lambda_i(A)$;

(iii) the eigenvalues of A on the imaginary axis are simple and the corresponding residues $\lim_{s \to s_0}(s - s_0)H(s)$, are Hermitian and nonnegative definite matrices.

Conversely, if $H(s)$ is positive real, then for any minimal realization of $H(s)$, there exists $P > 0$ which satisfies the passivity conditions (2.3.29). □

Extensions of the KYP Lemma to nonminimal realizations of $H(s)$ and to MIMO systems can be found in [2, 41, 107, 108]. In the next chapter, the KYP Lemma will be useful in the definitions of stability margins for nonlinear systems.

2.4 Feedback Passivity

2.4.1 Passivity: a tool for stabilization

The task of stabilization is the simplest when an output function $y = h(x)$ can be found such that the system

$$\begin{aligned} \dot{x} &= f(x) + g(x)u \\ y &= h(x) \end{aligned} \tag{2.4.1}$$

with u as the input and y as the output is passive. Then we know from Theorem 2.28 that stability is achieved if we close the feedback loop with $u = -y$. If, in addition, the system (2.4.1) is ZSD, the interconnection is GAS.

However, searching for an output $y = h(x)$ such that the system is passive with a positive definite storage function requires that the system be stable when $u = 0$. To remove this restriction, we include feedback as a means to achieve passivity. Instead of being stable, the uncontrolled system is assumed to be stabilizable. Therefore, we need to find an output $y = h(x)$ and a feedback transformation

$$u = \alpha(x) + \beta(x)v, \tag{2.4.2}$$

with $\beta(x)$ invertible, such that the system

$$\begin{aligned} \dot{x} &= f(x) + g(x)\alpha(x) + g(x)\beta(x)v \\ y &= h(x) \end{aligned} \tag{2.4.3}$$

is passive.

If a feedback transformation (2.4.2) can be found to render the system (2.4.3) passive, we call the original system (2.4.1) *feedback passive*. The selection of an output $y = h(x)$ and the construction of a passivating transformation (2.4.2) is referred to as *feedback passivation*. Under a ZSD assumption, asymptotic stability of the passive system (2.4.3) is simply achieved with the additional feedback $v = -\kappa y$, $\kappa > 0$.

As we will show next, the crucial limitation of the feedback passivation design is that the output must have two properties which cannot be modified by feedback. To identify these properties, we will use the characterization of passive systems given in Theorem 2.39. We first consider the linear systems.

2.4.2 Feedback passive linear systems

For a controllable and observable linear system

$$\begin{aligned} \dot{x} &= Ax + Bu \\ y &= Cx \end{aligned} , \tag{2.4.4}$$

where B and C have full rank, passivity is equivalent to the conditions (2.3.29): $PA + A^TP \leq 0$ and $B^TP = C$, where P is a positive definite matrix. It follows from $B^TP = C$ that the matrix

$$CB = B^TPB \tag{2.4.5}$$

is positive definite; hence the system (2.4.4) has relative degree one (see Appendix A). A linear change of coordinates

$$\begin{pmatrix} \xi_0 \\ y \end{pmatrix} = \begin{pmatrix} T \\ C \end{pmatrix} x \tag{2.4.6}$$

exists such that $TB = 0$. In the new coordinates (2.4.6), the system (2.3.28) is in normal form (see Appendix A)

$$\begin{aligned} \dot{\xi}_0 &= Q_{11}\xi_0 + Q_{12}y \\ \dot{y} &= Q_{21}\xi_0 + Q_{22}y + CBu \end{aligned} \tag{2.4.7}$$

Because CB is nonsingular, we can use the feedback transformation

$$u = -(CB)^{-1}(Q_{21}\xi_0 + Q_{22}y - v)$$

and reduce (2.4.4) to

$$\begin{aligned} \dot{\xi}_0 &= Q_{11}\xi_0 + Q_{12}y \\ \dot{y} &= v \end{aligned} \tag{2.4.8}$$

so that $y_i(s) = \frac{1}{s}v_i(s)$ where $\frac{1}{s}$ is an integrator – the simplest relative degree one transfer function.

The *normal form* (2.4.8) clearly shows that all the solutions which satisfy the constraint that the output be zero, $y(t) \equiv 0$, are defined by the *zero-dynamics* subsystem $\dot{\xi}_0 = Q_{11}\xi_0$. The eigenvalues of Q_{11} are, in fact, the zeros of the transfer function $C(sI - A)^{-1}B$ of the system (2.4.4). It is clear that the zero-dynamics subsystem in (2.4.8) remains unchanged by any feedback control $v(\xi_0, y)$ and the same is true for the relative degree. If the zero-dynamics subsystem is asymptotically stable, that is if the zeros are in the open left half plane, the system is said to be *minimum phase.* If the zero-dynamics subsystem is only Lyapunov stable, then the system is said to be *weakly minimum phase.*

We now return to the passivity conditions (2.3.29). Partitioning the matrix P according to the state partition (ξ_0, y), the passivity condition $B^T P = C$ yields

$$P_{12} = P_{21}^T = 0, \; P_{22} = (CB)^{-1} \tag{2.4.9}$$

and the first condition in (2.3.29) reduces to

$$P_{11}Q_{11} + Q_{11}^T P_{11} \leq 0 \tag{2.4.10}$$

This is a Lyapunov inequality for Q_{11} which shows that (2.4.7) is a weakly minimum phase system.

We see that, if the linear system (2.4.4) is passive, then it has relative degree one and is weakly minimum phase. Feedback passivation as a design tool is restricted by the fact that these two properties are invariant under the feedback transformation

$$u = Kx + Gv, \quad G \text{ nonsingular}, \tag{2.4.11}$$

The two structural properties, relative degree one and weak minimum phase, are not only necessary but also sufficient for a linear system to be feedback passive.

Proposition 2.42 *(Linear feedback passive systems)*
The linear system (2.4.4) where C has full rank, is feedback passive with a positive definite storage function $S(x) = x^T Px$ if and only if it has relative degree one and is weakly minimum phase.

Proof: The necessity was established in the discussion above. The sufficiency follows from the fact that the feedback

$$v = -2Q_{12}^T P_{11}\xi_0 + \bar{v} \tag{2.4.12}$$

transforms (2.4.8) into a passive system with the storage function

$$S(\xi_0, y) = \xi_0^T P_{11}\xi_0 + \frac{1}{2}y^T y \tag{2.4.13}$$

A straightforward calculation shows that $\dot{S} \leq \bar{v}^T y$. □

We know from Theorem 2.28 that ZSD passive systems are stabilizable. For linear systems the converse is also true as we now show following [92].

Proposition 2.43 *(Stabilizability and detectability)*
Under the assumptions of Proposition 2.42 a passive linear system is stabilizable if and only if it is detectable.

Proof: Because for linear systems detectability is equivalent to ZSD, by Theorem 2.28, passivity and detectability imply stabilizability. To prove the proposition we need to establish that the converse is also true. Using passivity conditions (2.3.29) we have already established that a storage function for the passive system (2.4.7) must be of the form

$$S(\xi_0, y) = \xi_0^T P_{11}\xi_0 + \frac{1}{2}y^T(CB)^{-1}y \tag{2.4.14}$$

The system $\dot{\xi}_0 = Q_{11}\xi_0$ is Lyapunov stable; we let $Q_{11} = diag\{Q_h, Q_c\}$, where Q_h is Hurwitz and Q_c is skew symmetric so that $Q_c + Q_c^T = 0$. The corresponding partitioned form of the system (2.4.7) is

$$\begin{aligned} \dot{\xi}_h &= Q_h\xi_h + G_h y \\ \dot{\xi}_c &= Q_c\xi_c + G_c y \\ \dot{y} &= D_h\xi_h + D_c\xi_c + Q_{22}y + CBu \end{aligned} \tag{2.4.15}$$

Because Q_h is Hurwitz and CB is nonsingular, stabilizability of (A, B, C) is equivalent to controllability of (Q_c, G_c). We now show that for passive systems this is equivalent to the observability of (D_c, Q_c) and hence, to detectability of (A, B, C).

In the new coordinates (ξ_h, ξ_c, y) the storage function (2.4.14) becomes

$$S = \xi_h^T P_h \xi + \frac{1}{2}\xi_c^T\xi_c + \frac{1}{2}y^T(CB)^{-1}y \tag{2.4.16}$$

Its derivative along the solutions of (2.4.15) is

$$\begin{aligned} \dot{S} = 2\xi_h^T P_h(Q_h\xi_h + G_h y) + \xi_c^T G_c y + y^T(CB)^{-1}D_h\xi_h + y^T(CB)^{-1}D_c\xi_c + \\ y^T(CB)^{-1}Q_{22}y + y^T u \end{aligned}$$

By passivity $\dot{S} \leq u^T y$, and hence, the two sign-indefinite terms which contain ξ_c must cancel out, that is

$$G_c^T = -(CB)^{-1}D_c \tag{2.4.17}$$

Because (Q_c, G_c) is controllable and $Q_c^T = -Q_c$, (2.4.17) implies that (D_c, Q_c^T) is observable, that is (A, B, C) is detectable. □

2.4.3 Feedback passive nonlinear systems

For input-affine nonlinear systems

$$\begin{aligned} \dot{x} &= f(x) + g(x)u \\ y &= h(x) \end{aligned} \tag{2.4.18}$$

we will proceed in full analogy with the linear case and assume that the matrices $g(0)$ and $\frac{\partial h}{\partial x}(0)$ have full rank. The nonlinear analog of the matrix CB is

$$\frac{\partial \dot{y}}{\partial u} = \frac{\partial h}{\partial x}\frac{\partial \dot{x}}{\partial u} = \frac{\partial h}{\partial x} g = L_g h \tag{2.4.19}$$

The system (2.4.18) has relative degree one at $x = 0$ if the matrix $L_g h(0)$ is invertible (see Appendix A).

Proposition 2.44 *(Relative degree of nonlinear passive systems)*
If the system (2.4.18) is passive with a C^2 storage function $S(x)$ then it has relative degree one at $x = 0$.

Proof: To derive the analog of the linear equation $B^T PB = CB$ we differentiate both sides of the passivity condition (2.3.27) and, upon the multiplication by $g(x)$, obtain

$$\frac{\partial}{\partial x}(g^T(x)\frac{\partial S}{\partial x}^T(x))g(x) = \frac{\partial h}{\partial x}(x)g(x) \tag{2.4.20}$$

At $x = 0$, $\frac{\partial S}{\partial x}(0) = 0$, and (2.4.20) becomes

$$g^T(0)\frac{\partial^2 S}{\partial x^2}(0)g(0) = L_g h(0)$$

The Hessian $\frac{\partial^2 S}{\partial x^2}$ of S at $x = 0$ is symmetric positive semidefinite and can be factored as $R^T R$. This yields

$$L_g h(0) = g^T(0)\ R^T R\ g(0) \tag{2.4.21}$$

which is the desired nonlinear analog of $B^T PB = CB$. However, $R^T R$ need not be positive definite and we need one additional condition which we obtain by differentiating (2.3.27):

$$\frac{\partial h}{\partial x}(0) = g^T(0)R^T R \tag{2.4.22}$$

Since by assumption $\frac{\partial h}{\partial x}(0)$ has rank m, the matrix $Rg(0)$ must have full rank m. With this we use (2.4.21) to conclude that $L_g h(0)$ is nonsingular. This means that the system (2.4.18) has relative degree one. □

From the proof of Proposition 2.44 we conclude that passivity of the system and full rank of $\frac{\partial h}{\partial x}(0)$ guarantee full rank of $g(0)$. This also excludes nonlinear systems which, because the rank of $L_gh(x)$ drops at $x = 0$, have no relative degree at $x = 0$. If we remove the rank assumption for $\frac{\partial h}{\partial x}(0)$, such systems may still be passive, but their relative degree may not be defined. For example, the system $\dot{x} = xu$ with output $y = x^2$ is passive with the storage function $S(x) = x^2$, but its relative degree at $x = 0$ is not defined.

If the system (2.1.2) has relative degree one at $x = 0$, we can define a local change of coordinates $(z, \xi) = (T(x), h(x))$ and rewrite (2.4.18) in the normal form

$$\begin{aligned} \dot{z} &= q(z,\xi) + \gamma(z,\xi)u \\ \dot{\xi} &= a(z,\xi) + b(z,\xi)u \\ y &= \xi \end{aligned} \tag{2.4.23}$$

where $b(z, \xi) = L_gh(x)$ is locally invertible near $x = 0$. As in the linear case, the zero dynamics are defined as the dynamics which satisfy the constraint $y(t) \equiv 0$, see Appendix A. For the system (2.4.23), the requirement $y \equiv 0$ is satisfied with the feedback law

$$u = -b^{-1}(z,0)a(z,0)$$

which is well-defined locally around $z = 0$. So, the zero-dynamics subsystem exists locally and is described by the differential equation

$$\dot{z} = q(z,0) - \gamma(z,0)b^{-1}(z,0)a(z,0) := f_{zd}(z) \tag{2.4.24}$$

Definition 2.45 *(Minimum phase and weak minimum phase)*
The system (2.4.18) is *minimum phase* if the equilibrium $z = 0$ of its zero-dynamics subsystem (2.4.24) is asymptotically stable. It is *weakly minimum phase* if it is Lyapunov stable and there exists a C^2 positive definite function $W(z)$ such that $L_{f_{zd}}W \leq 0$ in a neighborhood of $z = 0$. □

Proposition 2.46 (*Weak minimum phase of passive systems*)
If the system (2.4.18) is passive with a C^2 positive definite storage function $S(x)$ then it is weakly minimum phase.

Proof: By definition, the zero dynamics of the system (2.4.18) evolve in the manifold $\xi = h(x) = 0$. In this manifold, the second passivity condition $(L_gS)^T(x) = h(x)$ implies $L_gS = 0$, and, because $\dot{S} \leq u^Ty = 0$ we have

$$\dot{S} = L_fS + L_gSu = L_fS \leq 0$$

Thus, $S(x)$ is nonincreasing along the solutions in the manifold $h(x) = 0$ and the equilibrium $z = 0$ of (2.4.24) is stable. □

As in the linear case, the relative degree and the zero dynamics are invariant under the feedback transformation (2.4.2) because $L_gh(0)$ is simply multiplied by $b(0)$ and (2.4.24) is unchanged. So the relative degree one and the weak minimum phase conditions are necessary for feedback passivity. As in the linear case, they are also sufficient and the passivating transformation can be derived from the normal form (2.4.23).

To pursue the analogy with the linear case, we consider the special case when the z-coordinates can be selected such that $\gamma \equiv 0$ in (2.4.23). (A general case is covered in [15].) The normal form (2.4.23) then reduces to

$$\begin{aligned} \dot{z} &= q(z,\xi) \\ \dot{\xi} &= a(z,\xi) + b(z,\xi)u \\ y &= \xi \end{aligned} \tag{2.4.25}$$

and the zero-dynamics subsystem is $\dot{z} = q(z,0)$. We rewrite the first equation of (2.4.25) as

$$\dot{z} = q(z,0) + p(z,\xi)\xi \tag{2.4.26}$$

where $p(z,\xi)$ is smooth if $q(z,\xi)$ is smooth because the difference $\tilde{q}(z,\xi) = q(z,\xi) - q(z,0)$ vanishes at $\xi = 0$ and can be expressed as

$$\tilde{q}(z,\xi) = \int_0^1 \left(\frac{\partial \tilde{q}(z,\zeta)}{\partial \zeta}\right)\Bigg|_{\zeta = s\xi} \xi ds$$

Using (2.4.26), we proceed as in the linear case: if the system is weakly minimum phase, a C^2 positive definite function $W(z)$ exists such that

$$\dot{W}(z) = L_{q(z,0)}W + L_{p(z,\xi)}W\xi \leq L_{p(z,\xi)}W\xi$$

Therefore, with the feedback transformation

$$u(\xi,z) = b^{-1}(z,\xi)(-a(z,\xi) - (L_{p(z,\xi)}W)^T + v) \tag{2.4.27}$$

the positive definite function

$$S(z,\xi) = W(z) + \frac{1}{2}\xi^T\xi$$

satisfies $\dot{S} \leq y^T v$. So, the feedback transformation (2.4.27) renders the system (2.4.18) passive, as summarized in the following theorem.

Theorem 2.47 (*Feedback passivity*)
Assume that rank $\frac{\partial h}{\partial x}(0) = m$. Then the system (2.4.18) is feedback passive with a C^2 positive definite storage function $S(x)$ if and only if it has relative degree one at $x = 0$ and is weakly minimum phase. □

This theorem is of major interest for feedback passivation designs in Chapter 4. A brief example will serve as a preview.

Example 2.48 *(Feedback passivation design)*
By selecting the output $y = x_2$ for

$$\begin{aligned} \dot{x}_1 &= x_1^2 x_2 \\ \dot{x}_2 &= u \end{aligned} \tag{2.4.28}$$

we obtain a relative degree one system which is already in the normal form (2.4.25). Its zero-dynamics subsystem $\dot{x}_1 = 0$ is only stable, that is, (2.4.28) is weakly minimum phase. Feedback transformation (2.4.27) is

$$u = v + x_1^3$$

and renders the system

$$\begin{aligned} \dot{x}_1 &= x_1^2 x_2 \\ \dot{x}_2 &= -x_1^3 + v \\ y &= x_2 \end{aligned} \tag{2.4.29}$$

passive with the storage function $S(x) = \frac{1}{2}x_1^2 + \frac{1}{2}x_2^2$. Since $y(t) \equiv v(t) \equiv 0$ implies $x_1(t) = x_2(t) \equiv 0$, the additional output feedback $v = -y$ achieves GAS of $(x_1, x_2) = (0,0)$. Note that the original system 2.4.28 with $y = x_2$ is neither ZSO nor ZSD, but the transformed system (2.4.29) is ZSO. □

In the above example the feedback passivity property is global, while in Theorem 2.47 it is only local. Global results for feedback passivity depend on the existence of a global normal form (2.4.25). Existence conditions which are coordinate independent can be found in [15].

2.4.4 Output feedback passivity

We now briefly specialize our discussion to OFP systems. According to Definition 2.12, they are feedback passive with the feedback transformation restricted to the form

$$u = -\rho(h(x)) + v, \tag{2.4.30}$$

where ρ is a nonlinear function as in (2.2.3). The relative degree one and weak minimum phase conditions, which are necessary for feedback passivity, are also necessary for OFP. The following proposition provides an additional simple test.

Proposition 2.49 *(Additional test for output feedback passivity)*
If the system (2.4.18) is OFP with a C^2 positive definite storage function $S(x)$, then $L_g h(0)$ is symmetric positive definite.

Proof: For a passive system, it was proven in Proposition 2.44 that the matrix $L_g h(0)$ given by (2.4.21) is symmetric positive definite. This matrix remains unchanged by the output feedback transformation (2.4.30), so the condition is also necessary for output feedback passivity. □

Under what sufficient condition can the system (2.4.25) be rendered passive by output feedback? Because of the close relationship between passivity and stability, this problem is connected with the output feedback stabilization of nonlinear systems in the normal form (2.4.25). An example taken from [14] shows that the relative degree one and minimum phase conditions are not sufficient.

Example 2.50 *(Minimum phase does not imply OFP)*
The second order system

$$\begin{aligned} \dot{z} &= -z^3 + \xi \\ \dot{\xi} &= z + u \\ y &= \xi \end{aligned} \tag{2.4.31}$$

has relative degree one and is minimum phase since its zero-dynamics subsystem is $\dot{z} = -z^3$. So, (2.4.31) is feedback passive. We now prove that it is not OFP. The output feedback $u = -ky$ yields a closed-loop system whose Jacobian linearization at $(z, \xi) = (0, 0)$ has the characteristic polynomial

$$\lambda^2 + k\lambda - 1 = 0 \tag{2.4.32}$$

and is unstable for any $k > 0$. So the feedback $u = -ky + v$ cannot render the system passive, irrespective of the choice of k. □

To guarantee OFP we also require the minimum phase property of the Jacobian linearization.

Proposition 2.51 *(Local output feedback passivity)*
The system (2.4.18) is locally OFP with a quadratic positive definite storage function $S(x)$ if its Jacobian linearization at $x = 0$ is minimum phase and $L_g h(0)$ is symmetric positive definite. □

In this case, standard results of linear theory ensure stabilization of the linear approximation by high-gain output feedback. The associated quadratic Lyapunov function is a storage function for the original system and the high-gain output feedback renders the system passive.

2.5 Summary

The presentation of passivity concepts and results in this chapter has been geared to their subsequent use as design tools for feedback stabilization. The relationship of Lyapunov stability and passivity is one of the focal points, with the stress on the use of storage functions as Lyapunov functions. Because storage functions are allowed to be only positive semidefinite, rather than definite, the same assumption has been made about Lyapunov functions, and stability properties conditional to a set have been introduced. The stability analysis then relies on zero-state detectability properties.

The interplay of passivity and stability in feedback interconnections, which is of paramount importance for feedback stabilization designs, has been given a thorough treatment in Sections 2.2 and 2.3. In a feedback loop, the shortage of passivity in the plant to be controlled can be compensated for by the excess of passivity in the controller. To employ the concepts of shortage and excess of passivity as design tools, *output feedback passive* (OFP) and *input feedforward passive* (IFP) systems have been defined. As a special case, the classical absolute stability theorem has been proven using these concepts.

The chapter ends with *feedback passivity*, the property that a system can be made passive with state feedback. Recent passivity results have been presented which characterize the structural properties of feedback passive systems without throughput: the relative degree one and weak minimum phase. A full understanding of these properties is required in the rest of the book, and, in particular, in Chapters 4 and 6.

2.6 Notes and References

The students in the 1950's who, like one of the authors of this book, learned about passivity in a network synthesis course, and about absolute stability in a control theory course, were unsuspecting of the deep connection between the two concepts. This connection was revealed in the results of V.M. Popov, such as [87]. It stimulated a series of extensions by Yakubovich [121], Kalman [51], Naumov and Tsypkin [83], Sandberg [94], Zames [123] and many other authors. Written in the midst of that development, the monograph by Aizerman and Gantmacher [3], presents an eyewitness report on Lurie's pioneering results [70] and the impact of Popov's breakthrough [87]. Popov and circle stability criteria and various forms of the Positive Real Lemma (Kalman-Yakubovich-Popov Lemma) have since been used in many areas of control theory, especially in adaptive control [40, 61, 82].

Broader implications of passivity were analyzed by Popov in a series of paper, and the book [88]. These include the results on passivity of parallel and feedback interconnections of passive systems, playing the central role in this chapter. The book by Anderson and Vongpanitlerd [2] contains a presentation of the theory of linear passive systems, while the book by Desoer and Vidyasagar [18] treats dissipativity of input-output operators.

The starting point of our presentation is the state space approach presented in the 1972 paper by Willems [120]. This approach has been used by Hill and Moylan [37, 38] to establish conditions for stability of feedback interconnections of nonlinear dissipative systems, which motivated the concepts of excess and shortage of passivity, and of OFP and IFP systems presented in this chapter. Our treatment of these results reconciles the semidefiniteness of storage functions with the properties of Lyapunov functions needed to prove stability.

The characterization of dissipative nonlinear input-affine systems, which is a nonlinear generalization of the KYP Lemma, is due to Hill and Moylan [37]. Kokotović and Sussmann [59] have shown that feedback passive ("feedback positive real") linear systems are restricted by relative degree one and weak minimum phase requirements. General feedback passivity conditions for nonlinear systems have been derived by Byrnes, Isidori, and Willems [15].

Chapter 3

Stability Margins and Optimality

For stabilization of an unstable system, feedback is a necessity. With uncertainties in the operating environment, and in system components, feedback is needed to preserve stability and improve performance. However, feedback can also be dangerous. A tighter feedback loop, instead of achieving better performance, may cause instability. To guard against such dangers, the quantitative concepts of *gain* and *phase stability margins* were among the frequency domain tools of the classical Nyquist-Bode designs.

Although stability margins do not guarantee robustness, they do characterize certain basic robustness properties that every well-designed feedback system must possess. It will be shown in this chapter that optimal feedback systems satisfy this requirement because of their passivity properties.

The classical gain and phase margins, reviewed in Section 3.1, quantify the feedback loop's closeness to instability. *Gain margin* is the interval of gain values for which the loop will remain stable. *Phase margin* is an indicator of the amount of phase lag – and hence, of dynamic uncertainty – that the feedback loop can tolerate.

While the concept of gain margin extends to nonlinear feedback systems, the concept of phase margin does not. In Section 3.2 we interpret absolute stability as a stability margin and we define the notions of nonlinear *gain*, *sector* and *disk stability margins*. They are useful for input uncertainties which do not change the relative degree of the system. Such uncertainties include static nonlinearities, uncertain parameters and unmodeled dynamics of the type of pole-zero pairs. Dynamic uncertainties which change the system's relative degree are much more difficult to handle. We assume that they are faster than the rest of the system and treat them as singular perturbations.

Optimal control as a design tool for nonlinear systems is introduced in Section 3.3, where we present a connection between optimality and passivity established by Moylan [80] for nonlinear systems. In Section 3.4 these results are used to express stability margins achieved by optimal stabilization.

Optimal nonlinear control has a major handicap: it requires the solution of the complicated Hamilton-Jacobi-Bellman (HJB) partial differential equation. In Section 3.5 we follow the inverse path of Freeman and Kokotović [25, 26], which exploits the fact that for an optimal problem to be meaningful, it is not necessary to completely specify its cost functional. If a cost functional imposes a higher penalty for larger control effort in addition to a state cost term, it will result in desirable stability margins.

In Section 3.5 we employ the Artstein-Sontag control Lyapunov functions [4, 98] and Sontag's formula [101] to construct optimal value functions and optimal feedback laws for meaningful control problems.

3.1 Stability Margins for Linear Systems

3.1.1 Classical gain and phase margins

We begin with a review of the classical stability margins for the linear SISO system

$$(H)\quad \left\{ \begin{aligned} \dot{x} &= Ax + bu \\ y &= cx \end{aligned} \right. \tag{3.1.1}$$

also described by its transfer function

$$H(s) = c(sI - A)^{-1}b \tag{3.1.2}$$

In assuming that there is no throughput, $d = 0$, we have made the restriction to strictly proper transfer functions (the relative degree of H is at least one). In addition, throughout this chapter we assume that (A, b, c) is a minimal realization of the transfer function $H(s)$.

Classical gain and phase margins are equivalently defined on Nyquist and Bode plots of the transfer function $H(s)$. They describe the stability properties of $H(s)$ in the feedback loop with gain k, as in Figure 3.1. We will use the Nyquist plot of $H(s)$ which, in the complex plane, is the image of the imaginary axis under the mapping H, that is the curve

$$\Gamma \triangleq \{(a, jb) \mid a = Re\{H(j\omega)\},\ b = Im\{H(j\omega)\},\ \omega \in (-\infty, \infty)\} \tag{3.1.3}$$

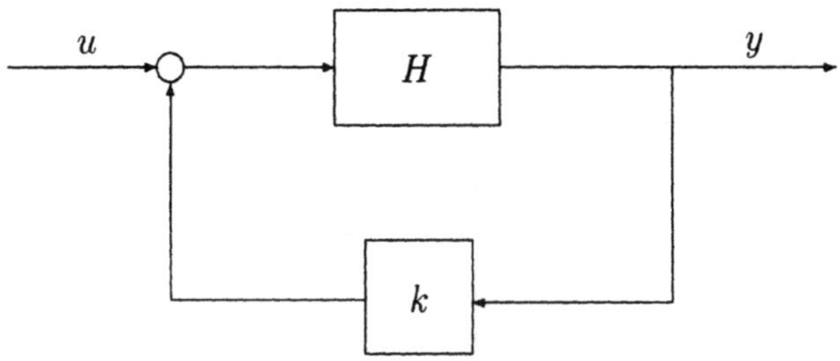

Figure 3.1: Simple static gain feedback.

For a proper rational transfer function H, which has no poles on the imaginary axis, the Nyquist plot is a closed, bounded curve. An example is the plot in Figure 3.2(a). In the case of poles on the imaginary axis, the Nyquist plot is unbounded, as in Figure 3.2(b). We imagine that unbounded plots connect at infinity.

For the feedback system in Figure 3.1, the absence of right half plane poles of $\frac{H(s)}{1+kH(s)}$ is deduced from the relative position of the point $-\frac{1}{k}$ with respect to the Nyquist curve.

Proposition 3.1 *(Nyquist criterion)*
Suppose that the Nyquist plot of H is bounded and let μ be the number of poles of H in the open right half-plane. If the Nyquist curve of $H(s)$ encircles the point $(-\frac{1}{k}, j0)$ in the counterclockwise direction μ times when ω passes from $-\infty$ to $+\infty$, then the feedback interconnection with the constant gain k is GAS. □

The Nyquist criterion is necessary and sufficient for asymptotic stability. If the Nyquist curve of H passes through the point $(-\frac{1}{k}, j0)$, the closed-loop system has a pole on the imaginary axis, and hence, is not asymptotically stable. When H has one or several poles on the imaginary axis, the Nyquist criterion still applies, with each pole on the imaginary axis circumvented by a small half-circle in the right half-plane.

The Nyquist criterion defines a gain margin:

- *gain margin* is an interval $(\alpha, \beta) \subset I\!R$ such that for each constant $\kappa \in (\alpha, \beta)$, the point $(-\frac{1}{\kappa}, j0)$ satisfies the encirclement condition of the Nyquist criterion.

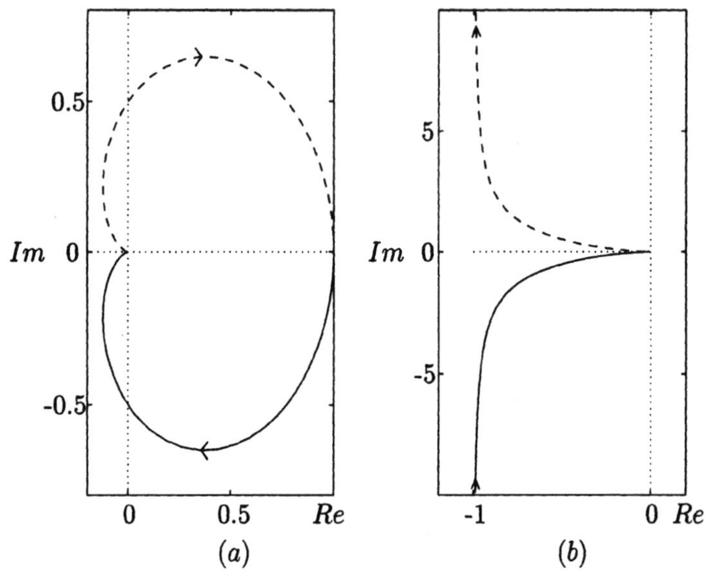

Figure 3.2: Nyquist plots for $H(s) = \frac{1}{(s+q)(s+1)}$, $q = 1$ in (a) and $q = 0$ in (b). The intersection with the real axis $\frac{1}{q}$ for plot (b) is at infinity.

In both plots in Figure 3.2 the gain margin is $(0, \infty)$.

Phase margin is introduced to guard against the effects of unmodeled dynamics which cause phase delays. The worst case is a pure time delay element $e^{-s\tau}$ in the series with $H(s)$. The addition of such an element causes the rotation of each point in the Nyquist curve by the angle $-\tau\omega$. Motivated by this consideration, phase margin is defined as follows:

- *phase margin* ϕ_k for a nominal gain $k > 0$ is the minimal rotation of the Nyquist curve that causes it to pass through the point $(-\frac{1}{k}, j0)$.

In general, phase margin depends on the nominal gain k. We see from the plots in Figure 3.2 that the closer the point $(-\frac{1}{k}, j0)$ gets to the origin, the smaller is the angle for which the Nyquist curve can be rotated without the encirclement of that point. In these two plots the phase margin decreases when k increases. For example, in the plot 3.2(b), if $k = 1$ the phase margin is 51.8°, and if $k = 20$ the phase margin is only 12.8°. However, this is not always the case and a general relation between phase and gain margins does not exist.

3.1.2 Sector and disk margins

The absolute stability conditions (Proposition 2.38) define a stability margin because they guarantee that the feedback loop of $H(s)$ with static nonlinearity $\varphi(\cdot)$ remains asymptotically stable as long as the nonlinearity belongs to a sector (α, β), that is as long as $\alpha y^2 < y\varphi(y) \leq \beta y^2$, $\forall y \in I\!R$.

Definition 3.2 *(Sector margin)*
H has a *sector margin* (α, β) if the feedback interconnection of H with a static nonlinearity $\varphi(\cdot)$ is GAS for any locally Lipschitz nonlinearity φ in the sector (α, β).

□

A special case of the sector nonlinearity $\varphi(y)$ is the linear function κy which belongs to the sector (α, β) whenever $\kappa \in (\alpha, \beta)$. So, if $H(s)$ has a sector margin (α, β), it also has a gain margin (α, β). In 1949 Aizerman [3] made a conjecture that the converse is also true. This conjecture was shown to be incorrect in [86] and in many other counter-examples. One of them, taken from [119], is particularly instructive.

Example 3.3 *(Gain margin versus sector margin)*
Consider the feedback interconnection of the transfer function $H(s) = \frac{s+1}{s^2}$ with a static nonlinearity $\varphi(\cdot)$, described by

$$\begin{aligned} \dot{x}_1 &= x_2 \\ \dot{x}_2 &= -\varphi(x_1 + x_2) \end{aligned} \tag{3.1.4}$$

Clearly, H has a gain margin $(0, \infty)$, that is, the feedback loop of $H(s)$ with $\varphi(y) = \kappa y$, is GAS for any gain $0 < \kappa < \infty$. Next consider the nonlinearity depicted in Figure 3.3 and defined by

$$\varphi(y) = \begin{cases} \dfrac{y}{(e+1)e}, & \text{for } y \leq 1 \\ \dfrac{e^{-y}}{(e^y + 1)}, & \text{for } y \geq 1 \end{cases} \tag{3.1.5}$$

In this case, the solution of (3.1.4) with initial conditions $x_1(0) = \frac{e-1}{e}$, $x_2(0) = \frac{1}{e}$ satisfies $x_2(t) = e^{-(x_1(t)+x_2(t))}$ for all $t \geq 0$. This proves that the solution $x_1(t)$ is increasing for all t. Clearly, the closed-loop system is not asymptotically stable. In fact, it can be shown that $x_1(t)$ grows unbounded. □

Gain and sector margins characterize the class of *static* uncertainties which the feedback loop can tolerate without losing asymptotic stability. Phase margin pertains to *dynamic* uncertainties, but, as a frequency domain concept, cannot be directly generalized to nonlinear systems.

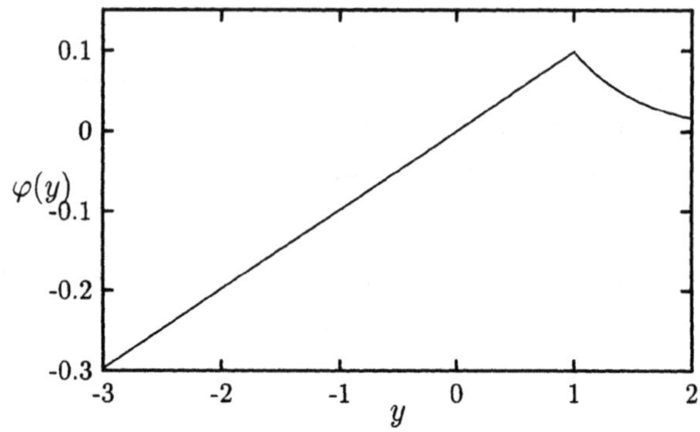

Figure 3.3: Nonlinear function $\varphi(y)$.

We now introduce a *disk margin* as an indicator of the feedback loop's robustness to dynamic uncertainties. For $\alpha < \beta$, we denote by $D(\alpha, \beta)$ the open disk in the complex plane with its center on the real axis and its boundary intersecting the real axis at the points $(-\frac{1}{\alpha}, j0)$ and $(-\frac{1}{\beta}, j0)$ when $\alpha\beta > 0$. When $\alpha\beta < 0$, $D(\alpha, \beta)$ denotes the complement of the closed disk with its center on the real axis and its boundary intersecting the real axis at the points $(-\frac{1}{\alpha}, j0)$ and $(-\frac{1}{\beta}, j0)$. When $\alpha = 0$, $D(0, \beta)$ denotes the open half-plane to the left of the line $Re\{a + jb\} = -\frac{1}{\beta}$. In all these cases we call $D(\alpha, \beta)$ a disk.

Definition 3.4 *(Disk margin)*
Let μ be the number of poles of $H(s)$ in the open right half-plane. We say that H has a disk margin $D(\alpha, \beta)$ if the Nyquist curve of $H(s)$ does not intersect the disk $D(\alpha, \beta)$ and encircles it μ times in the counterclockwise sense.

□

How are different margins related to each other? Let us consider the case $0 < \alpha < \beta$ in Figure 3.4. If H has a disk margin $D(\alpha, \beta)$, then it has a gain margin of (α, β), since for any $k \in (\alpha, \beta)$ the point $(-\frac{1}{k}, j0)$ is in the interior of the disk and the encirclement condition is satisfied. For a phase margin we first need to specify a nominal gain $k^* > 0$ such that $(-\frac{1}{k^*}, j0) \in D(\alpha, \beta)$. Then phase margin is not smaller than ϕ_k in Figure 3.4.

The following result from [9] establishes a connection between passivity and disk margin.

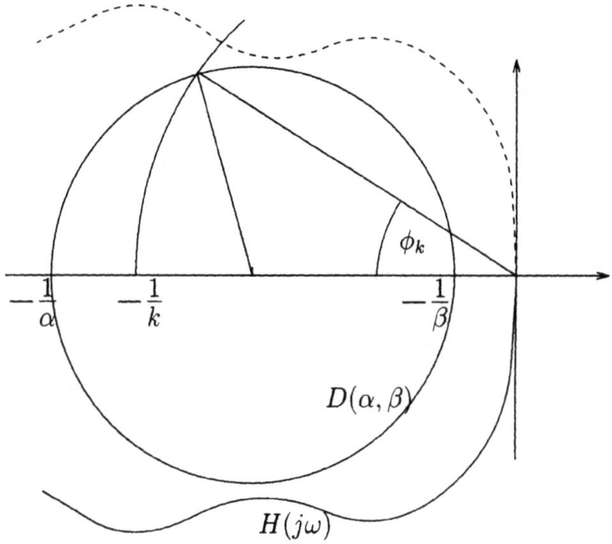

Figure 3.4: Phase and gain margins for systems with $D(\alpha, \beta)$ disk margin.

Proposition 3.5 *(Disk margin and positive realness)*
For $\beta > 0$ the following holds:

(i) If $H(s)$ has a disk margin $D(\alpha, \beta)$, then the transfer function

$$\bar{H}(s) = \frac{H(s) + \frac{1}{\beta}}{\alpha H(s) + 1} \tag{3.1.6}$$

is positive real.

(ii) If the Nyquist curve of $H(s)$ does not intersect $D(\alpha, \beta)$ but encircles it counterclockwise fewer times than the number of poles of $H(s)$ in the open right half-plane, then the transfer function $\bar{H}(s)$ in (3.1.6) is not positive real. □

This theorem allows us to reformulate Proposition 2.38 for linear systems as the well known circle criterion [9, 83, 94, 123].

Proposition 3.6 *(Circle criterion)*
If H has a disk margin $D(\alpha, \beta)$, with $\beta > 0$, then the feedback interconnection of H and the static nonlinearity $\varphi(\cdot)$ is GAS for any nonlinearity in the sector (α, β). □

Thus a disk margin $D(\alpha, \beta)$ implies a sector margin (α, β). However, the converse is not true, as shown by the following example.

Example 3.7 *(Sector margin versus disk margin)*
The system

$$\begin{array}{rcl} \dot{x}_1 & = & -x_1 + x_2 \\ \dot{x}_2 & = & -x_2 + u \\ y & = & x_1 \end{array} \tag{3.1.7}$$

has a sector margin $(0, \infty)$, because for any nonlinearity φ in the sector $(0, \infty)$, the feedback system with $u = -\varphi(y)$,

$$\begin{array}{rcl} \dot{x}_1 & = & -x_1 + x_2 \\ \dot{x}_2 & = & -x_2 - \varphi(x_1) \end{array} \tag{3.1.8}$$

is GAS. This is proven with the Lyapunov function

$$V(x_1, x_2) = \frac{1}{2}x_1^2 + \int_0^{x_1} \varphi(s)\, ds + \frac{1}{2}x_2^2 \tag{3.1.9}$$

Its time-derivative for (3.1.8) is negative definite: $\dot{V} = -x_1^2 + x_1x_2 - x_2^2 - x_1\phi(x_1)$.

In spite of its sector margin $(0, \infty)$, the system (3.1.7) does not have a disk margin $D(\alpha, \infty)$ for any α. This can be verified on the Nyquist plot of its transfer function $H(s) = \frac{1}{(s+1)^2}$ in Figure 3.2(a).

This example also shows that a sector margin does not imply a phase margin. It is clear from the Nyquist plot in Figure 3.2(a) that the phase margin decreases to zero when the nominal gain k is increased. On the other hand, the sector margin remains $(0, \infty)$ for all $k > 0$.

□

To summarize: a system with a disk margin $D(\alpha, \beta)$ has both gain and sector margins (α, β), and a phase margin $\phi_k(\alpha, \beta)$. This shows that disk margin guarantees stronger robustness properties than the other three margins. Furthermore, disk margin will allow us to characterize the class of dynamic uncertainties which do not destabilize the feedback loop. This is done in the next section.

3.1.3 Disk margin and output feedback passivity

When $\beta = \infty$, the disk boundary passes through the origin, and a disk margin is $D(\alpha, \infty)$, denoted simply by $D(\alpha)$. This stability margin is equivalent to the

OFP(-α) property. In one direction this follows immediately from Proposition 3.5. When we let $\beta \to \infty$, then a disk margin $D(\alpha)$ for $H(s)$ implies that

$$\bar{W}(s) = \frac{H(s)}{\alpha H(s) + 1} \tag{3.1.10}$$

is positive real. By KYP Lemma (Theorem 2.41), any minimal realization of $\bar{H}(s)$ is passive. This means that the feedback interconnection of H and a scalar gain α is passive, that is, H is OFP($-\alpha$). The following proposition shows that the converse is also true.

Proposition 3.8 *($D(\alpha)$ is OFP($-\alpha$))*
If H is OFP($-\alpha$) then it has a disk margin $D(\alpha)$.

Proof: By assumption, $\bar{H}(s)$ in (3.1.10) is positive real. By applying the KYP Lemma to the following state space representation of $\bar{H}(s)$:

$$\begin{aligned} \dot{x} &= (A - \alpha bc)x + bu \\ y &= cx \end{aligned} \tag{3.1.11}$$

we obtain a positive definite matrix P such that

$$\begin{array}{l} (A - \alpha bc)^T P + P(A - \alpha bc) \leq 0 \\ Pb = c^T \end{array} \tag{3.1.12}$$

Adding and subtracting $j\omega P$ from the right hand side of the inequality in (3.1.12) and multiplying by -1 we get

$$(-j\omega I - A^T)P + P(j\omega I - A) + \alpha c^T b^T P + \alpha Pbc \geq 0 \tag{3.1.13}$$

Next, multipling both sides of (3.1.13) by $b^T(-j\omega I - A^T)^{-1}$ from the left and by $(j\omega I - A)^{-1}b$ from the right, and substituting $Pb = c^T$, we obtain

$$\begin{array}{c} c(j\omega I - A)^{-1}b + b^T(-j\omega I - A^T)^{-1}c^T + \\ +2\alpha b^T(-j\omega I - A^T)^{-1}c^T c(j\omega I - A)^{-1}b \geq 0 \end{array}$$

Noting that $c(j\omega I - A)^{-1}b = H(j\omega)$ we rewrite the above inequality as

$$H(j\omega) + H(-j\omega) + 2\alpha H(j\omega)H(-j\omega) \geq 0 \tag{3.1.14}$$

If $\alpha > 0$, we divide the inequality (3.1.14) by 2α and rewrite it as

$$\left(\frac{1}{2\alpha} + H(-j\omega)\right)\left(\frac{1}{2\alpha} + H(j\omega)\right) \geq \frac{1}{4\alpha^2}$$

or equivalently

$$\left|\frac{1}{2\alpha} + H(j\omega)\right| \geq \frac{1}{2\alpha}$$

Therefore the Nyquist curve of $H(s)$ does not intersect the disk $D(\alpha)$. Analogously, if $\alpha < 0$, we divide (3.1.14) by 2α and reverse the inequality sign to obtain

$$\left|\frac{1}{2\alpha} + H(j\omega)\right| \leq \frac{1}{2|\alpha|}$$

Again, the Nyquist curve of $H(s)$ does not intersect $D(\alpha)$. Finally, because $\frac{H}{1+\alpha H}$ is positive real, it follows from Proposition 3.5, part (ii), that the number of encirclements of the disk by the Nyquist curve of $H(s)$ is equal to the number of the poles of $H(s)$ in the right half-plane. □

With $\alpha = 0$, from (3.1.14) we recover the positive realness property that, if the linear system H is passive, the Nyquist curve of its transfer function lies in the closed right half plane: its real part is nonnegative. Finally, disk margin $D(0, \beta)$ is an IFP property.

Proposition 3.9 *($D(0,\beta)$ is IFP($-\frac{1}{\beta}$))*
H has a disk margin $D(0, \beta)$ if and only if H is IFP($-\frac{1}{\beta}$).

Proof: This property is a direct consequence of the fact that H is IFP($-\frac{1}{\beta}$) if and only if $H' = H + \frac{1}{\beta}$ is passive. It is clear that $H'(s)$ has a disk margin $D(0, \infty)$, that is the Nyquist curve of $H'(s)$ is in the closed right half plane, if and only if $H(s) = H'(s) - \frac{1}{\beta}$ has a disk margin $D(0, \beta)$ because the subtraction of $\frac{1}{\beta}$ just shifts the Nyquist curve by $-\frac{1}{\beta}$.

□

The following example illustrates the Nyquist plot of an IFP system, and will be helpful in the proof of the subsequent theorem.

Example 3.10 *(Nyquist plot of an IFP system)*
For $p = 1$ the Nyquist curve of the transfer function

$$G(s) = \frac{p}{(s+1)^2}$$

is given in Figure 3.2(a). This transfer function has a disk margin $D(0, \frac{8}{p})$ so that the system is IFP($-\frac{p}{8}$). This is verified in Figure 3.2(a) because the Nyquist curve of $G(s)$ lies to the right of the vertical line passing through the minimal value of $Re\{G(j\omega)\}$. This minimal value is equal to $-\frac{p}{8}$ at $\omega = \sqrt{3}$. The imaginary part at $\omega = \sqrt{3}$ is $Im\{G(j\sqrt{3})\} = -\frac{p\sqrt{3}}{8}$.

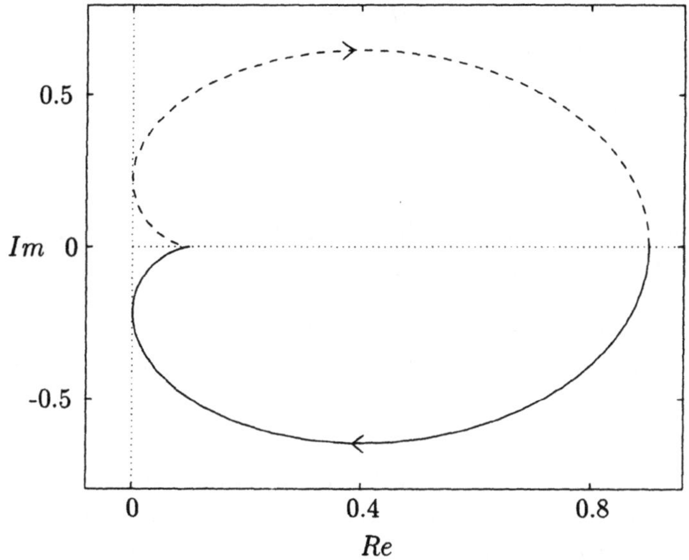

Figure 3.5: Nyquist plot for $G(s) = \frac{1}{8} + \frac{1}{(s+1)^2}$.

Note now that the Nyquist plot of $G(s)$ augmented by a throughput term $r > 0$,

$$G'(s) = r + \frac{p}{(s+1)^2}$$

is translated to the right. By selecting $r = \frac{p}{8}$, as shown in Figure 3.5 for $p = 1$, we make $Re\{G'(j\omega)\}$ nonnegative, that is, we make the transfer function $G'(s)$ positive real. By changing p we can make the graph touch the imaginary axis at any symmetric pair of purely imaginary points.

By increasing r the shift is further to the right and $r = \nu + \frac{p}{8}$ renders $Re\{G'(j\omega)\} \geq \nu$. Because $G(s)$ has no poles in the right half plane this means that $G'(s)$ has a disk margin $D(0, -\frac{1}{\nu})$ and, by Proposition 3.9, any minimal realization of $G'(s)$ is IFP(ν). □

The equivalence between a disk margin $D(\alpha)$ and the OFP($-\alpha$) property provides us with a characterization of the dynamic uncertainties which do not cause the loss of stability. Such a characterization will be essential for our definition of a disk margin for nonlinear systems.

Theorem 3.11 *(Disk margin and IFP uncertainties)*
For linear systems, the following statements are equivalent:

(i) H_1 has a disk margin $D(\alpha)$;

(ii) H_1 is OFP$(-\alpha)$;

(iii) The feedback loop formed of H_1 and any linear system H_2 which is GAS and IFP(ν), with $\nu > \alpha$, is GAS.

Proof: (i) $\Rightarrow$ (ii) follows from Proposition 3.5 with $\beta = \infty$. (ii) $\Rightarrow$ (iii) is an application of the interconnection Theorem 2.34. What remains to be proven is (iii) $\Rightarrow$ (i).

We prove that H_1 has a disk margin $D(\alpha)$ by contradiction. First, if the Nyquist curve of $H_1(s)$ does not enter the disk, but the number of encirclements is not equal to μ, then, by Nyquist criterion, the feedback interconnection of H_1 and k is unstable for any $k > \alpha$. This is because the Nyquist curve does not encircle the point $-\frac{1}{k}$ μ times. Hence, since k is an IFP(k) system and $k > \alpha$, we have a contradiction.

The second case is when the Nyquist curve of $H_1(s)$ intersects the disk $D(\alpha)$. Assume that there exists $\omega_1 > 0$ such that $H_1(j\omega_1) := a + jb \in D(\alpha)$ with $b < 0$ (the case $b > 0$ is treated below; if $b = 0$ we can always find another point inside the disk with $b \neq 0$). This implies that

$$(\frac{1}{2\alpha} + a)^2 + b^2 < \frac{1}{4\alpha^2}$$

and thus $\alpha < \frac{-a}{a^2+b^2}$.

Let $G(s)$ be a positive real transfer function with its poles in the open left half-plane and satisfying the condition $G(j\omega_1) = \frac{jb}{a^2+b^2}$. Such a function is provided by Example 3.10:

$$G(s) = \frac{p}{8} + \frac{p\omega_1^2}{(\sqrt{3}s + \omega_1)^2}, \qquad p = \frac{8}{\sqrt{3}} \frac{-b}{a^2 + b^2} > 0$$

Then

$$H_2(s) := \frac{-a}{a^2 + b^2} + G(s)$$

satisfies $H_2(j\omega_1) = -\frac{a-jb}{a^2+b^2}$ and defines a GAS system H_2 which is IFP$(\frac{-a}{a^2+b^2})$. Because $\alpha < \frac{-a}{a^2+b^2}$, H_2 is IFP(ν), with $\nu > \alpha$.

However, the feedback interconnection of H_1 and H_2 is not GAS because

$$1 + H_1(j\omega_1)H_2(j\omega_1) = 0 \tag{3.1.15}$$

We conclude that the closed-loop system has poles on the imaginary axis, which contradicts the asymptotic stability of the interconnection.

The case when $H_1(j\omega_1) = a + jb \in D(\alpha)$ with $b > 0$ is handled in a similar way with

$$H_2(s) = \frac{-a}{a^2+b^2} + \frac{p}{8} + \frac{ps^2}{(\frac{1}{\sqrt{3}}s + \omega_1)^2}, \quad p = 8\sqrt{3}\frac{b}{a^2+b^2} > 0$$

□

3.2 Input Uncertainties

3.2.1 Static and dynamic uncertainties

For linear systems the stability margins discussed in Section 3.1 delineate types of uncertainties with which the feedback loop retains asymptotic stability. We now extend this analysis to a wider class of nonlinear feedback systems shown in Figure 3.6 where u and y are of the same dimension and Δ represents modeling uncertainty. In the nominal case Δ is identity, and the feedback

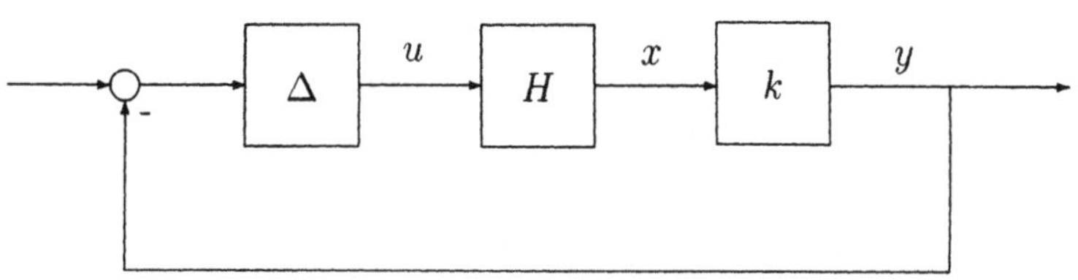

Figure 3.6: Nonlinear feedback loop with the control law $k(x)$ and input uncertainty Δ.

loop consists of the (nominal) nonlinear plant H in the feedback loop with the nominal control $u = -k(x) =: -y$. We denote the nominal system by (H, k) and the perturbed system by (H, k, Δ).

The block-diagram in Figure 3.6 restricts modeling uncertainties to be at the input. This is a common physical situation, in particular when simplified models of actuators are used for the design.

As we shall see, our disk margin will guarantee robustness with respect to the input uncertainties which do not change the relative degree of the nominal model. This restricts the relative degree of Δ to be 0. Uncertainties which cause a change in the relative degree are more severe. For general nonlinear systems with such uncertainties we can at most preserve the desired stability properties in a certain region of attraction. For fast unmodeled dynamics, which can be characterized as singular perturbations, we will be able to give estimates of that region.

The input uncertainties Δ which do not change the relative degree can be static or dynamic. The two most common static uncertainties are

- *unknown static nonlinearity* $\varphi(\cdot)$ which belongs to a known sector (α, β), including, as a special case, the unknown static gain,
- *unknown parameters* belonging to known intervals in which the relative degree remains the same.

It is important to clarify the above restriction on parametric uncertainty.

Example 3.12 *(Parametric uncertainty)*
In the following three systems

$$\frac{1}{s+q_1} \tag{3.2.1}$$

$$\frac{s+q_2}{(s+1)(s+2)} \tag{3.2.2}$$

$$\begin{aligned} \dot{x}_1 &= f(x_1, x_2) + q_3 u \\ \dot{x}_2 &= u, \qquad y = x_1 \end{aligned} \tag{3.2.3}$$

the unknown parameter is denoted by $q_i, i = 1, 2, 3$. The admissible intervals of parameter uncertainties do not include $q_3 = 0$, because then the relative degree changes from one to two, even though the dynamic order of the system remains two. No such restriction is imposed on q_1 and q_2, because even when at $q_2 = 2$ the input-output description reduces to $\frac{1}{s+1}$, the relative degree remains the same. Likewise, no finite variation of q_1 can change the relative degree of (3.2.1).

It should also be pointed out that the value $q_3 = 0$ must not be used for the nominal model, because for any variation of q_3 the relative degree will drop from two to one.

□

In linear systems a *dynamic* uncertainty which does not change the relative degree is due to neglected pole-zero pairs. For example, in the system (3.2.2), if q_2 is known to be close to 2, the designer may choose to treat $\frac{1}{s+1}$ as the nominal plant and to neglect the dynamics $\frac{s+q_2}{s+2}$ which thus becomes the input unmodeled dynamics Δ.

In some cases, parametric uncertainty can be represented as an input dynamic uncertainty Δ. As an illustration, we consider again (3.2.1). Instead of q_1 which is unknown, we let an estimate $\hat{q}_1$ be used in the nominal plant H. Then the difference Δ between the actual plant and the nominal plant becomes

$$\Delta = \frac{s + \hat{q}_1}{s + q_1} \tag{3.2.4}$$

In this way a parametric uncertainty $q_1 - \hat{q}_1$ is converted into a dynamic uncertainty which does not change the relative degree of the nominal plant, as in the case (3.2.2). The fact that the actual plants in the two cases are of different dynamic order is of no consequence for the stability analysis. All that matters is that the control design for the nominal system possesses sufficient stability margin, which tolerates Δ as an input uncertainty.

When Δ represents an uncertainty which changes the relative degree, the concepts of gain, sector, and disk margins are no longer applicable, except when Δ has relative degree one and the nominal system is passive.

It is common practice to neglect the dynamics of the devices which are much faster than the rest of the system. In this case we have to deal with *fast unmodeled dynamics.* The separation of time scales into slow and fast allows the design to be performed on the nominal slow model. This has been justified by the theory of singular perturbations [57]. A standard singular perturbation form is

$$\begin{aligned} \dot{x} &= f_c(x, z, u), \quad x \in I\!R^{n_1} \\ \mu\dot{z} &= q_c(x, z, u), \quad z \in I\!R^{n_2} \end{aligned} \tag{3.2.5}$$

where $\mu > 0$ is the singular perturbation parameter. In the nominal model we set $\mu = 0$ and obtain

$$\dot{x} = f_c(x, h(x, u), u) \tag{3.2.6}$$

where $h(x, u)$ satisfies $q_c(x, h(x, u), u) = 0$, that is $z = h(x, u)$ is a root of $q_c(x, z, u) = 0$. Thus the order of the nominal slow model (3.2.6) is n_1, while that of the actual system (3.2.5) is $n_1 + n_2$. In general, such an increase in model order leads to an increase in the relative degree.

A fundamental property of the singular perturbation model is that it possesses two time scales: the slow time scale of the x-dynamics, and a fast time scale of the z-dynamics. The separation of time scales is parameterized by μ: with smaller μ, the z-state is faster, as can be seen from the fact that $\dot{z}$ is proportional to $\frac{1}{\mu}$. Hence the term *fast unmodeled dynamics.*

3.2.2 Stability margins for nonlinear feedback systems

To deal with uncertainties which do not change the relative degree we extend the concept of stability margins to nonlinear feedback systems. The extension of the definitions of gain and sector margins is straightforward.

Definition 3.13 *(Gain margin)*
The nonlinear feedback system (H, k) is said to have a gain margin (α, β) if the perturbed closed-loop system (H, k, Δ) is GAS for any Δ which is of the form $diag\{\kappa_1, \ldots, \kappa_m\}$ with constants $\kappa_i \in (\alpha, \beta)$, $i = 1, \ldots, m$. □

Definition 3.14 *(Sector margin)*
The nonlinear feedback system (H, k) is said to have a sector margin (α, β) if the perturbed closed-loop system (H, k, Δ) is GAS for any Δ which is of the form $diag\{\varphi_1(\cdot), \ldots, \varphi_m(\cdot)\}$ where $\varphi_i(\cdot)$'s are locally Lipschitz static nonlinearities which belong to the sector (α, β). □

Phase margin, which is defined in the frequency domain, cannot be extended to the nonlinear case. In contrast, disk margin, which is also defined in the frequency domain, can be extended to nonlinear systems using the characterization given in Theorem 3.11.

Definition 3.15 *(Disk margin)*
The nonlinear feedback system (H, k) is said to have a disk margin $D(\alpha)$ if the closed-loop system (H, k, Δ) is GAS for any Δ which is GAS and IFP(ν), $\nu > \alpha$, with a radially unbounded storage function. □

When (H, k) is a SISO linear system, the above definition of disk margin coincides with Definition 3.4 which defines the notion of disk margin in terms of the Nyquist curve of the transfer function. This is guaranteed by the equivalence of (i) and (iii) in Theorem 3.11. Note that the above assumptions on Δ are such that Theorem 2.34 guarantees a $D(\alpha)$ disk margin for any ZSD, OFP($-\alpha$) nonlinear feedback system.

A nonlinear system having a disk margin $D(\alpha)$ also has gain and sector margins (α, ∞). This is so because constant gain and static nonlinearity are IFP uncertainties with void state space.

3.2.3 Stability with fast unmodeled dynamics

Do wider stability margins imply improved robustness with respect to fast unmodeled dynamics? Unfortunately, this is not always the case and judicious trade-offs may be required. For example, an increase in the nominal gain may increase stability margins, but it may also increase the bandwidth thus leading to higher danger of instability caused by fast unmodeled dynamics.

Example 3.16 *(Trade-off between two types of robustness)*
For $k > 1$ the nominal system

$$H(s) = \frac{1}{s-1}$$

is stabilized with the control law $u = -ky$. By choosing larger k we increase

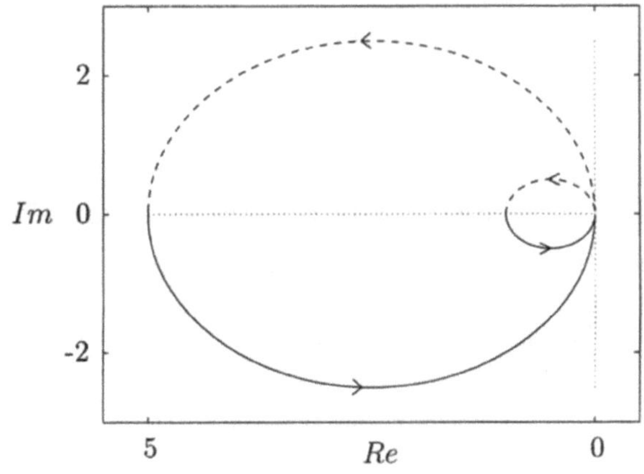

Figure 3.7: Nyquist plot of $H(s) = \frac{k}{s-1}$ with $k = 1$ and $k = 5$.

the disk margin of the system as shown in Figure 3.7 where the smaller circle corresponds to $k = 1$ and the larger circle corresponds to $k = 5$. This tells us that with larger nominal gain k, the feedback system can tolerate larger uncertainty. However, this is true only for uncertainties which *do not change the relative degree*. With $\Delta(s) = \frac{100}{(s+10)^2}$, which has relative degree two, the perturbed systems is unstable for $k > 16.6$. □

The representation of fast unmodeled dynamics in the standard singular perturbation form (3.2.5) is natural for many physical plants and will be illustrated by a robotic example from [104].

Example 3.17 *(Single-link manipulator)*
For the single link with joint flexibility, shown in Figure 3.8, actuator M delivers a torque τ_m to the motor shaft which is coupled, via the gear transmission

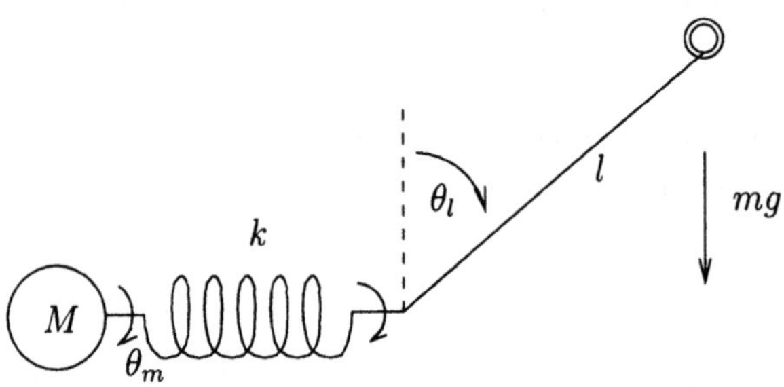

Figure 3.8: Single-link manipulator with joint flexibility.

with ratio n, to the link of length l, mass m, and moment of inertia $\frac{1}{3}ml^2$. When the flexibility is modeled by a linear torsional spring with stiffness k, the equations of motion are

$$\frac{1}{3}ml^2\ddot{\theta}_l + B_l\dot{\theta}_l + \frac{mgl}{2}\sin\theta_l + \zeta = 0 \tag{3.2.7}$$

$$J_m\ddot{\theta}_m + B_m\dot{\theta}_m + \frac{1}{n}\zeta = \tau_m \tag{3.2.8}$$

$$\zeta = k(\theta_l - \frac{1}{n}\theta_m) \tag{3.2.9}$$

Introducing the notation

$$a_1 = -\frac{3B_l}{ml^2},\ a_2 = -\frac{3g}{2l},\quad a_3 = \frac{B_m}{J_m} - \frac{3B_l}{ml^2},\quad a_4 = -\frac{B_m}{J_m}$$

$$A_1 = -\frac{3}{ml^2},\quad A_2 = A_1 - \frac{1}{nJ_m},\quad b = \frac{1}{nJ_m}$$

we rewrite the equations of motion in terms of θ_l, ζ, and $\frac{1}{k}$ as a small parameter:

$$\begin{aligned}\ddot{\theta}_l &= a_1\dot{\theta}_l + a_2\sin\theta_l + A_1\zeta \\ \tfrac{1}{k}\ddot{\zeta} &= a_4\tfrac{1}{k}\dot{\zeta} + A_2\zeta + a_3\dot{\theta}_l + a_2\sin\theta_l + b\tau_m\end{aligned}$$

A common actuator for this application is a DC-motor. Its torque is $\tau_m = k_mI$, where k_m is a motor constant and I is the armature current governed by

$$\frac{L}{R}\dot{I} = -I - \frac{\beta}{R}\dot{\theta}_m + \frac{1}{R}v \tag{3.2.10}$$

with R and L being the armature resistance and inductance, and β the speed voltage constant. The control input is the armature voltage v. In (3.2.10) the time constant $\frac{L}{R}$ is exhibited as another small parameter. We can represent $\frac{1}{k}$ and $\frac{L}{R}$ as functions of a single small parameter:

$$\frac{1}{\sqrt{k}} = c\frac{L}{R} = \mu$$

with $\sqrt{k}$ rather than k, because $\sqrt{k}$ is proportional to the natural frequency of the flexible mode. Using the state variables

$$x_1 = \theta_l,\ x_2 = \dot{\theta}_l,\ z_1 = \zeta,\ z_2 = \frac{1}{\sqrt{k}}\dot{\zeta},\ z_3 = I$$

it is easy to verify that the above equations constitute a fifth order singularly perturbed system in the standard form (3.2.5):

$$\begin{aligned}
\dot{x}_1 &= x_2 \\
\dot{x}_2 &= a_1 x_2 + a_2 \sin x_1 + A_1 z_1 \\
\mu \dot{z}_1 &= z_2 \\
\mu \dot{z}_2 &= a_3 x_2 + a_2 \sin x_1 + \mu a_4 z_2 + A_2 z_1 + b k_m z_3 \\
\mu \dot{z}_3 &= a_5 z_3 + a_6 \mu z_2 + a_6 x_2 + u
\end{aligned}$$

where $a_5 = -c$, $a_6 = -\frac{c\beta n}{R}$, and $u = \frac{c}{R}v$ is the control input. In the nominal model we neglect the fast unmodeled dynamics by letting $\mu = 0$, that is $\frac{1}{\sqrt{k}} = 0$ and $\frac{L}{R} = 0$. The nominal slow model is the second order system

$$\begin{aligned}
\dot{x}_1 &= x_2 \\
\dot{x}_2 &= \tilde{a}_1 x_2 + \tilde{a}_2 \sin x_1 + \tilde{a}_3 u,
\end{aligned}$$

where $\tilde{a}_1 = a_1 + A_1(bk_m\frac{\beta n}{R} - a_3 A_2^{-1})$, $\tilde{a}_2 = (1 - A_2^{-1}A_1)a_2$, and $\tilde{a}_3 = -A_1 A_2^{-1} b k_m$. It represents the single link manipulator with a rigid joint driven by an ideal DC-motor. The armature current transients and the flexible mode are the fast unmodeled dynamics. □

In the above example, the perturbation block Δ with input u and output z_1 is a dynamic system with relative degree three, it is not passive, and hence, cannot be handled by our stability margins. This situation is typical of fast unmodeled dynamics, for which we need a different robustness indicator. A sufficient time-scale separation between "fast" unmodeled dynamics and the "slow" nominal model validates a design based on the nominal model. For this purpose, we extend a stability result [57] for the system

$$\dot{x} = f_c(x, z, u), \quad x \in \mathbb{R}^{n_x} \tag{3.2.11}$$

$$\mu \dot{z} = q_c(x, z, u), \quad z \in \mathbb{R}^{n_z} \tag{3.2.12}$$

When we let the nominal feedback control law be $u = -k(x)$ and denote

$$f_c(x, z, -k(x)) =: f(x, z), \quad q_c(x, z, -k(x)) =: q(x, z),$$

we obtain the standard singular perturbation form

$$\dot{x} = f(x, z), \quad x \in I\!R^{n_x} \tag{3.2.13}$$
$$\mu\dot{z} = q(x, z), \quad z \in I\!R^{n_z} \tag{3.2.14}$$

where, without loss of generality, we assume that $f(0,0) = 0$ and $q(0,0) = 0$. For this system the following stability result is proven in Appendix B.

Theorem 3.18 *(Robustness with respect to fast unmodeled dynamics)*
Let the following assumptions be satisfied:

(i) The equation

$$0 = q(x, z)$$

obtained by setting $\mu = 0$ in (3.2.14) has a unique C^2 solution $z = \bar{z}(x)$

(i) The equilibrium $x = 0$ of the reduced (slow) model

$$\dot{x} = f(x, \bar{z}(x)) \tag{3.2.15}$$

is GAS and LES.

(iii) For any *fixed* $x \in I\!R^{n_x}$ the equilibrium $z_e = \bar{z}(x)$ of the subsystem (3.2.14) is GAS and LES.

Then for every two compact sets $C_x \in \mathbb{R}^{n_x}$ and $C_z \in \mathbb{R}^{n_z}$ there exists $\mu^* > 0$ such that for all $0 < \mu \le \mu^*$ the equilibrium $(x, z) = (0, 0)$ of the system (3.2.13), (3.2.14) is asymptotically stable and its region of attraction contains $C_x \times C_z$. □

We refer to this form of asymptotic stability as "semiglobal in μ" because a larger size of the region of attraction requires a smaller singular perturbation parameter μ, that is, a wider time-scale separation between the nominal model and the fast unmodeled dynamics.

3.3 Optimality, Stability, and Passivity

3.3.1 Optimal stabilizing control

We now introduce optimal control as a design tool which guarantees stability margins. Of the two types of optimality conditions, Pontryagin-type necessary conditions ("Maximum Principle") and Bellman-type sufficient conditions ("Dynamic Programming"), the latter is more suitable for feedback design over infinite time intervals [1]. This will be our approach to the problem of finding a feedback control $u(x)$ for the system

$$\dot{x} = f(x) + g(x)u, \tag{3.3.1}$$

with the following properties:

(i) $u(x)$ achieves asymptotic stability of the equilibrium $x = 0$

(ii) $u(x)$ minimizes the cost functional

$$J = \int_0^\infty (l(x) + u^T R(x)u)\, dt \tag{3.3.2}$$

where $l(x) \geq 0$ and $R(x) > 0$ for all x.

For a given feedback control $u(x)$, the value of J, if finite, is a function of the initial state $x(0)$: $J(x(0))$, or simply $J(x)$. When J is at its minimum, $J(x)$ is called the *optimal value function*. Preparatory for our use of the optimal value function $J(x)$ as a Lyapunov function, we denote it by $V(x)$. When we want to stress that $u(x)$ is optimal, we denote it by $u^*(x)$. The functions $V(x)$ and $u^*(x)$ are related to each other via the following optimality condition.

Theorem 3.19 *(Optimality and stability)*
Suppose that there exists a C^1 positive semidefinite function $V(x)$ which satisfies the Hamilton-Jacobi-Bellman equation

$$l(x) + L_f V(x) - \frac{1}{4} L_g V(x) R^{-1}(x)(L_g V(x))^T = 0, \quad V(0) = 0 \tag{3.3.3}$$

such that the feedback control

$$u^*(x) = -\frac{1}{2} R^{-1}(x)(L_g V)^T(x) \tag{3.3.4}$$

achieves asymptotic stability of the equilibrium $x = 0$. Then $u^*(x)$ is the *optimal stabilizing control* which minimizes the cost (3.3.2) over all u guaranteeing $\lim_{t\to\infty} x(t) = 0$, and $V(x)$ is the optimal value function.

Proof: Substituting

$$v = u + \frac{1}{2}R^{-1}(x)(L_gV(x))^T$$

into (3.3.2) and using the HJB-identity we get the following chain of equalities:

$$\begin{aligned} J &= \int_0^\infty (l + v^TRv - v^T(L_gV)^T + \frac{1}{4}L_gVR^{-1}(L_gV)^T)\, dt \\ &= \int_0^\infty (-L_fV + \frac{1}{2}L_gVR^{-1}(L_gV)^T - L_gVv)\, dt + \int_0^\infty v^TR(x)v\, dt \\ &= -\int_0^\infty \frac{\partial V}{\partial x}(f + gu)\, dt + \int_0^\infty v^TR(x)v\, dt = -\int_0^\infty \frac{dV}{dt} + \int_0^\infty v^TR(x)v\, dt \\ &= V(x(0)) - \lim_{T\to\infty} V(x(T)) + \int_0^\infty v^TR(x)v\, dt \end{aligned}$$

Because we minimize (3.3.2) only over those u which achieve $\lim_{t\to\infty} x(t) = 0$, the above limit of $V(x(T))$ is zero and we obtain

$$J = V(x(0)) + \int_0^\infty v^TR(x)v\, dt$$

Clearly, the minimum of J is $V(x(0))$. It is reached for $v(t) \equiv 0$ which proves that $u^*(x)$ given by (3.3.4) is optimal and that $V(x)$ is the optimal value function.

□

Example 3.20 *(Optimal stabilization)*
For the optimal stabilization of the system

$$\dot{x} = x^2 + u$$

with the cost functional

$$J = \int_0^\infty (x^2 + u^2)\, dt \tag{3.3.5}$$

we need to find a positive semidefinite solution of the HJB equation

$$x^2 + \frac{\partial V}{\partial x}x^2 - \frac{1}{4}\left(\frac{\partial V}{\partial x}\right)^2 = 0, \quad V(0) = 0$$

Solving it first as the quadratic equation in $\frac{\partial V}{\partial x}$ we get

$$\frac{\partial V}{\partial x} = 2x^2 + 2x\sqrt{x^2+1}$$

where the positive sign is required for the optimal value function to be positive semidefinite:

$$V(x) = \frac{2}{3}(x^3 + (x^2+1)^{\frac{3}{2}} - 1) \tag{3.3.6}$$

It can be checked that $V(x)$ is positive definite and radially unbounded. The control law

$$u^*(x) = -\frac{1}{2}\frac{\partial V}{\partial x} = -x^2 - x\sqrt{x^2+1} \tag{3.3.7}$$

achieves GAS of the resulting feedback system

$$\dot{x} = -x\sqrt{x^2+1}$$

and hence, is the optimal stabilizing control for (3.3.5). □

In the statement of Theorem 3.19 we have assumed the existence of a positive semidefinite solution $V(x)$ of the HJB equation. For the LQR-problem the HJB equation (3.3.3) can be solved with the help of an algebraic Ricatti equation whose properties are well known. For further reference we quote a basic version of this well known result.

Proposition 3.21 *(LQR-problem)*
For optimal stabilization of the linear system

$$\dot{x} = Ax + Bu \tag{3.3.8}$$

with respect to the cost functional

$$J = \int_0^\infty (x^T C^T C x + u^T R u)dt, \quad R > 0$$

consider the Ricatti equation

$$PA + A^T P - PBR^{-1}B^T P + C^T C = 0 \tag{3.3.9}$$

If (A, B) is controllable and (A, C) is observable, then (3.3.9) has a unique positive definite solution P^*, the optimal value function is $V(x) = x^T P^* x$, and the optimal stabilizing control is

$$u^*(x) = -R^{-1}B^T P^* x$$

If (A, B) is stabilizable and (A, C) is detectable then P^* is positive semidefinite. □

A proof of this result can be found in any standard text, such as [1]. For our further discussion, the semidefiniteness of $l(x) = x^T C^T C x$ is of interest because it shows the significance of an observability property. It is intuitive that "the detectability in the cost" of the unstable part of the system is necessary for an optimal control to be stabilizing. A scalar example will illustrate some of the issues involved.

Example 3.22 *(Optimal control and "detectability in the cost")*
For the linear system

$$\dot{x} = x + u$$

and the cost functional

$$J = \int_0^\infty u^2 \, dt \tag{3.3.10}$$

we have $A = 1$, $B = 1$, $C = 0$, $R = 1$. The Ricatti equation and its solutions P_1 and P_2 are

$$2P - P^2 = 0, \;\; P_1 = 0, \;\; P_2 = 2 \tag{3.3.11}$$

It can also be directly checked that the solutions of the HJB equation

$$x\frac{\partial V}{\partial x} - \frac{1}{4}(\frac{\partial V}{\partial x})^2 = 0, \qquad V(0) = 0$$

are $V_1(x) = 0$ and $V_2(x) = 2x^2$, that is $V_1(x) = P_1x^2$, $V_2(x) = P_2x^2$. The smaller of the two, $V_1(x)$, gives the minimum of the cost functional, but the control law $u(x) = 0$ is not stabilizing. The reason is that $l(x) = 0$ and the instability of $\dot{x} = x$ is not detected in the cost functional.

According to Theorem 3.19, in which the minimization of J is performed only over the set of stabilizing controls, $V_2(x) = 2x^2$ is the optimal value function and $u(x) = -2x$ is the optimal stabilizing control.

The assumptions of Theorem 3.19 can be interpreted as incorporating a detectability condition. This can be illustrated by letting the cost functional J in (3.3.10) be the limit, as $\epsilon \to 0$, of the augmented cost functional

$$J^\epsilon = \int_0^\infty (\epsilon^2 x^2 + u^2) dt$$

in which the state is observable. The corresponding Ricatti equation, and its solutions P_1^ϵ and P_2^ϵ are

$$2P - P^2 + \epsilon^2 = 0, \;\; P_1^\epsilon = 1 - \sqrt{1+\epsilon}, \; P_2^\epsilon = 1 + \sqrt{1+\epsilon}$$

The HJB solutions $V_1^\epsilon(x) = (1 - \sqrt{1+\epsilon})x^2$ and $V_2^\epsilon(x) = (1 + \sqrt{1+\epsilon})x^2$ converge, as $\epsilon \to 0$, to $V_1(x) = 0$ and $V_2(x) = 2x^2$, respectively. This reveals that $V_1(x) = 0$ is the limit of $V_1^\epsilon(x)$ which, for $\epsilon > 0$, is negative definite while J^ϵ must be nonnegative. Hence $V_1^\epsilon(x)$ cannot be a value function, let alone an optimal value function. The optimal value function for J^ϵ is $V_2^\epsilon(x)$ and Theorem 3.19 identifies its limit $V_2(x)$ as the optimal value for J. □

In our presentation thus far we have not stated the most detailed conditions for optimality, because our approach will be to avoid the often intractable task of solving the HJB equation (3.3.3). Instead, we will employ Theorem 3.19 only as a test of optimality for an already designed stabilizing control law.

3.3.2 Optimality and passivity

In the special case $R(x) = I$, that is when (3.3.2) becomes

$$J = \int_0^\infty (l(x) + u^T u)\, dt \tag{3.3.12}$$

the property that the system (3.3.1) is stabilized with a feedback control which minimizes (3.3.12) is closely related to a passivity property. The following result is a variant of Theorem 4 in [81].

Theorem 3.23 *(Optimality and passivity)*
The control law $u = -k(x)$ is optimal stabilizing for the cost functional (3.3.12) if and only if the system

$$\begin{aligned} \dot{x} &= f(x) + g(x)u \\ y &= k(x) \end{aligned} \tag{3.3.13}$$

is ZSD and OFP(-$\frac{1}{2}$) with a C^1 storage function $S(x)$.

Proof: The control law $u = -k(x)$ is optimal stabilizing for (3.3.12) if

(i) it achieves asymptotic stability of $x = 0$ for (3.3.13), and

(ii) there exists a C^1, positive semidefinite, function $V(x)$ such that

$$\begin{aligned} k(x) &= \tfrac{1}{2}(L_gV)^T \\ l(x) &= \tfrac{1}{4}L_gV(L_gV)^T - L_fV \geq 0 \end{aligned} \tag{3.3.14}$$

To verify that condition (ii) is equivalent to the OFP(-$\frac{1}{2}$) property, we note that with $S(x) = \frac{1}{2}V(x)$ the equalities (3.3.14) become

$$\begin{aligned} L_gS &= k^T \\ L_fS &= -l + \tfrac{1}{2}k^T k \end{aligned}$$

This means that the system (3.3.13) satisfies Theorem 2.39 with $\nu = 0, \rho = -\frac{1}{2}$ and any q such that $q^T q = 2l$. So, (ii) is satisfied if and only if the system (3.3.13) is OFP(-$\frac{1}{2}$).

In view of (i) the equilibrium $x = 0$ of

$$\dot{x} = f(x) - g(x)k(x) \tag{3.3.15}$$

is asymptotically stable. In particular, near $x = 0$, the solutions of $\dot{x} = f(x)$ that satisfy $y = k(x) \equiv 0$ converge to zero. Hence the system (3.3.13) is ZSD. So, $u = -k(x)$ being an optimal stabilizing control implies the OFP(-$\frac{1}{2}$) and ZSD properties of (3.3.13).

Conversely, by Theorem 2.33, these two properties imply that the equilibrium $x = 0$ for (3.3.13) with any feedback control $u = -\kappa y$, $\kappa > \frac{1}{2}$, is asymptotically stable. Therefore (i) is satisfied, which shows that OFP(-$\frac{1}{2}$) and ZSD imply optimal stabilization. □

Example 3.24 *(Optimality and passivity)*
From Example 3.20 we know that for the system

$$\dot{x} = x^2 + u$$

and the cost functional $J = \int_0^\infty (x^2 + u^2)\, dt$ the optimal stabilizing control law is $u = -x^2 - x\sqrt{x^2+1}$. Now Theorem 3.23 implies that the system

$$\begin{aligned} \dot{x} &= x^2 + u \\ y &= x^2 + x\sqrt{x^2+1} \end{aligned} \tag{3.3.16}$$

is OFP($-\frac{1}{2}$). This is verified by taking the time-derivative of the storage function $S(x) = \frac{1}{2}V(x) = \frac{1}{3}(x^3 + (x^2+1)^{\frac{3}{2}} - 1)$. We get

$$\dot{S} = \frac{1}{2}\frac{\partial V}{\partial x}(x^2 + u) = (x^2 + x\sqrt{x^2+1})(x^2 + u) = yx^2 + yu$$

From the expression for y in (3.3.16) we see that if $x < 0$, then $y < 0$, and hence, $yx^2 < 0$. Otherwise $yx^2 \le \frac{1}{2}y^2$, which can be verified by a simple calculation. In either case we obtain

$$\dot{S}(x) \le \frac{1}{2}y^2 + yu$$

which proves that (3.3.16) is OFP($-\frac{1}{2}$). The ZSD property is immediate because $y = 0$ implies $x = 0$.

□

In Section 2.4 we have given structural conditions for output feedback passivity. We now use Theorem 3.23 to show how these conditions apply to optimal stabilization. The violation of any one of these conditions excludes the possibility for a given stabilizing feedback $u = -k(x)$ to be optimal for any functional of the form (3.3.12).

Proposition 3.25 *(Structural conditions for optimality)*
If $u = -k(x)$ is optimal stabilizing for (3.3.12) and if $\frac{\partial k}{\partial x}(0)$ has full rank, then

the system (3.3.13) has relative degree one, is weakly minimum phase, and $L_g k(0)$ is symmetric positive definite.

Conversely, if the system (3.3.13) has relative degree one, its Jacobian linearization at $x = 0$ is minimum phase, and $L_g k(0)$ is symmetric positive definite, then there exists $\epsilon^* > 0$ such that for all $\epsilon \in (0, \epsilon^*]$, the feedback $u = -\frac{1}{\epsilon}k(x)$ is optimal stabilizing for (3.3.12). □

Example 3.26 *(Structural obstacle to optimality)*
For the linear system

$$\begin{aligned} \dot{x}_1 &= x_2 \\ \dot{x}_2 &= -x_2 + u \end{aligned}$$

the linear stabilizing control law $u = -x_1$ has a gain margin $(0, \infty)$. This means that for any $\kappa > 0$ the control $u = -\kappa x_1$ is also stabilizing. However, for the output $y = x_1$ the relative degree is two, so the stabilizing control $u = -x_1$ cannot be optimal with respect to any cost of the form (3.3.12). □

For our future use we examine when the optimality and stability properties are global. This is certainly the case when the optimal control u^* achieves GAS and the optimal value function V is positive definite and radially unbounded. Alternative assumptions, needed when V is only positive semidefinite, are discussed in the following two examples.

Example 3.27 *(Optimality with a global invariant manifold)*
For the nonlinear system

$$\begin{aligned} \dot{x}_1 &= -x_1^3 + x_1 u \\ \dot{x}_2 &= u \end{aligned} \tag{3.3.17}$$

and the cost functional

$$J = \int_0^\infty (x_2^2 + u^2)\, dt \tag{3.3.18}$$

the solution to the HJB equation $V = x_2^2$ is only positive semidefinite. The corresponding control is $u = -\frac{1}{2}\frac{\partial V}{\partial x} = -x_2$. Because in the set $\{x : V(x) = 0\}$ the closed-loop system reduces to $\dot{x}_1 = -x_1^3$, from Theorems 2.24 and 2.21 we conclude that $x = 0$ is asymptotically stable, and hence, $u = -x_2$ is the optimal stabilizing control.

To examine the global behavior, we use the "bounded-input bounded-state" property of $\dot{x}_1 = -x_1^3 + x_1 u$, see Example 2.25. Furthermore, in the closed-loop system $x_2 = -u = e^{-t}x_2(0)$. It follows that all solutions are bounded and, by Theorem 2.21, the origin is GAS. □

In the above example the optimal stabilization is achieved globally despite the unobservability of x_1 in the cost functional. This was so because of the strong stability property of the x-subsystem: bounded x_2 produces bounded x_1 and, moreover, if x_2 converges to 0, so does x_1. In the following example, the situation where the unobservable subsystem does not possess this strong stability property, but the properties are global thanks to the existence of a radially unbounded value function.

Example 3.28 *(Optimality with positive semidefinite radially unbounded V)* The problem of minimizing (3.3.18) for the system

$$\begin{aligned} \dot{x}_1 &= -x_1^3 + x_1^3 u \\ \dot{x}_2 &= u \end{aligned} \tag{3.3.19}$$

results in the optimal value function $V(x) = x_2^2$ and the control law $u = -x_2$ which are the same as in Example 3.27 and asymptotic stability of $x = 0$ is established in the same way, with exponential convergence of x_2. However, in this case the x_1-subsystem is not bounded-input bounded-state. In fact, whenever $x_2(0) > 1$, the solutions of the closed-loop system for sufficiently large $x_1(0)$ escape to infinity in finite time.

Even though the Jacobian linearization of (3.3.19) is not stabilizable, we can achieve global asymptotic stability and retain the exponential convergence of x_2 if we use a cost which penalizes x_1 only when it is far from 0 as in

$$J = \int_0^\infty \left(2\varphi(x_1)x_1^3 + (\varphi(x_1) + x_2)^2 + u^2\right)\, dt \tag{3.3.20}$$

where $\varphi(x_1) = 0$ for $|x_1| \leq 1$, $x_1 - 1$ for $x_1 > 1$, $x_1 + 1$ for $x_1 < -1$. This renders x_1 is unobservable in the cost when $|x_1| \leq 1$. The solution of the HJB equation

$$V = \varphi^2(x_1) + x_2^2 \tag{3.3.21}$$

is C^1 positive semidefinite and radially unbounded. The corresponding control law

$$u = -\varphi(x_1)x_1^3 - x_2$$

is equal to $-x_2$ in a neighborhood of $x = 0$ and thus achieves asymptotic stability of the closed-loop system and exponential convergence of x_2. Moreover, because V is radially unbounded and satisfies

$$\dot{V} = -2\varphi(x_1)x_1^3 - 2(\varphi(x_1) + x_2)^2 \leq 0$$

all the solutions are bounded. Since $\dot{V} \equiv 0 \Rightarrow |x_1| \leq 1 \Rightarrow x_2 = 0 \Rightarrow x_1 \to 0$, by Theorem 2.21, $x = 0$ is GAS. □

The above two examples represent alternative means for achieving global properties of optimal feedback systems. The approach which uses radially unbounded optimal value functions is more suitable for our designs and is adopted in the following definition.

Definition 3.29 *(Optimal globally stabilizing control)*
The control law

$$u^*(x) = -\frac{1}{2}R^{-1}(x)(L_gV)^T(x) \tag{3.3.22}$$

is optimal globally stabilizing if

(i) it achieves global asymptotic stability of $x = 0$ for the system (3.5.1).

(ii) V is a C^1, positive semidefinite, radially unbounded function which satisfies the Hamilton-Jacobi-Bellman equation (3.3.3).

□

With this definition, we obtain the following global version of Theorem 3.23.

Theorem 3.30 *(Global optimality and passivity)*
The control law $u = -k(x)$ is optimal globally stabilizing for the cost functional (3.3.12) if and only if the system

$$\begin{aligned} \dot{x} &= f(x) + g(x)u \\ y &= k(x) \end{aligned} \tag{3.3.23}$$

is ZSD and OFP(-$\frac{1}{2}$) with a C^1, radially unbounded storage function $S(x)$. □

3.4 Stability Margins of Optimal Systems

3.4.1 Disk margin for $R(x) = I$

Theorems 2.34 and 3.30 show that optimal stabilization for a cost functional guarantees a disk stability margin.

Proposition 3.31 *(Disk margin of optimal stabilization)*
If $u = -k(x)$ is optimal globally stabilizing for

$$J = \int_0^\infty (l(x) + u^T R(x)u)\, dt \tag{3.4.1}$$

then $u = -k(x)$ achieves a disk margin $D(\frac{1}{2})$. □

A well known special case is that the LQR-design for linear systems guarantees the disk margin $D(\frac{1}{2})$ and hence, a gain margin $(\frac{1}{2}, \infty)$ and a phase margin $\pm 60°$.

The constant $\frac{1}{2}$ in the above statements is relative to the nominal feedback $k(x)$. Disk margin, and therefore gain, phase, and sector margins, can be increased by rescaling the control law using the scaling lemma (Lemma 2.17).

Proposition 3.32 *(Scaling and high gain)*
If the control law $u = -k(x)$ is optimal globally stabilizing for the cost functional (3.4.1), then the feedback law $u = -\frac{1}{\epsilon}k(x)$, $\epsilon \leq 1$, has the disk margin $D(\frac{\epsilon}{2})$. □

When $\epsilon \to 0$, the disk margin tends to $D(0)$ which means that the gain and sector margins tend to $(0, \infty)$, and the phase margin tends to $90°$. Thus, as $\epsilon \to 0$, the stability margins of optimal stabilization designs tend to the stability margins of a passive system. However, there is a caveat: when ϵ is small, the loop gain with the control $u = -\frac{1}{\epsilon}k(x)$ is very high. In general, this reduces the robustness to unmodeled dynamics which change the relative degree of the system as shown in Example 3.16. Thus ϵ is a design parameter which reflects a trade-off between different types of robustness.

3.4.2 Sector margin for diagonal $R(x) \neq I$

By employing the connection between optimality and passivity, we have shown in Proposition 3.31 that an optimal stabilizing feedback law for a cost functional (3.4.1), where $R(x) = I$, achieves a disk margin. Does a similar property hold when $R(x) \neq I$? The answer is negative: for a more general cost functional

$$J = \int_0^\infty (l(x) + u^T R(x) u)\, dt \tag{3.4.2}$$

the connection with passivity established in Theorem 3.23 no longer holds.

Example 3.33 *(Lack of passivity when $R(x) \neq I$)*
For $a > 0$ we consider the system

$$\begin{aligned} \dot{x}_1 &= -ax_1 + \tfrac{1}{4}R^{-1}(x)(x_1 + x_2) + u \\ \dot{x}_2 &= \tfrac{1}{4}R^{-1}(x)(x_1 + x_2) + u \end{aligned} \tag{3.4.3}$$

and the cost functional

$$J = \int_0^\infty (ax_1^2 + R(x)u^2)\, dt,$$

with

$$R(x) = \frac{1}{1 - \frac{1}{2}\sigma(\frac{2+2a^2}{a}x_2(x_1 - \frac{1}{1+a^2}x_2))} > 0, \tag{3.4.4}$$

where $\sigma(\cdot)$ saturates at one. By direct substitution it can be verified that the positive definite solution of the HJB equation (3.3.3) is

$$V(x) = \frac{1}{2}x_1^2 + \frac{1}{2}x_2^2$$

and that the corresponding control

$$u(x) = -\frac{1}{2}R^{-1}(x)(x_1 + x_2) \tag{3.4.5}$$

is stabilizing because

$$\dot{V} = -ax_1^2 - \frac{1}{4}R^{-1}(x)(x_1 + x_2)^2 < 0, \quad \text{for all } x \neq 0$$

Hence $u(x)$ in (3.4.5) is an optimal globally stabilizing control.

However, Theorem 3.23 does not apply because the system

$$\begin{aligned} \dot{x}_1 &= -ax_1 + \tfrac{1}{4}R^{-1}(x)(x_1 + x_2) + u \\ \dot{x}_2 &= \tfrac{1}{4}R^{-1}(x)(x_1 + x_2) + u \\ y &= \tfrac{1}{2}R^{-1}(x)(x_1 + x_2) \end{aligned} \tag{3.4.6}$$

is not OFP(-$\frac{1}{2}$). We show this by proving the equivalent statement that the system

$$\begin{aligned} \dot{x}_1 &= -ax_1 + u \\ \dot{x}_2 &= u \\ y &= \tfrac{1}{2}R^{-1}(x)(x_1 + x_2) \end{aligned} \tag{3.4.7}$$

is not passive.

For $x_1(0) = \frac{a}{1+a^2}$, $x_2(0) = 0$, and $u(t) = \cos t$ the solution of (3.4.7) is

$$\begin{aligned} x_1(t) &= \tfrac{1}{1+a^2}\sin t + \tfrac{a}{1+a^2}\cos t \\ x_2(t) &= \sin t \end{aligned}$$

Along this solution, $R^{-1}(x(t)) = 1 - \frac{1}{2}\sin(2t)$ and

$$\int_0^T u(t)y(t)\,dt = \int_0^T \frac{1}{2}\left(1 - \frac{1}{2}\sin(2t)\right)\left(\frac{1}{2}\frac{a^2+2}{a^2+1}\sin(2t) + \frac{a}{1+a^2}\cos^2 t\right)\,dt$$

For $T = 2n\pi, \;\; n = 1, 2, \ldots$ we are left with

$$\int_0^{2n\pi} u(t)y(t) = \int_0^{2n\pi} \frac{1}{2}\frac{a}{1+a^2}\cos^2 t\,dt - \int_0^{2n\pi} \frac{1}{8}\frac{a^2+2}{a^2+1}\sin^2(2t)\,dt \tag{3.4.8}$$

For $a > 2+\sqrt{2}$, the right hand side of (3.4.8) is negative and converges to $-\infty$ as $n \to \infty$. Thus the system (3.4.7) with $a > 2+\sqrt{2}$ is not passive. This shows that, when $R(x)$ is not constant, the connection between optimality and passivity no longer holds.

□

In the absence of a disk margin, a sector margin exists when $R(x)$ is a diagonal matrix.

Proposition 3.34 *(Sector margin of optimal stabilizing control)*
If the control law $u = -k(x)$ is optimal globally stabilizing for a cost functional (3.4.2) with

$$R(x) = diag\{r_1(x), \ldots, r_m(x)\}, \tag{3.4.9}$$

then it achieves a sector margin $(\frac{1}{2}, \infty)$.

Proof: By assumption, the optimal stabilizing feedback $u = -k(x)$ is of the form

$$k(x) = \frac{1}{2}R^{-1}(x)(L_gV(x))^T \tag{3.4.10}$$

where the optimal value function V is radially unbounded. Moreover, along the solutions of the closed-loop system

$$\dot{x} = f(x) - g(x)k(x) =: F(x), \tag{3.4.11}$$

the time-derivative of V is

$$\dot{V} = L_FV(x) = L_fV(x) - L_gVk(x) = -l(x) - \frac{1}{4}(L_gV)R^{-1}(L_gV)^T(x) \leq 0$$

When u is replaced by $\varphi(u)$, where $\varphi = diag\{\varphi_1, \ldots, \varphi_m\}$, with φ_i in the sector $(\frac{1}{2}, \infty)$, the closed-loop system becomes

$$\dot{x} = f(x) + g(x)\varphi(-k(x)) = f(x) - g(x)\varphi(k(x)) =: \tilde{F}(x) \tag{3.4.12}$$

and the time-derivative of V is

$$\dot{V} = L_{\tilde{F}}V = L_FV + L_gV(\varphi(k(x)) - k(x)) = -l(x) + L_gV(x)(\varphi(k(x)) - \frac{1}{2}k(x))$$

Using (3.4.9),(3.4.10), $l(x) \geq 0$, and $\varphi(k(x)) = diag\{\varphi_1(k_1(x)), \ldots, \varphi_m(k_m(x))\}$, we obtain

$$\begin{aligned} \dot{V} &= L\tilde{F}V \leq -2k(x)^TR(x)(\varphi(k(x)) - \frac{1}{2}k(x)) \\ &= -2\sum_{i=1}^{m}[r_i(x)k_i(x)(\varphi_i(k_i(x)) - \frac{1}{2}k_i(x))] \leq 0 \end{aligned}$$

Now, because $s\varphi(s) > \frac{1}{2}s^2$, for all $s \neq 0$, we obtain that $\dot{V}(x) = 0$ implies $k(x) = \varphi(k(x)) = 0$. Thus the solutions of (3.4.12) converge to the set E where $k(x) = 0$.

The GAS of the system (3.4.12) is established as follows. Because V is radially unbounded, the solutions of the two systems (3.4.11) and (3.4.12) are bounded and converge to the same invariant set E where $k(x) = \varphi(k(x)) = 0$, which means that

$$\forall x \in E : F(x) = \tilde{F}(x) = f(x)$$

Because the equilibrium $x = 0$ of the system $\dot{x} = F(x)$ is GAS, the solutions of $\dot{x} = F(x)$ which remain in E for all t converge to 0. Then the same must hold for the solutions of $\dot{x} = \tilde{F}(x)$ which remain in E. By Theorem 2.21 this proves global attractivity of $x = 0$ for the system $\dot{x} = \tilde{F}(x)$. Stability follows from Theorem 2.24 because $Z = \{x|V(x) = 0\} \subset \{x|\dot{V}(x) = 0\}$, and hence, $x = 0$ is asymptotically stable conditionally to Z.

□

In the above proof, the assumption that $R(x)$ is diagonal is crucial for the negativity of

$$-k(x)^T R(x)(\varphi(k(x)) - \frac{1}{2}k(x)) \tag{3.4.13}$$

With R nondiagonal, the negativity of (3.4.13) can be violated even with a constant positive definite matrix R and with linear gains $\phi_i(s) = \alpha_i s,\ \alpha > \frac{1}{2}$. For linear multivariable systems, it is known from [1, 64] that an LQR design with nondiagonal R may result in an arbitrary small gain margin.

To summarize, optimal stabilization of the system $\dot{x} = f(x) + g(x)u$ for the cost functional

$$J = \int_0^\infty (l(x) + u^T R(x)u)\, dt$$

- achieves a disk margin $D(\frac{1}{2})$ if $R(x) = I$,
- achieves a sector margin $(\frac{1}{2}, \infty)$ if $R(x)$ is diagonal,
- but does not guarantee any stability margin for a general $R(x)$.

3.4.3 Achieving a disk margin by domination

Although for a general positive definite $R(x)$ an optimal globally stabilizing control

$$u(x) = -\frac{1}{2}R^{-1}(x)(L_g V(x))^T \tag{3.4.14}$$

does not achieve a desired stability margin, it can still be used as a starting point for a *domination redesign* in which the control is rendered optimal for a cost with $R(x) = I$ and achieves a disk margin $D(\frac{1}{2})$.

We define a continuous *dominating function* $\gamma : \mathbb{R}^+ \to \mathbb{R}^+$ which satisfies the two conditions

$$\gamma(V(x))I \geq R^{-1}(x), \quad \forall x \in \mathbb{R}^n \tag{3.4.15}$$

$$\lim_{t\to\infty} \int_0^t \gamma(s)ds = +\infty \tag{3.4.16}$$

Such a function always exists if $V(x)$ is radially unbounded. One possible choice is

$$\gamma(s) = a + \sup_{\{x:V(x)\leq s\}} \lambda_{\max}(R^{-1}(x)), \quad a > 0$$

with $\lambda_{\max}$ denoting the largest eigenvalue. The redesigned optimal value function

$$\tilde{V}(x) := \int_0^{V(x)} \gamma(s)ds$$

inherits the properties of $V(x)$: it is C^1, positive semidefinite (because $\gamma(s) > 0$ for all s, $\tilde{V} = 0$ if and only if $V = 0$), and radially unbounded.

To show that the redesigned control law

$$\tilde{u}(x) = \frac{1}{2}(L_g\tilde{V}(x))^T = -\frac{1}{2}\gamma(V(x))(L_gV(x))^T \tag{3.4.17}$$

achieves GAS, we use (3.4.15) to obtain

$$\begin{aligned}\dot{\tilde{V}} &= \gamma(V)L_fV - \frac{1}{2}\gamma^2(V)L_gV(L_gV)^T \\ &\leq \gamma(V)(-l - \frac{1}{4}L_gV\, R^{-1}(L_gV)^T) \leq 0\end{aligned}$$

Boundedness of solutions follows because V is radially unbounded. To prove GAS, we examine the set E where $\dot{\tilde{V}} = 0$. In E we have $L_gV(x) = 0$ so that $\tilde{u}(x) = 0$ and hence, $u(x) = 0$. Because $u(x)$ is optimal stabilizing, the solutions of $\dot{x} = f(x) + g(x)u(x)$ contained in E converge to the origin. But, since in E the two closed-loop systems corresponding to $u(x)$ and $\tilde{u}(x)$ coincide, we conclude that the redesigned feedback $\tilde{u}(x)$ achieves GAS.

To prove optimality, we define the state cost as

$$\tilde{l}(x) := -L_f\tilde{V} + \frac{1}{4}(L_g\tilde{V})(L_g\tilde{V})^T$$

By construction $\tilde{V}$ is a radially unbounded, positive semidefinite solution of the Hamilton-Jacobi-Bellman equation and $\tilde{l}(x)$ is positive semidefinite because

$$\begin{aligned} -\tilde{l} &= L_f\tilde{V} - \frac{1}{4}(L_g\tilde{V})(L_g\tilde{V})^T \\ &= \gamma(V)(L_fV - \frac{1}{4}\gamma(V)(L_gV)(L_gV)^T) \\ &\leq \gamma(V)(L_fV - \frac{1}{4}(L_gV)R^{-1}(L_gV)^T) = -\gamma(V)l \leq 0 \end{aligned}$$

Thus the control law (3.4.17) minimizes the modified cost functional

$$\tilde{J} = \int_0^\infty (\tilde{l}(x) + u^Tu)\, dt \tag{3.4.18}$$

with $\tilde{l}(x) \geq 0$. We arrive at the following conclusion.

Proposition 3.35 *(Dominating optimal control)*
Assume that $u = -\frac{1}{2}R^{-1}(L_gV)^T$ is optimal globally stabilizing with respect to the cost (3.3.2). Then, for any dominating function γ satisfying (3.4.15) and (3.4.16), the redesigned control law $\tilde{u} = -\frac{1}{2}\gamma(V)(L_gV)^T$ is optimal globally stabilizing for the modified cost functional (3.4.18) and hence, achieves a disk margin $D(\frac{1}{2})$. □

The redesign in Proposition 3.35 improves the stability margins of the closed-loop system, but it often does so at the expense of an increased control effort, as we now illustrate.

Example 3.36 *(Domination increases control effort)*
For the system

$$\begin{aligned} \dot{x}_1 &= x_2 + x_1^2x_2^2 \\ \dot{x}_2 &= -x_1 + u \end{aligned}$$

the time-derivative of $V = \frac{1}{2}x^Tx$ is $\dot{V} = x_2(u + x_1^3x_2)$. The control law

$$u = -2(1 + \max(0, x_1^3))x_2 \tag{3.4.19}$$

renders $\dot{V}$ negative semidefinite

$$\dot{V} = -(2 + |x_1^3|)x_2^2 \leq -2x_2^2$$

and, because $x_2 \equiv 0 \Rightarrow x_1 = 0$, the equilibrium $(x_1, x_2) = (0, 0)$ is GAS. Defining

$$R^{-1}(x) := 2(1 + \max(0, x_1^3))$$

it is easy to verify that the control law (3.4.19) minimizes the cost

$$J = \int_0^\infty [(1 + \max(0, x_1^3))x_2^2 + R(x)u^2]\, dt$$

A sector margin $(\frac{1}{2}, \infty)$ is therefore guaranteed by Proposition 3.34. In order to achieve a disk margin, we use the dominating function $\gamma(s) = 2(1 + s^{3/2})$. The redesigned control law

$$u = -2(1 + (x_1^2 + x_2^2)^{\frac{3}{2}})x_2 \tag{3.4.20}$$

results in

$$\dot{V} = -(2 + 2(x_1^2 + x_2^2)^{\frac{3}{2}} - x_1^3)x_2^2 \leq -2x_2^2$$

and achieves GAS.

Comparing the two control laws, (3.4.19) and (3.4.20), we observe that with the redesign the control effort has increased at every point, even in the directions where $u = 0$ would suffice for stabilization.

□

The increased control effort is not necessarily wasted, nor is the domination tantamount to high-gain feedback. In the above example the extra effort is used to enhance the negativity of $\dot{V}$ at each point. However, this effort is never used to cancel a beneficial nonlinearity. Furthermore, while the control law (3.4.19) makes use of a detailed knowledge of the nonlinearity $x_1^2x_2^2$, the redesigned control law (3.4.20) is optimal globally stabilizing even when the nonlinearity $x_1^2x_2^2$ is replaced by any nonlinearity $\phi(x_1, x_2)x_2^2$ such that

$$|x_1\phi(x_1, x_2)| \leq (x_1^2 + x_2^2)^{\frac{3}{2}}$$

This means that the system with dominating feedback can tolerate more uncertainty.

An indirect consequence of Proposition 3.35 is that the structural conditions for feedback passivity are necessary for optimal stabilization, not only when $R(x) = I$, but also with a general cost functional (3.3.2).

Proposition 3.37 *(Structural conditions for optimality when $R(x) \neq I$)*
If the control law $u(x) = -k(x)$ is optimal stabilizing for $J = \int_0^\infty (l(x) + u^T R(x)u)\, dt$ and if $\frac{\partial k}{\partial x}(0)$ has full rank, then the system

$$\begin{aligned} \dot{x} &= f(x) + g(x)u, \\ y &= k(x) \end{aligned} \tag{3.4.21}$$

has relative degree one at $x = 0$ and is weakly minimum phase.

Proof: If $u(x) = -k(x)$ is optimal stabilizing, it is of the form

$$u(x) = -\frac{1}{2}R^{-1}(x)(L_gV)^T(x)$$

and, using a domination redesign, there exists a control of the form

$$\tilde{u}(x) = -\tilde{k}(x) = -\frac{1}{2}\gamma(V(x))(L_gV)^T(x)$$

which is optimal stabilizing for a modified cost functional where $R(x) = I$. By Proposition 3.25, the system

$$\begin{aligned} \dot{x} &= f(x) + g(x)u \\ y &= \tilde{k}(x) \end{aligned} \tag{3.4.22}$$

has relative degree one and is locally weakly minimum phase. Noting that

$$\tilde{k}(x) = 0 \Leftrightarrow L_gV(x) = 0 \Leftrightarrow k(x) = 0,$$

we conclude that the systems (3.4.21) and (3.4.22) have the same zero dynamics. Therefore the system (3.4.21) is weakly minimum phase. To prove the relative degree condition, we observe that

$$L_gk(0) = \frac{1}{2}R^{-1}(0)\gamma^{-1}(V(0))L_g\tilde{k}(0)$$

Because the system (3.4.22) is OFP($-\frac{1}{2}$), the matrix $L_g\tilde{k}(0)$ is symmetric positive definite. So $L_gk(0)$ is nonsingular, that is, the system (3.4.21) has relative degree one.

□

3.5 Inverse Optimal Design

3.5.1 Inverse optimality

Optimal stabilization guarantees several desirable properties for the closed-loop system, including stability margins. In a direct approach we would have to solve the HJB equation which in general is not a feasible task. On the other hand, the robustness achieved as a result of the optimality is largely independent of the particular choice of functions $l(x) \geq 0$ and $R(x) > 0$. This motivated Freeman and Kokotović [25, 26] to pursue the development of design methods which solve the *inverse problem of optimal stabilization*. In the inverse approach, a stabilizing feedback is designed first and then shown to be optimal for a cost functional of the form $J = \int_0^\infty (l(x) + u^T R(x)u)\, dt$. The problem is *inverse* because the functions $l(x)$ and $R(x)$ are a posteriori determined by the stabilizing feedback, rather than a priori chosen by the designer.

A stabilizing control law $u(x)$ solves an inverse optimal problem for the system

$$\dot{x} = f(x) + g(x)u \tag{3.5.1}$$

if it can be expressed as

$$u = -k(x) = -\frac{1}{2}R^{-1}(x)(L_gV(x))^T, \quad R(x) > 0, \tag{3.5.2}$$

where $V(x)$ is a positive semidefinite function, such that the negative semidefiniteness of $\dot{V}$ is achieved with the control $u = -\frac{1}{2}k(x)$, that is

$$\dot{V} = L_fV(x) - \frac{1}{2}L_gV(x)k(x) \leq 0 \tag{3.5.3}$$

When the function $-l(x)$ is set to be the right-hand side of (3.5.3):

$$l(x) := -L_fV(x) + \frac{1}{2}L_gV(x)k(x) \geq 0 \tag{3.5.4}$$

then $V(x)$ is a solution of the HJB equation

$$l(x) + L_fV(x) - \frac{1}{4}(L_gV(x))R^{-1}(x)(L_gV(x))^T = 0 \tag{3.5.5}$$

Hence, consistent with Definition 3.29 we will say that the control law $u^*(x)$ is an *inverse optimal* (globally) stabilizing control law for the system (3.5.1) if

(i) it achieves (global) asymptotic stability of $x = 0$ for the system (3.5.1).

(ii) it is of the form

$$u^*(x) = -\frac{1}{2}R^{-1}(x)L_gV(x)$$

where $V(x)$ is (radially unbounded) positive semidefinite function such that

$$\dot{V}\mid_{u=\frac{1}{2}u^*(x)} \triangleq L_fV + \frac{1}{2}L_gVu^* \leq 0$$

The design methods presented in subsequent chapters solve in a systematic way global inverse optimal stabilization problems for important classes of nonlinear systems. The main task of these design methods is the construction of positive (semi)definite functions whose time-derivatives can be rendered negative semidefinite by feedback control. In the inverse approach, such functions become optimal value functions.

Some designs of stabilizing control laws employ cancellation and do not have satisfactory stability margins, let alone optimality properties. The inverse optimal approach is a constructive alternative to such designs, which achieves desired stability margins. Let us clarify this important issue.

Example 3.38 *(Nonrobustness of cancellation designs)*
For the scalar system

$$\dot{x} = x^2 + u, \tag{3.5.6}$$

one possible design is to let u cancel x^2 in (3.5.6) and add a stabilizing term. This is accomplished with the the feedback linearizing control law

$$u_l(x) = -x^2 - x \tag{3.5.7}$$

which results in what appears to be a desirable closed-loop system $\dot{x} = -x$. However, because of the cancellation, this feedback linearizing control law does not have any stability margin: with a slightly perturbed feedback $(1+\epsilon)u_l(x)$, the closed-loop system

$$\dot{x} = -(1+\epsilon)x - \epsilon x^2 \tag{3.5.8}$$

has solutions which escape to infinity in finite time for any $\epsilon \neq 0$.

Let us instead use the optimal feedback $u^*(x) = -x^2 - x\sqrt{x^2+1}$ designed in Example 3.20. This control law has two desirable properties.

- For $x < 0$, it recognizes the beneficial effect of the nonlinearity x^2 to enhance the negativity of $\dot{V}$. For large negative x, the control is negligible: as $x \to -\infty$, it converges to $\frac{1}{2}$.

- Instead of cancelling the destabilizing term x^2 for $x > 0$, the optimal control $u^*(x)$ dominates it and, by doing so, achieves a disk margin $(\frac{1}{2}, \infty)$.

The benefit of optimal stabilization is graphically illustrated in Figure 3.9. The graph of any stabilizing control law $u(x)$ must lie outside the shaded region; because at a given point x, the negativity of $\dot{V}(x)$ and hence, the pointwise gain margin, increase with the distance of $u(x)$ from the parabola $-x^2$. The feedback linearizing control law $u_l(x)$ has two major drawbacks: first, for $x < -1$, its graph is in the third quadrant, which shows that the control effort is wasted to cancel a beneficial nonlinearity; second, for $|x|$ large, its graph approaches the parabola $-x^2$, that is the control law loses its stability

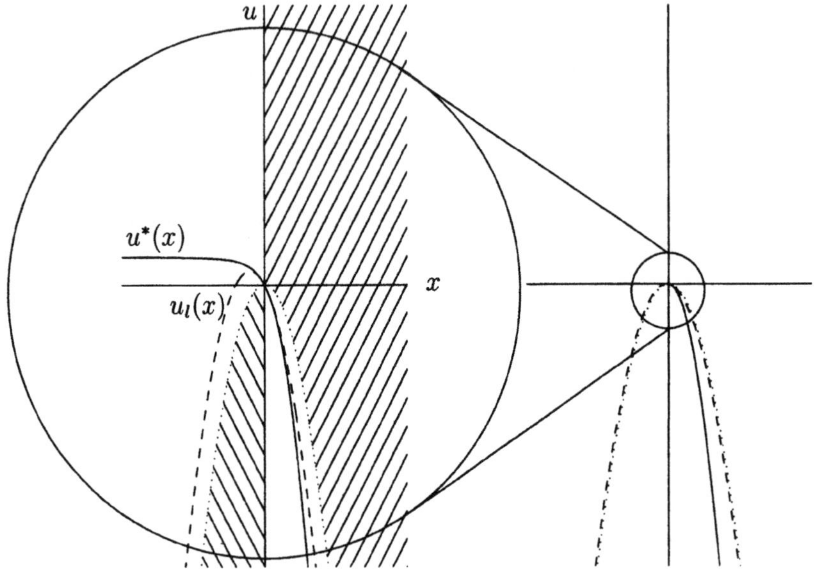

Figure 3.9: Control laws $u_l(x)$ and $u^*(x)$ in Example 3.38.

margin. The optimal control law $u^*(x)$ never wastes the effort because its graph is entirely in the second and fourth quadrants. The stabilizing effect of $u^*(x)$ and its stability margin are superior to those of $u_l(x)$ because the distance of its graph from the parabola $-x^2$ is larger at every point x. Finally, the optimality property guarantees that even the graph of $\frac{1}{2}u^*(x)$ stays away from the parabola $-x^2$ for all $x \neq 0$. □

After a Lyapunov function has been constructed, instead of cancelling nonlinearities, a stabilizing feedback can be constructed to be in the inverse optimal form (3.5.2). We will now examine situations in which this design task can be solved in a systematic way.

3.5.2 Damping control for stable systems

In many applications the equilibrium $x = 0$ of the uncontrolled part

$$\dot{x} = f(x), \quad f(0) = 0 \tag{3.5.9}$$

of the system

$$\dot{x} = f(x) + g(x)u \tag{3.5.10}$$

is stable and the task of the control is to provide additional damping which will render $x = 0$ *asymptotically* stable. If a radially unbounded Lyapunov function $V(x)$ is known such that $L_f V \leq 0$ for all $x \in I\!R^n$, then it is tempting to employ $V(x)$ as a Lyapunov function for the whole system (3.5.10). In view of $L_f V \leq 0$, the time-derivative of $V(x)$ for (3.5.10) satisfies

$$\dot{V} \leq L_g V u$$

This shows that $\dot{V}$ can be made more negative with the control law

$$u = -\kappa (L_g V)^T, \quad \kappa > 0 \tag{3.5.11}$$

We use the terminology "damping control" because (3.5.11) can be viewed as additional damping which dissipates the "system energy" $V(x)$. This type of control law, known as Jurdjevic-Quinn control [49], has also been used in [60] and [44]. We deduce from Theorem 3.19 that, if the control law (3.5.11) achieves GAS of $x = 0$, then it also solves the global optimal stabilization problem for the cost functional

$$J = \int_0^\infty (l(x) + \frac{2}{\kappa} u^T u) \, dt$$

with the state cost given by

$$l(x) = -L_f V + \frac{\kappa}{2} L_g V (L_g V)^T \geq 0 \tag{3.5.12}$$

The optimal value function is $V(x)$. We have thus made use of the inverse optimality idea to make the Lyapunov function for (3.5.9) an optimal value function for (3.5.10).

The connection with passivity is clear: the system

$$\begin{array}{rcl} \dot{x} & = & f(x) + g(x)u \\ y & = & (L_g V)^T (x) \end{array} \tag{3.5.13}$$

is passive when $L_f V \leq 0$ because $\dot{V} = L_f V + L_g V u \leq y^T u$. Furthermore, for the output $y = (L_g V)^T (x)$, the control law (3.5.11) is the simplest output feedback $u = -\kappa y$ which guarantees GAS if the system is ZSD. Hence, the control law (3.5.11) achieves a disk margin $D(0)$.

However, the damping control (3.5.11) has a limitation. It stems from the fact that $V(x)$ is chosen for the uncontrolled system (3.5.9) in complete disregard of the flexibilities that may be offered by the control term $g(x)u$ in (3.5.10). A simple example will show how this may lead to unnecessary degradation of performance.

Example 3.39 *(Performance limitation of damping control)*
The uncontrolled part of the system

$$\begin{aligned} \dot{x}_1 &= x_2 \\ \dot{x}_2 &= -\epsilon x_2 + u, \ \epsilon > 0, \end{aligned} \tag{3.5.14}$$

is stable. For this part, a Lyapunov function $V = x^T P x$ is obtained from the condition $L_f V \leq 0$, that is

$$PA + A^T P \leq 0 \tag{3.5.15}$$

This condition imposes the constraint $p_{12} \leq \epsilon p_{22}$. The damping control law (3.5.11) is

$$u = -2kL_gV = -2kB^T Px = -2\tilde{k}(\frac{p_{12}}{p_{22}}x_1 + x_2), \ 0 < \frac{p_{12}}{p_{22}} \leq \epsilon$$

where the gain $\tilde{k} = kp_{22} > 0$ can be freely chosen. Because of the constraint $p_{12} \leq \epsilon p_{22}$, the closed-loop system

$$\begin{aligned} \dot{x}_1 &= x_2 \\ \dot{x}_2 &= -2\tilde{k}\tfrac{p_{12}}{p_{22}}x_1 - (\epsilon + 2\tilde{k})x_2 \end{aligned}$$

has one real eigenvalue in the interval $(-\epsilon, 0)$ regardless of the choice of $\tilde{k}$. For ϵ small this results in an unacceptably slow response of the system. In this case, the damping control, although optimal, "overlooked" the possibility to achieve a faster response. □

3.5.3 CLF for inverse optimal control

Performance limitation in Example 3.39 is not due to the inverse optimality approach, but rather to our choice of the optimal value function $V(x)$ which imposed the constraint $p_{12} \leq \epsilon p_{22}$. This constraint is due to the choice of V as a Lyapunov function for $\dot{x} = f(x)$ and dictated by the requirement $L_f V \leq 0$. It is clear, therefore, that even when the uncontrolled part is stable, our choice of a Lyapunov function should not be based only on the properties of $\dot{x} = f(x)$, but it should also include the flexibility provided by the control term $g(x)u$.

Example 3.40 *(Removing $L_f V \leq 0$)*
We now investigate when $V = x^T P x$ can be an optimal value function for (3.5.14) without imposing the condition $L_f V \leq 0$, that is (3.5.15). With $R(x) = \frac{1}{k}$, $k > 0$, an optimal stabilizing control corresponding to $V = x^T P x$ is

$$u^*(x) = -\frac{k}{2}L_gV(x) = -kB^T Px = -k(p_{22}x_2 + p_{12}x_1)$$

The constraints on p_{12} and p_{22} are now imposed by the condition for optimal stabilization

$$\dot{V}\Big|_{u=\frac{1}{2}u*(x)} = L_f V + \frac{1}{2} L_g V u^* \leq 0 \tag{3.5.16}$$

Evaluating $\dot{V}$ along the solutions of (3.5.14) shows that for any choice of $p_{22} > 0$ and $p_{12} > 0$, the inequality (3.5.16) is satisfied with k sufficiently large. The constraint $p_{12} \leq \epsilon p_{22}$ has disappeared and the choice of the optimal value function $V = x^T P x$ can be made to achieve an arbitrarily fast response for the closed-loop system. □

The flexibility in the choice of an optimal value function $V(x)$ comes from the fact that, by substituting the control law $u = \frac{k}{2}u^*(x)$ into the inequality (3.5.16), we have relaxed the constraint (3.5.15): the inequality $x^T(PA + A^T P)x < 0$ must hold only when $x^T PB = 0$, that is in the directions of the state space where the column vectors of the matrix B are tangent to the level sets of V. To characterize the analogous property for nonlinear systems

$$\dot{x} = f(x) + g(x)u, \tag{3.5.17}$$

we employ the concept of a "control Lyapunov function" (CLF) of Artstein [4] and Sontag [98].

Definition 3.41 (*Control Lyapunov function*)
A smooth, positive definite, and radially unbounded function $V(x)$ is called a control Lyapunov function (CLF) for the system $\dot{x} = f(x) + g(x)u$ if for all $x \neq 0$,

$$L_g V(x) = 0 \Rightarrow L_f V(x) < 0 \tag{3.5.18}$$

□

By definition, any Lyapunov function whose time-derivative can be rendered negative definite is a CLF. In Chapters 4, 5, and 6, we develop systematic methods for construction of Lyapunov functions which can be used as CLF's. The importance of the CLF concept in the framework of inverse optimality is that, when a CLF is known, an inverse optimal stabilizing control law can be selected from a choice of explicit expressions such as those in [26]. Then the CLF becomes an optimal value function.

A particular optimal stabilizing control law, derived from a CLF, is given by Sontag's formula [100],

$$u_S(x) = \begin{cases} -\left(c_0 + \frac{a(x)+\sqrt{a^2(x)+(b^T(x)b(x))^2}}{b^T(x)b(x)}\right) b(x) & , \quad b(x) \neq 0 \\ 0 & , \quad b(x) = 0 \end{cases} \tag{3.5.19}$$

where $L_fV(x) = a(x)$ and $(L_gV(x))^T = b(x)$. The control law (3.5.19) achieves negative definiteness of $\dot{V}$ for the closed-loop system since for $x \neq 0$

$$\dot{V} = a(x) - p(x)b^T(x)b(x) = -\sqrt{a^2(x) + (b^T(x)b(x))^2} - c_0 b^T(x)b(x) < 0 \tag{3.5.20}$$

where

$$p(x) = \begin{cases} c_0 + \frac{a(x)+\sqrt{a^2(x)+(b^T(x)b(x))^2}}{b^T(x)b(x)} & , \quad b(x) \neq 0 \\ c_0 & , \quad b(x) = 0 \end{cases} \tag{3.5.21}$$

It is easy to see that $c_0 > 0$ is not required for the negative definiteness of V since, away from $x = 0$, $a(x)$ and $b(x)$ never vanish together because of (3.5.18).

To analyze the continuity properties of the control law (3.5.19), we consider separately the open set

$$\Omega = \{x \mid b(x) \neq 0 \text{ or } a(x) < 0\}$$

and its complement $\Omega^c = \mathbb{R}^n \backslash \Omega$. Inside Ω, the control law (3.5.19) is a smooth function of x if a and b are smooth, because

$$\frac{a + \sqrt{a^2 + (b^T b)^2}}{b^T b} b$$

as a function of $a \in \mathbb{R}$ and $b \in \mathbb{R}^m$ is analytic when $b \neq 0$ or $a < 0$.

When V is a CLF, the set Ω is the whole state space except for the origin, because of the strict inequality in (3.5.18). Then the set Ω^c is just the origin $x = 0$. The control law (3.5.19) is continuous at $x = 0$ if and only if the CLF satisfies the *small control property*: for each $\epsilon > 0$, we can find $\delta(\epsilon) > 0$ such that, if $0 < \| x \| < \delta$, there exists u which satisfies $L_fV(x) + (L_gV)^T(x)u < 0$ and $\| u \| < \epsilon$.

The small control property is a mild assumption on V. If Ω^c were to include points other than the origin, which happens when the inequality in (3.5.18) is not strict, the continuity of the control law (3.5.19) would require the small control property at *every point* of Ω^c. This is a restrictive assumption, as illustrated in the following example, which also explains why the CLF concept is defined only with a strict inequality.

Example 3.42 *(Strict inequality in the CLF condition)*
For the system

$$\begin{aligned} \dot{x}_1 &= x_2 \\ \dot{x}_2 &= u \end{aligned} \tag{3.5.22}$$

and the Lyapunov function $V = \frac{1}{2}x_1^2 + \frac{1}{2}x_2^2$, the inequality (3.5.18) is not strict because $L_gV = x_2 = 0$ implies $L_fV = x_1x_2 = 0$. The set Ω^c is the axis $x_2 = 0$ and the formula (3.5.19) yields

$$u_S(x) = -c_0x_2 - x_1 - sgn(x_2)\sqrt{x_1^2 + x_2^2}$$

which is discontinuous in Ω^c. □

It is often desirable to guarantee at least Lipschitz continuity of the control law at $x = 0$ in addition to its smoothness elsewhere. If there exists a stabilizing feedback $\bar{u}(x)$, which is Lipschitz continuous at the origin and achieves negative definiteness of $\dot{V}$, we say that the CLF $V(x)$ satisfies a *Lipschitz control property*. Under this additional assumption, the same Lipschitz property holds for the control law $u_S(x)$.

Proposition 3.43 *(Lipschitz continuity of Sontag's formula)*
Assume that $V(x)$ is a CLF with the Lipschitz control property for the nonlinear system (3.5.17). Then the control law given by Sontag's formula (3.5.19) is Lipschitz continuous at $x = 0$.

Proof: Let $\bar{u}(x)$ be a stabilizing control for (3.5.17) and K_u be a constant such that, for $\| x \| < \delta$, with $\delta > 0$, we have

$$\| \bar{u}(x) \| \leq K_u \| x \| \tag{3.5.23}$$

and

$$a(x) + b^T(x)\bar{u}(x) < 0 \quad \text{for } x \neq 0. \tag{3.5.24}$$

We restrict our attention to the open ball B_δ of radius δ centered at $x = 0$ and prove that the control law $u_s(x)$ given by (3.5.19) is Lipschitz in B_δ. Because V is smooth and $\frac{\partial V}{\partial x}(0) = 0$, there exists a constant $K_b > 0$ such that $\|b(x)\| < K_b\|x\|$ in B_δ.

We now distinguish the cases $a(x) > 0$ and $a(x) \leq 0$. If $a(x) > 0$, the inequality (3.5.24) implies

$$a(x) < -b(x)\bar{u}(x), \quad \forall x \neq 0$$

and we have $|a(x)| < |b(x)\bar{u}(x)| \leq \|b(x)\|K_u\|x\|$. From (3.5.19), we conclude

$$|b^T u_S| = a + \sqrt{a^2 + (b^Tb)^2} + c_0b^Tb \leq 2a + (1 + c_0)b^Tb$$

which implies that

$$0 < \frac{a + \sqrt{a^2 + (b^Tb)^2}}{b^Tb} + c_0 \leq \frac{2K_u\|x\|}{\|b\|} + 1 + c_0 \tag{3.5.25}$$

Thus, when $a(x) > 0$

$$\|u_S(x)\| \leq (c_0 + \frac{a + \sqrt{a^2 + (b^Tb)^2}}{b^Tb})\|b\| \leq (2K_u + K_b(c_0 + 1))\|x\|$$

In the case $a(x) \leq 0$, we have $0 \leq a + \sqrt{a^2 + (b^Tb)^2} \leq b^Tb$ which implies

$$c_0 \leq c_0 + \frac{a + \sqrt{a^2 + (b^Tb)^2}}{b^Tb} \leq 1 + c_0 \tag{3.5.26}$$

Thus

$$\|u_S(x)\| \leq (c_0 + \frac{a + \sqrt{a^2 + (b^Tb)^2}}{b^Tb})\|b\| \leq (1 + c_0)K_b\|x\|$$

which proves that $u_S(x)$ is Lipschitz continuous at the origin. □

In view of this proposition, the control law $u_S(x)$ in (3.5.19) with any $c_0 \geq 0$ is globally stabilizing, smooth away from the origin and Lipschitz continuous at the origin. Moreover, $u_S(x)$ is in the form $-\frac{1}{2}R^{-1}(x)(L_gV(x))^T$ where by construction

$$R(x) = \frac{1}{2}p^{-1}(x)I > 0 \tag{3.5.27}$$

which means that $u_S(x)$ is an optimal globally stabilizing control law. The parameter $c_0 \geq 0$ is not present in the original Sontag's formula, but a choice $c_0 > 0$ may be needed to ensure the strict positivity of $p(x)$. This in turn guarantees that $R(x)$ is bounded on compact sets. From the bounds (3.5.25) and (3.5.26) we obtain a further characterization of $R(x)$:

$$\begin{aligned} 0 < \|R(x)\| &\leq \frac{1}{c_0 + 1} \quad \text{if } a(x) > 0 \\ \frac{1}{2c_0} \leq \|R(x)\| &\leq \frac{1}{c_0 + 1} \quad \text{if } a(x) \leq 0 \end{aligned}$$

The above inequalities show that $R(x)$ may be small when $a(x)$ is positive, which reflects the fact that the cost on the control is small at those points where a large effort is necessary to achieve the negativity of $\dot{V}$.

To prove that (3.5.19) is optimal stabilizing, it remains to show that $\dot{V} \leq 0$ is satisfied with the control law $\frac{1}{2}u_S(x)$. This is verified by adding $\frac{1}{2}(L_gV)^Tu_S(x)$ to the right-hand side of (3.5.20) which yields

$$\begin{aligned} \dot{V}|_{\frac{u_s(x)}{2}} &= -\sqrt{a^2(x) + (b^T(x)b(x))^2} - c_0b^T(x)b(x) + \frac{1}{2}p(x)b^T(x)b(x) \\ &= -\frac{1}{2}p(x)b^T(x)b(x) \leq 0 \end{aligned}$$

Proposition 3.44 *(Optimal stabilizing control from a CLF)*
The control law (3.5.19) is optimal stabilizing for the cost functional

$$J = \int_0^\infty (\frac{1}{2}p(x)b^T(x)b(x) + \frac{1}{2p(x)}u^T u)\, dt \tag{3.5.28}$$

where $p(x)$ is defined by (3.5.21). □

A consequence of the optimality is that the control law (3.5.19) has a sector margin $(\frac{1}{2}, \infty)$. In general, a disk margin $D(\frac{1}{2})$ is not guaranteed because $R(x)$ in (3.5.28) is diagonal but not constant. However, as an application of Proposition 3.35, the control law (3.5.19) may serve as the starting point of a domination redesign which, at the expense of an increased control effort, achieves a disk margin $D(\frac{1}{2})$. Because the domination redesign of Proposition 3.35 results in a smooth feedback, this redesign can also be used to smoothen the control law (3.5.19) at the origin.

Example 3.45 *(Design with CLF)*
In Example 3.20, we have explicitly solved the HJB equation to achieve optimal stabilization of the system

$$\dot{x} = x^2 + u,$$

for the cost functional

$$J = \int_0^\infty (x^2 + u^2)\, dt, \tag{3.5.29}$$

We have found the optimal stabilizing control

$$u^*(x) = -x^2 - x\sqrt{x^2+1} \tag{3.5.30}$$

and the optimal value function

$$V(x) = \frac{2}{3}(x^3 + (x^2+1)^{\frac{3}{2}} - 1) \tag{3.5.31}$$

We now reconsider the same system with the CLF approach. For scalar stabilizable systems, $V(x) = \frac{1}{2}x^2$ is always a CLF, from which we immediately get $L_fV(x) = x^3$ and $L_gV(x) = x$. The formula (3.5.19) with $c_0 = 0$ yields

$$u_S(x) = -\frac{x^3 + \sqrt{x^6 + x^4}}{x} = -x^2 - x\sqrt{x^2+1} \tag{3.5.32}$$

which is the same as the optimal control law $u^*(x)$ in (3.5.30). By Proposition 3.44, the control law $u_S(x)$ is optimal for the cost (3.5.28) with $p(x) = x + \sqrt{x^2+1}$. It can be observed that $p(x)x$ is the gradient of the optimal value function (3.5.31). This fact is particular to the scalar case and explains why $u_S(x)$ is also optimal with respect to the simpler cost (3.5.29) where $R(x) = I$. □

In the subsequent chapters, we will delineate several classes of systems for which the construction of a CLF is systematic. The construction of a CLF is usually performed together with the construction of a stabilizing feedback, but it can be of interest to separate the two tasks. In particular, Propositions 3.44 and 3.35 can be applied to the constructed CLF in order to obtain an optimal stabilizing feedback which achieves desirable stability margins. An illustration is the class of feedback linearizable systems [24]. For simplicity, we limit our attention to the single input nonlinear system

$$\dot{x} = f(x) + g(x)u$$

which is feedback linearizable if there exists a global change of coordinates $z = T(x)$ such that, in the new coordinates, the system has the normal form

$$\begin{array}{rcl} \dot{z}_1 & = & z_2 \\ & \vdots & \\ \dot{z}_{n-1} & = & z_n \\ \dot{z}_n & = & \alpha(z) + \beta(z)u \end{array} \tag{3.5.33}$$

with $\beta(z)$ globally invertible. Feedback linearization can be used for stabilization since the feedback

$$u_l(z) = \beta^{-1}(z)(-\alpha(z) - c^T z) \tag{3.5.34}$$

renders the closed-loop system linear and GAS provided that the polynomial $c_1 s + \ldots + c_n s^n$ is Hurwitz. However, because of the cancellations, the control law (3.5.34) in general does not have stability margins, as already illustrated in Example 3.38.

Instead of pursuing feedback linearization (3.5.34), we use the normal form (3.5.33) only to construct a CLF with which we then design an optimal stabilizing control. Because the nonlinear system (3.5.33) can be transformed by feedback into a chain of integrators, a CLF is obtained for the nonlinear system (3.5.33) by constructing a CLF for a chain of integrators. This can be performed in many different ways. For instance, a quadratic CLF $z^T P z$ can be chosen to satisfy the Ricatti inequality

$$A^T P + PA - PBB^T P < 0 \tag{3.5.35}$$

For a linear system, a quadratic CLF has always the Lipschitz control property since a linear feedback can be used to achieve the negative definiteness of $\dot{V}$. The quadratic CLF $V = z^T P z$ for a chain of integrators is also a CLF for the system (3.5.33) and has the Lipschitz control property. An optimal stabilizing control is then obtained with the formula (3.5.19). This control law is smooth away from the origin, and Lipschitz continuous at the origin.

Example 3.46 *(Inverse optimal design for a feedback linearizable system)* The system

$$\begin{aligned} \dot{x}_1 &= x_2 + x_1^2 \\ \dot{x}_2 &= u \end{aligned} \tag{3.5.36}$$

is feedback linearizable and, in the linearizing coordinates $(z_1, z_2) = (x_1, x_2 + x_1^2)$, it takes the normal form

$$\begin{aligned} \dot{z}_1 &= z_2 \\ \dot{z}_2 &= 2z_1 z_2 + u \end{aligned} \tag{3.5.37}$$

A stabilizing nonlinear control law based on feedback linearization is

$$u_l(z) = -2z_1 z_2 - k_1 z_1 - k_2 z_2, \quad k_1 > 0, k_2 > 0.$$

It cancels the term $2z_1 z_2$. To avoid the cancellation and achieve a sector margin for the feedback system, we use the linearizing coordinates only to construct a CLF. With this CLF we then design an optimal stabilizing control. With

$$A = \begin{pmatrix} 0 & 1 \\ 0 & 0 \end{pmatrix}, \quad B = \begin{pmatrix} 0 \\ 1 \end{pmatrix}, \quad P = \begin{pmatrix} 1 & c \\ c & 1 \end{pmatrix},$$

the Ricatti inequality (3.5.35) is satisfied for any $c \in (0, 1)$. Then $V = z^T P z$ is a CLF for $\dot{z}_1 = z_2$, $\dot{z}_2 = u$ and hence, it is also a CLF for the nonlinear system (3.5.37). Proposition 3.44 yields the optimal stabilizing control law

$$u = -2z_1 z_2 - \frac{(z_1 + cz_2)z_1 + \sqrt{(2z_1 z_2(z_2 + cz_1) + z_2(z_1 + cz_2))^2 + (cz_1 + z_2)^4}}{cz_1 + z_2}$$

As in Example 3.38, this optimal control law has two desirable properties not present in the feedback linearizing design: it recognizes the beneficial effect of $L_f V$, when $L_f V < 0$, and dominates $L_f V$ instead of cancelling it, when $L_f V > 0$. □

3.6 Summary

While stability margins do not guarantee robustness with respect to all types of uncertainties, they are sine-qua-non properties of well designed control systems. When input uncertainties are static nonlinearities, the Nyquist curve is required to stay outside a disk in the complex plane. In our terminology, the system is required to possess a disk margin, a notion which we have extended to nonlinear systems with the help of passivity properties.

In both linear and nonlinear systems a disk margin guards against two types of input uncertainties: static nonlinearities and dynamic uncertainties which do not change the relative degree of the system. This relative degree restriction may not appear significant, but, unfortunately, it does eliminate many realistic unmodeled dynamics. If the unmodeled dynamics evolve in a time scale significantly faster than the system, they can be treated as singular perturbations. The stability properties are then preserved in a region whose size increases with the increase of the separation of the time scales. Even though we do not characterize this semiglobal property as a margin, it is a robustness property.

We have next examined the stability margins of optimal feedback systems using the connection between optimality and passivity (Theorem 3.23). We have first shown that with a purely quadratic control penalty ($R(x) = I$) in the cost functional, a nonlinear optimal stabilizing control guarantees a disk margin, which, in the special case of the LQR design implies the familiar gain and phase margins of $(\frac{1}{2}, \infty)$ and $60°$, respectively. With $R(x) = diag\{r_1(x), \ldots, r_m(x)\}$ a sector margin is achieved. Our redesign strengthens this property and achieves a disk margin by dominating the original optimal value function by a larger one, which, in general, requires larger control effort.

Optimal control methods requiring the solution of the Hamilton-Jacobi-Bellman equation are impractical. We have instead, taken an inverse path. As the remaining chapters in this book will show, our design methods first construct Lyapunov functions for various classes of nonlinear systems. We then follow the inverse path by interpreting the constructed Lyapunov functions as optimal value functions for some meaningful cost functionals.

For systems which are open-loop stable, a well known inverse optimal control is the damping control, also called "L_gV-control." In Chapters 5 and 6, our forwarding procedure will recursively extend this inverse optimal design to feedforward systems, which, in general, are open loop unstable.

For more general situations, we derive an inverse optimal control from Control Lyapunov functions which are constructed by methods in remaining chapters.

3.7 Notes and References

Our disk margin is motivated by the property that the systems whose Nyquist curve does not intersect a disk remain stable in feedback interconnections with either static or dynamic "conic uncertainties." Following the work of Lurie [70] and Popov [88], this was shown by Zames [123] in the operator theoretic framework, and by Hill and Moylan [37, 38] in the state space framework.

In the 1971 edition of [1], Anderson and Moore have shown that the linear optimal regulator design results in a feedback system with the Nyquist curve outside the disk $D(\frac{1}{2})$, that is with a disk stability margin. Multivariable generalization of the gain and phase margins were given by Safonov [91], Lehtomaki, Sandell, and Athans [64] and Grimble and Owens [31], among others.

We have defined nonlinear gain, sector, and disk stability margins by specifying the class of uncertainties, in series with the plant, that the feedback systems must tolerate. A gain margin introduced by Sontag [102] deals with nonlinear additive uncertainty. The small gain stability margins, which are implicit in the recent global stability results by Krstić, Sun, and Kokotović [62] and Praly and Wang [89], can be an alternative to our passivity based margins.

The connection between optimality and passivity established by Kalman [52] for linear systems, and by Moylan [80] for nonlinear systems, has been exploited by Glad [29, 28] and Tsitsiklis and Athans [114] to prove certain robustness properties of nonlinear optimal systems. Recent developement of the inverse optimality approach is due to Freeman and Kokotović [25, 26]. The two specific "inverse optimal" control laws considered in this chapter are the *damping control*, due to Jurdjevic and Quinn [49], Jacobson [44], and Krasovsky [60], and the control law given by *Sontag's formula* [100] which uses Artstein-Sontag's *control Lyapunov function* [4, 101]. Other explicit formulae can be found in [25, 26].

Chapter 4

Cascade Designs

With this chapter we begin the presentation of feedback stabilization designs which exploit structural properties of nonlinear systems. In Section 4.1 we introduce a class of *cascade structures* formed of two subsystems, with the subsystem states z and ξ, as illustrated in Figure 4.1.

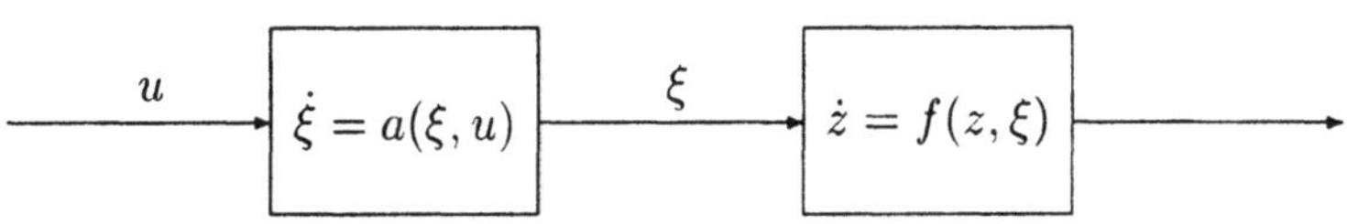

Figure 4.1: A cascade system.

The first characteristic of the cascade is that the control u enters only the ξ-subsystem. A further characterization specifies the properties of the z-subsystem and how they can be changed by the interconnection, which may act either as a control input or as an external disturbance.

In *partial-state* feedback designs presented in Section 4.2, only the ξ-subsystem state is used for feedback. The problem is to stabilize the ξ-subsystem without destroying the GAS property of the z-subsystem. In this case the interconnection with the ξ-subsystem acts as a disturbance on the z-subsystem.

In *full-state* feedback passivation designs presented in Section 4.3, the interconnection term plays an active role and the GAS assumption for the z-subsystem is relaxed to a stabilizability assumption. In this case ξ is treated as the input of the z-subsystem. A detailed case study of a translational platform stabilized by a rotating actuator (TORA) is presented in Section 4.4, as an illustration of several cascade and passivation designs.

Our design goal is either *global* or *semiglobal* stabilization. For semiglobal stabilization a control law is designed to guarantee that a *prescribed compact set* belongs to the region of attraction of the equilibrium $(z, \xi) = (0, 0)$.

A hidden danger in the deceptively simple cascade structure of Figure 4.1 is the intricate *peaking phenomenon.* An attempt to force ξ to rapidly converge to zero in order to preserve the stability properties of the z-subsystem may instead cause explosive forms of instability. Unexpectedly, the peaking phenomenon emerges as a fundamental structural obstacle not only to the solution of global, but also semiglobal stabilization problem. In Section 4.5 we characterize the class of nonpeaking cascades in which the peaking obstacle can be avoided.

4.1 Cascade Systems

4.1.1 TORA system

Cascade structures often reflect configurations of major system components, especially when each of these components constitutes a dynamical subsystem. A typical example, which will be our case study in Section 4.4, is the TORA system[1] in Figure 4.2, where a translational platform of mass M is stabilized by an eccentric rotating mass m.

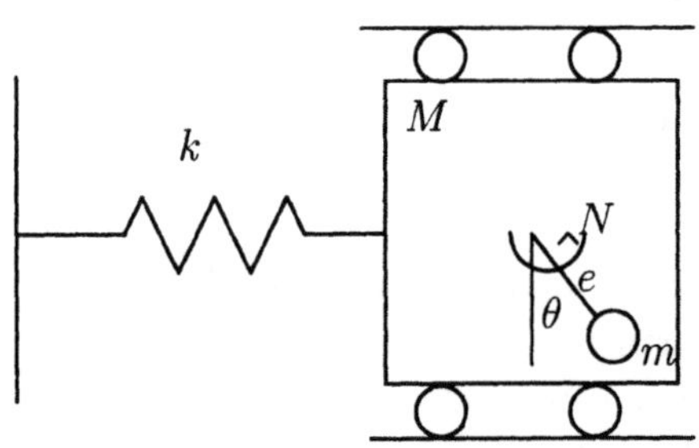

Figure 4.2: TORA system configuration.

Even without a detailed model, the TORA subsystems are physically recognizable. The controlling subsystem is the rotating mass which acts upon the second subsystem – the translational platform. The rotating mass qualifies as the ξ-subsystem because it is acted upon by the control torque directly. The

[1] TORA = Translational Oscillator with Rotating Actuator. This case study was suggested to the authors by Professor Dennis Bernstein who has built such a system in his laboratory at the University of Michigan, Ann Arbor.

platform qualifies as the z-subsystem, which, disregarding the rotating mass and friction, is a conservative mass-spring system.

4.1.2 Types of cascades

For a complete description of a cascade system, it is not sufficient to identify its subsystems and their stability properties. It is also necessary to characterize the nature of the *interconnection* of the subsystems. In the TORA system, the important interconnection term is the force of the rotating mass which acts upon the platform. This force can add damping to the oscillations of the platform, but it can also act as a destabilizing disturbance. When an interconnection term acts as a disturbance, its growth as a function of z is a critical factor which determines what is achievable with feedback design. We will return to this issue in Section 4.2. At this point we only stress the importance of the *nonlinear growth* properties of the interconnection terms.

In the simplest cascade we consider, the controlling subsystem is linear

$$\begin{array}{rcll} \dot{z} &=& \tilde{f}(z,\xi), & z \in I\!R^{n_z} \\ \dot{\xi} &=& A\xi + Bu, & \xi \in I\!R^{n_\xi} \end{array}$$

where $\tilde{f}(z,\xi)$ is C^1 and $\tilde{f}(0,0) = 0$, so that the equilibrium is at $(z,\xi) = (0,0)$. The stability assumption for the z-subsystem will be that the equilibrium $z = 0$ of $\dot{z} = \tilde{f}(z,0)$ is either globally stable (GS) or globally asymptotically stable (GAS). The ξ-subsystem is assumed to be stabilizable.

For a further characterization of the cascade, we need to specify the properties of the interconnection term

$$\psi(z,\xi) := \tilde{f}(z,\xi) - \tilde{f}(z,0) \tag{4.1.1}$$

so that the cascade can be rewritten as

$$\begin{array}{rcll} \dot{z} &=& f(z) + \psi(z,\xi), & f(z) := \tilde{f}(z,0) \\ \dot{\xi} &=& A\xi + Bu & \end{array} \tag{4.1.2}$$

When $\dot{z} = f(z)$ is GAS and the growth of $\|\psi(z,\xi)\|$ is linear in $\|z\|$, we will show that, for stabilization of the cascade, it is not important how ξ enters the interconnection term. However, if $\|\psi\|$ grows with $\|z\|$ faster than linear, then the nature of its dependence on ξ becomes critical. To analyze this more complex case, we will factor out of $\psi(z,\xi)$ a linear function,

$$\psi(z,\xi) = \tilde{\psi}(z,\xi)\, C\xi \tag{4.1.3}$$

and treat $y = C\xi$ as an "output" of the ξ-subsystem. For a given $\psi(z, \xi)$, many such factorizations are possible, a flexibility useful in some of the cascade designs. The cascade form

$$\begin{aligned} \dot{z} &= f(z) + \tilde{\psi}(z, \xi)y, \\ \dot{\xi} &= A\xi + Bu, \\ y &= C\xi \end{aligned} \tag{4.1.4}$$

is useful because it also exhibits the *input-output* properties of the ξ-subsystem, which are important for our designs.

The partially linear cascade is sometimes the result of an "input-output" linearization of a nonlinear system, achieved by a preliminary nonlinear change of coordinates, and a feedback transformation, as shown in Appendix A.

The most general nonlinear cascades to be considered in this chapter are of the form

$$\begin{aligned} \dot{z} &= f(z, \xi) + \psi(z, \xi), \\ \dot{\xi} &= a(z, \xi, u) \end{aligned} \tag{4.1.5}$$

This configuration is informative if the structural properties of the cascade are retained. For the z-subsystem, this means that the stability properties of $\dot{z} = f(z, \xi)$ must be uniform in ξ. For the ξ-subsystem, it is required that a feedback control exists which achieves global asymptotic stability of $\xi = 0$, uniformly in z. Under these conditions, the behavior of the cascade (4.1.5) is qualitatively the same as if f were independent of ξ, and a were independent of z. We will therefore concentrate on the cascades with $f(z)$ and $a(\xi, u)$, and illustrate more general situations (4.1.5) by examples.

4.2 Partial-State Feedback Designs

4.2.1 Local stabilization

In some cases the stabilization of the ξ-subsystem ensures the stabilization of the entire cascade. Such partial-state feedback designs are of interest because of their simplicity.

During the stabilization of the ξ-subsystem in the cascade

$$\begin{aligned} \dot{z} &= f(z) + \psi(z, \xi), \\ \dot{\xi} &= a(\xi, u) \end{aligned} \tag{4.2.1}$$

the interconnection term ψ acts as a disturbance which must be driven to zero without destabilizing the z-subsystem. A potentially destabilizing effect of ψ is not an obstacle to achieving *local* asymptotic stability [100].

Proposition 4.1 *(Asymptotic stability)*
If $z = 0$ is an asymptotically stable equilibrium of $\dot{z} = f(z)$, then any partial-state feedback control $u = k(\xi)$ which renders the ξ-subsystem equilibrium $\xi = 0$ asymptotically stable, also achieves asymptotic stability of $(z, \xi) = (0, 0)$. Furthermore, if $\dot{z} = f(z)$ and $\dot{\xi} = a(\xi, k(\xi))$ are both GAS, then, as $t \to \infty$, every solution $(z(t), \xi(t))$ either converges to $(z, \xi) = (0, 0)$ or is unbounded.

Proof: Let $U(\xi)$ be a Lyapunov function for the subsystem $\dot{\xi} = a(\xi, k(\xi))$. Then $V(z, \xi) = U(\xi)$ is a positive semidefinite Lyapunov function for the whole cascade. Stability of $(z, \xi) = (0, 0)$ follows from Theorem 2.24, because $(z, \xi) = (0, 0)$ is asymptotically stable conditionally to the set $\{(z, \xi) | V(z, \xi) = 0\} = \{(z, \xi) | \xi = 0\}$. Let Ω_z be the region of attraction of $z = 0$ for $\dot{z} = f(z)$ and Ω_ξ be the region of attraction of $\xi = 0$ for $\dot{\xi} = a(\xi, k(\xi))$. Because the equilibrium $(z, \xi) = (0, 0)$ is stable, it has a neighborhood Ω such that every solution $(z(t), \xi(t))$ starting in Ω is bounded and remains inside $\Omega_z \times \Omega_\xi$ for all $t \geq 0$. As $t \to \infty$, $\xi(t) \to 0$, and, by Theorem 2.21, $(z(t), \xi(t))$ converges to the largest invariant set of $\dot{z} = f(z)$ in $\Omega_z \times \{0\}$, which is the equilibrium $(z, \xi) = (0, 0)$. This proves asymptotic stability. Finally, if $\Omega_z \times \Omega_\xi = \mathbb{R}^{n_z} \times \mathbb{R}^{n_\xi}$, the attractivity argument applies to every bounded solution. This means that the solutions which do not converge to $(z, \xi) = (0, 0)$ are unbounded. □

The usefulness of a local stability property depends on the size of the region of attraction, which, in turn, is determined by the choice of $k(\xi)$.

Example 4.2 *(Semiglobal region of attraction)*
For the system

$$\begin{aligned} \dot{z} &= -z + \xi z^2 \\ \dot{\xi} &= u \end{aligned} \tag{4.2.2}$$

a linear feedback $u = -k\xi$, $k > 0$, achieves asymptotic stability of $(z, \xi) = (0, 0)$. The region of attraction can be estimated with the Lyapunov function $V = z^2 + \xi^2$. Its time-derivative

$$\dot{V} = -2(z^2 + k\xi^2 - \xi z^3) = -\begin{bmatrix} z & \xi \end{bmatrix} \begin{bmatrix} 2 & -z^2 \\ -z^2 & 2k \end{bmatrix} \begin{bmatrix} z \\ \xi \end{bmatrix} \tag{4.2.3}$$

is negative for $z^2 < 2\sqrt{k}$. An estimate of the region of attraction is the largest set $V = c$ in which $\dot{V} < 0$. This shows that with feedback gain $k > \frac{c^2}{4}$ we can guarantee any prescribed c, which means that asymptotic stability is *semiglobal*. The price paid is that feedback gain k grows as c^2. □

Semiglobal stabilizability allows the designer to achieve any desired region of attraction, but it also involves trade-offs with robustness, because the expanded system bandwidth reduces its robustness to noise and unmodeled dynamics. It is important to stress that semiglobal stabilizability does not imply global stabilizability. The system (4.2.2) will again serve as an illustration.

Example 4.3 *(Obstacle to global stabilization with partial-state feedback)*
We now show that global stabilization of the system (4.2.2) cannot be achieved with partial-state feedback $u = k(\xi)$. Worse yet: the solutions from some initial conditions escape to infinity in finite time. To see this we let $z = \frac{1}{\sigma}$, which transforms the nonlinear equation $\dot{z} = -z + \xi z^2$ into $\dot{\sigma} = \sigma - \xi$. Using its explicit solution $\sigma(t)$ and returning to $z(t)$ we obtain

$$z(t) = \frac{e^{-t}}{\frac{1}{z(0)} - \int_0^t e^{-\tau}\xi(\tau)d\tau}$$

It is clear that starting with

$$z(0) > \left(\int_0^\infty e^{-\tau}\xi(\tau)d\tau\right)^{-1} \tag{4.2.4}$$

the denominator will be zero at some finite time $t_e > 0$ and, hence, $z(t)$ escapes to infinity as $t \to t_e$. If we restrict u to be a function of ξ only, the right hand side of the inequality (4.2.4) will be bounded and independent of $z(0)$. Thus, for any $\xi(0)$ we can find $z(0)$ such that $z(t)$ escapes to infinity in finite time.
□

In the system 4.2.2, even an arbitrarily fast exponential decay of ξ is unable to prevent the destabilization of the z-subsystem. This is due to the quadratic growth in z of the interconnection term ξz^2. We will show later that global stabilization of the same system is possible with full-state feedback.

4.2.2 Growth restrictions for global stabilization

The task of global stabilization of the cascade (4.2.1) by partial-state feedback $u = k(\xi)$ not only requires that we make stability and stabilizability assumptions about the subsystems, but it also imposes a severe linear growth restriction on the interconnection term $\psi(z,\xi)$. In the last section of this chapter, we will see that, if the growth of $\psi(z,\xi)$ in z is faster than linear, a structural obstacle to both global and semiglobal stabilization is a "peaking phenomenon". Of the three assumptions we now make, Assumptions 4.4 and 4.6 are the stability and stabilizability requirements, and Assumption 4.5 is the interconnection growth restriction.

Assumption 4.4 *(Subsystem stability/stabilizability)*
In the cascade (4.2.1) the equilibrium $z = 0$ of $\dot{z} = f(z)$ is GAS and there exists a C^1 partial-state feedback control $u = k(\xi)$ such that the equilibrium $\xi = 0$ of $\dot{\xi} = a(\xi, k(\xi))$ is GAS.

□

Assumption 4.5 *(Interconnection growth restriction)*
The function $\psi(z, \xi)$ has *linear growth* in z, that is, there exist two class-$\mathcal{K}$ functions $\gamma_1(\cdot)$ and $\gamma_2(\cdot)$, differentiable at $\xi = 0$, such that

$$\| \psi(z,\xi) \| \leq \gamma_1(\| \xi \|) \| z \| + \gamma_2(\| \xi \|) \tag{4.2.5}$$

□

Assumption 4.6 *(Local exponential stabilizability of the ξ-subsystem)*
The Jacobian linearization (A, B) of $\dot{\xi} = a(\xi, u)$ at $\xi = 0$ is stabilizable. □

Theorem 4.7 *(Global stabilization with partial-state feedback)*
Suppose that Assumptions 4.5 and 4.6 hold and let $u = k(\xi)$ be any C^1 partial-state feedback such that the equilibrium $\xi = 0$ of $\dot{\xi} = a(\xi, k(\xi))$ is GAS and LES. If there exists a positive semidefinite radially unbounded function $W(z)$ and positive constants c and M such that for $\|z\| > M$

(i) $L_f W(z) \leq 0$;

(ii) $\| \frac{\partial W}{\partial z} \| \| z \| \leq c\, W(z)$

then the feedback $u = k(\xi)$ guarantees boundedness of all the solutions of (4.2.1). If, in addition, $\dot{z} = f(z)$ is GAS, then the feedback $u = k(\xi)$ achieves GAS of the equilibrium $(z, \xi) = (0, 0)$.

Proof: Let $(z(0), \xi(0))$ be an arbitrary initial condition. For $\|z\| > M$, the sequence of inequalities below follows from (i), (ii) and Assumption 4.5:

$$\begin{aligned} \dot{W} &= L_f W + L_\psi W \leq L_\psi W \leq \| \frac{\partial W}{\partial z} \| \| \psi \| \\ &\leq \| \frac{\partial W}{\partial z} \| (\gamma_1(\| \xi \|) + \gamma_2(\| \xi \|) \| z \|) \end{aligned}$$

Because the equilibrium $\xi = 0$ of $\dot{\xi} = a(\xi, k(\xi))$ is LES, we know that $\| \xi(t) \|$ converges to zero exponentially fast. This implies that there exist a positive constant α and a function $\gamma(\cdot) \in \mathcal{K}$, such that

$$\begin{aligned} \dot{W}(z(t)) &\leq \| \frac{\partial W}{\partial z} \| (\gamma(\|\xi(0)\|)e^{-\alpha t} + \gamma(\|\xi(0)\|)e^{-\alpha t} \| z(t) \|) \\ &\leq \| \frac{\partial W}{\partial z} \| \| z(t) \| \gamma(\|\xi(0)\|)e^{-\alpha t}, \quad \text{for } \| z(t) \| \geq 1 \end{aligned}$$

Using (ii), we obtain the estimate

$$\dot{W} \leq K_1(\|\xi(0)\|)e^{-\alpha t}W \tag{4.2.6}$$

for some $K_1 \in \mathcal{K}$ and for $\|z(t)\| > \max\{1, M\}$. This estimate proves the boundedness of $W(z(t))$ because

$$W(z(t)) \leq W(z(0))e^{\int_0^T K_1(\|\xi(0)\|)\, e^{-\alpha s} ds} \leq K(\|\xi(0)\|)W(z(0)) \tag{4.2.7}$$

for some $K \in \mathcal{K}$.

Because $W(z)$ is radially unbounded, the boundedness of $W(z(t))$ implies the boundedness of $\|z(t)\|$. If $\dot{z} = f(z)$ is GAS, global asymptotic stability of the equilibrium $(z, \xi) = (0, 0)$ follows from Proposition 4.1. □

Condition (ii) of Theorem 4.7 is a growth restriction imposed on $W(z)$ as a Lyapunov function for $\dot{z} = f(z)$, which can be interpreted as a polynomial growth condition.

Proposition 4.8 *(Polynomial $W(z)$)*
If $W(z)$ is a polynomial function which is positive semidefinite and radially unbounded, then it satisfies the growth condition (ii) of Theorem 4.7.

Proof: Choose $c = 4N^*$ where N^* is the degree of the polynomial $W(z)$. Pick any $z_c \in S(0,1)$ where $S(0,1) := \{z \in I\!R^{n_z} \mid \|z\| = 1\}$. First we show that for every z_c there exists $\mu(z_c)$ such that

$$\lambda\|\frac{\partial W}{\partial z}(\lambda z_c)\| < cW(\lambda z_c) \text{ for } \lambda \geq \mu(z_c) > 0 \tag{4.2.8}$$

Assume that $z_c = e_1 = (1, 0, \ldots, 0)^T$. Then $W(\lambda z_c) = P(\lambda)$ with P a polynomial in λ. Let $a_N\lambda^N$ be the highest-order term of P (clearly $N \leq N^*$). Because a_N must be positive, for λ sufficiently large, we have

$$W(\lambda z_c) = P(\lambda) > \frac{a_N\lambda^N}{2} \tag{4.2.9}$$

$$\|\frac{\partial W}{\partial z}(\lambda z_c)\| = |\, P'(\lambda)\, | < 2Na_N\lambda^{N-1} \tag{4.2.10}$$

From (4.2.9) and (4.2.10) it follows that

$$\lambda\|\frac{\partial W}{\partial z}(\lambda z_c)\| \leq 2Na_N\lambda^N < 4N\, W(\lambda z_c) \tag{4.2.11}$$

which proves (4.2.8) for $z_c = e_1$ since $4N \leq c$.

For any $z_c \in S(0,1)$, there exists an orthonormal matrix T such that $z_c = Te_1$. Defining $\tilde{z} = T^{-1}z$, we obtain a new polynomial $\tilde{W}$ in $\tilde{z}$:

$$\tilde{W}(\tilde{z}) = W(z) = W(T\tilde{z})$$

Due to linearity of the transformation, $\tilde{W}(\tilde{z})$ is a positive semidefinite, radially unbounded, polynomial function of degree N^*. Moreover,

$$\|\frac{\partial W}{\partial z}(z)\| \leq \|\frac{\partial \tilde{W}}{\partial \tilde{z}}(T^{-1}z)\|\|T^{-1}\| = \|\frac{\partial \tilde{W}}{\partial \tilde{z}}(\tilde{z})\|$$

In particular, for $z = \lambda z_c$ we obtain

$$\begin{array}{c} W(\lambda z) = W(\lambda Te_1) = \tilde{W}(\lambda e_1) \\ \|\frac{\partial W}{\partial z}(\lambda z_c)\| \leq \|\frac{\partial \tilde{W}}{\partial \tilde{z}}(\lambda e_1)\| \end{array}$$

Since the inequality (4.2.11) applies to $\tilde{W}(\lambda e_1)$, we conclude that

$$\lambda\|\frac{\partial W}{\partial z}(\lambda z_c)\| \leq \lambda\|\frac{\partial \tilde{W}}{\partial \tilde{z}}(\lambda e_1)\| < c\tilde{W}(\lambda e_1) = cW(\lambda z_c)$$

for $\lambda > \mu(z_c)$, which establishes (4.2.8) for any z_c.

Because W and $\frac{\partial W}{\partial z}$ are continuous and the inequality (4.2.8) is strict, then each $z_c \in S(0,1)$ has an open neighborhood $\mathcal{O}(z_c)$ in $S(0,1)$ such that

$$z \in \mathcal{O}(z_c) \Rightarrow \lambda\|\frac{\partial W}{\partial z}(\lambda z)\| < cW(\lambda z) \text{ for } \lambda \geq \mu(z)$$

The union of the neighborhoods $(\mathcal{O}(z_c))_{z_c \in S(0,1)}$ provides an open covering of $S(0,1)$. By compactness of the unit sphere, there exists a finite number of points $(z_{ci})_{i\in I} \subset S(0,1)$ such that $\cup_{i\in I}\mathcal{O}(z_{ci})$ is still an open covering of $S(0,1)$. As a consequence, we can choose a constant μ as the maximum of $\mu(z_{ci})$, $i \in I$ and the condition (b) of Theorem 4.7 is satisfied for $\|z\| > \mu$.

□

The growth restriction (4.2.5) on the interconnection $\psi(z,\xi)$ and exponential convergence of ξ are not sufficient to prevent destabilization of the z-subsystem. The stability properties of the z-subsystem have been strengthened by the $W(z)$-growth condition (ii) of Theorem 4.7 which cannot be dropped.

Example 4.9 *(Linear growth of ψ is insufficient for global stabilization)*
The system

$$\begin{array}{rcl} \dot{z}_1 &=& -z_1 + z_2\xi \\ \dot{z}_2 &=& -z_2 + z_1^2 z_2 \\ \dot{\xi} &=& u \end{array} \qquad (4.2.12)$$

satisfies Assumptions 4.4, 4.5, and 4.6 because the interconnection term $\psi = [z_2\xi, 0]^T$ is linear in z and $\dot{\xi} = u$ is controllable. Global asymptotic stability of the z-subsystem

$$\begin{aligned} \dot{z}_1 &= -z_1 \\ \dot{z}_2 &= (-1 + z_1^2)z_2 \end{aligned} \tag{4.2.13}$$

is established with $W(z) = z_1^2 + z_2^2 e^{z_1^2}$ which yields $\dot{W}(z) = -2W(z)$.

$W(z)$ is radially unbounded and satisfies condition (i) of Theorem 4.7. However, it does not satisfy condition (ii). We now prove that the system (4.2.12) cannot be globally stabilized by any C^1 partial-state feedback $u = k(\xi)$.

Let $\xi(0) > 0$, so that $\xi(t) \geq 0$ for all $t \geq 0$. Because $k(\xi)$ is C^1, there exists a constant $K > 0$ such that $\dot{\xi}(t) \geq -K\xi(t)$. Let $z_2(0) > 0$, so that, as long as $z_1^2(t) \geq K + 2$, we have $\dot{z}_2(t) \geq (K+1)z_2(t)$. Combining both estimates we obtain that, if $z_1^2(t) \geq K + 2$, then

$$\frac{d}{dt}(z_2\xi) \geq (K+1)z_2\xi - Kz_2\xi = z_2\xi \tag{4.2.14}$$

Choosing $z_2(0)\xi(0) > z_1(0) > \sqrt{K+2}$, we have $\dot{z}_1(0) > 0$. But $\dot{z}_1(t)$ is itself increasing because

$$\ddot{z}_1(t) = \frac{d}{dt}(z_2\xi - z_1) \geq z_1(t) \geq 0$$

We conclude that (4.2.14) holds for all $t \geq 0$. Because $\xi(t)$ converges to zero, this proves that $z_2(t)$ grows unbounded. □

The unboundedness in (4.2.12) is due to the nonlinear growth of the term $z_1^2 z_2$ in $\dot{z} = f(z)$. Because of this, $W(z) = z_1^2 + z_2^2 e^{z_1^2}$ did not satisfy the polynomial growth condition (ii) of Theorem 4.7. In the absence of such a Lyapunov function for $\dot{z} = f(z)$, further restrictions need to be imposed on both f and ψ, as in the following result proved by Sussmann and Kokotović [105].

Proposition 4.10 *(Global stabilization with linear growth)*
Suppose that Assumptions 4.4 and 4.6 hold. Let $u = k(\xi)$ be any C^1 control law which achieves GAS and LES of the equilibrium $\xi = 0$ of $\dot{\xi} = a(\xi, u)$ and denote by $\tilde{A}$ the Jacobian of $a(\xi, k(\xi))$ at $\xi = 0$. If there exist constants α and β such that

$$Re\{\lambda(\tilde{A})\} < -\alpha, \quad \| f(z) \| \leq \alpha \| z \|, \quad \| \psi(z,\xi) \| \leq \beta \| \xi \|, \tag{4.2.15}$$

for all (z, ξ), then $u = k(\xi)$ achieves GAS of the equilibrium $(z, \xi) = (0, 0)$ of (4.2.1). □

4.2.3 ISS condition for global stabilization

Instead of relying on the exponential decay of ξ, we can strengthen the *input-to-state* properties of the z-subsystem

$$\dot{z} = f(z) + \psi(z, \xi) \tag{4.2.16}$$

by requiring that for any input $\xi(t)$ which converges to zero, the corresponding solution $z(t)$ of (4.2.16) be bounded. By Proposition 4.1, this "converging input - bounded state" property is sufficient for global asymptotic stability of $(z, \xi) = (0, 0)$ if $\dot{z} = f(z)$ is GAS. For a more specific result, we assume that $\dot{z} = f(z)$ is globally exponentially stable (GES).

Proposition 4.11 (*GES of* $\dot{z} = f(z)$ and linear growth of ψ)
If Assumption 4.5 holds and if the system $\dot{z} = f(z)$ is GES, with a Lyapunov function $W(z)$ which satisfies

$$\alpha_1 \|z\|^2 \leq W(z) \leq \alpha_2 \|z\|^2, \quad \left\|\frac{\partial W}{\partial z}\right\| \leq \alpha_3 \|z\|$$

$$L_f W(z) \leq -\alpha_4 W(z), \qquad \alpha_i > 0, \; i = 1, \ldots, 4$$

then the solutions $z(t)$ of (4.2.16) are bounded and converge to zero for any $\xi(t)$ which converges to zero. Furthermore, any $u = k(\xi)$ which satisfies Assumption 4.4 for the cascade (4.2.1) achieves GAS of its equilibrium $(z, \xi) = (0, 0)$

Proof: Along the solutions of (4.2.16) we have

$$\dot{W}(z) \leq -\alpha_4 W(z) + \alpha_3 \|z\| \|\psi(z, \xi)\|$$

For $\|z\| \geq 1$, Assumption 4.5 implies $\|\psi\| \leq \gamma(\|\xi\|)\|z\|$ for some $\gamma \in \mathcal{K}$, so that

$$\dot{W}(z) \leq (-\alpha_4 + \frac{\alpha_3}{\alpha_1}\gamma(\|\xi\|))\, W(z)$$

This proves that $W(z(t))$ exists for all $t \geq 0$. Moreover, because $\xi(t)$ converges to zero, there exists a finite period after which $\dot{W}(z) \leq -\frac{1}{2}\alpha_4 W(z)$. This proves that $z(t)$ is bounded and converges exponentially to zero. □

The "converging input - bounded state" property is often difficult to verify and it is more practical to employ the stronger input-to-state stability (ISS) condition introduced by Sontag [99].

Definition 4.12 (*Input-to-state stability*)
The system $\dot{z} = \tilde{f}(z, \xi)$ is *input-to-state stable* (ISS) if there exist functions $\beta \in \mathcal{KL}$ and $\gamma \in \mathcal{K}$ such that for each bounded input $\xi(.)$ and each initial condition $z(0)$, the solution $z(t)$ exists for all $t \geq 0$ and is bounded by

$$\|z(t)\| \leq \beta(\|z(0)\|, t) + \gamma(\sup_{0\leq\tau\leq t} \|\xi(\tau)\|) \tag{4.2.17}$$

□

In a recent result by Sontag and Wang [102], the ISS property is characterized by the existence of an ISS-Lyapunov function introduced in [99].

Theorem 4.13 (*Characterization of ISS*)
The system $\dot{z} = \tilde{f}(z, \xi)$ is ISS if and only if there exists a C^1 positive definite radially unbounded function $W(z)$ such that

$$\|z\| \geq \chi_1(\|\xi\|) \Rightarrow \frac{\partial W}{\partial z}\tilde{f}(z, \xi) \leq -\chi_2(\| z \|) \tag{4.2.18}$$

where χ_1 and χ_2 are two class $\mathcal{K}$ functions. Such a $W(z)$ is called an ISS-Lyapunov function. □

An application to the cascade (4.2.1) is immediate.

Corollary 4.14 (*Global stabilization with ISS property*)
If the system $\dot{z} = f(z) + \psi(z, \xi)$ is ISS, then, under Assumption 4.4, the feedback $u = k(\xi)$ achieves GAS of the equilibrium $(z, \xi) = (0, 0)$ of the cascade (4.2.1). □

In the presence of the ISS property no growth assumption on the interconnection or exponential stability of the ξ-subsystem are needed to establish boundedness.

Example 4.15 (*ISS property – global stabilization*)
With the ISS-Lyapunov function $W(z) = \frac{z^2}{2}$, it is readily verified that the z-subsystem in the nonlinear cascade

$$\begin{aligned} \dot{z} &= -z^3 + z^2\xi \\ \dot{\xi} &= \xi^2 u \end{aligned} \tag{4.2.19}$$

has the desired ISS property. This is because

$$\dot{W} = -z^4 + \xi z^3 \leq -\frac{1}{4}z^4 + \frac{1}{4}\xi^4 \tag{4.2.20}$$

satisfies (4.2.18). Thus, if $\xi(t)$ is a bounded input, the solution $z(t)$ is bounded for all $t \geq 0$. For large z, the stabilizing term $-z^3$ in (4.2.19) dominates the destabilizing perturbation $z^2\xi$ and the linear feedback $u = -\xi$ achieves GAS of the cascade, even though the convergence of ξ to zero is not exponential. □

4.2.4 Stability margins: partial-state feedback

When a partial-state feedback $u = k(\xi)$ achieves GAS of the equilibrium $(z, \xi) = (0, 0)$ of the cascade

$$\begin{aligned} \dot{z} &= f(z) + \psi(z, \xi) \\ \dot{\xi} &= a(\xi, k(\xi)), \end{aligned} \tag{4.2.21}$$

the underlying geometry is that all the solutions converge to the manifold $\xi = 0$ which is invariant because $\xi = 0 \Rightarrow \dot{\xi} = 0$. The system (4.2.21) reduced to this manifold is the GAS z-subsystem $\dot{z} = f(z)$.

We have seen, however, that the convergence to the manifold $\xi = 0$ does not guarantee boundedness because $z(t)$ may grow unbounded while $\xi(t) \to 0$. To guarantee the *boundedness* of $z(t)$ we have introduced additional assumptions, such as LES of ξ in Section 4.2.2, or the ISS assumption in Section 4.2.3. An important consequence is that, if a control law $u = k(\xi)$ achieves GAS/LES of the subsystem $\dot{\xi} = a(\xi, u)$ with a certain stability margin, then the same stability margin is guaranteed for the entire system. This speaks in favor of partial-state feedback designs with which it is easier to achieve stability margins at the subsystem level.

Stability margins for the ξ-subsystem can be guaranteed by a stabilizing control law $u = k(\xi)$ which minimizes a cost functional of the form

$$J(\xi, u) = \int_0^\infty (l(\xi) + u^T R(\xi) u) dt, \quad l(\xi) \geq 0 \tag{4.2.22}$$

where $R(\xi) > 0$ is diagonal. We know from Chapter 3 that such an optimal control law achieves a sector margin $(\frac{1}{2}, \infty)$ and, if $R(\xi) = I$, a disk margin $D(\frac{1}{2})$, that is, $u = k(\xi)$ preserves GAS of $\xi = 0$ in the presence of any IFP($\frac{1}{2}$) input uncertainty.

To deduce the stability margins for the whole cascade from the stability margins of the ξ-subsystem, we must distinguish between Proposition 4.11 and Corollary 4.14, which require only GAS of $\dot{\xi} = a(\xi, k(\xi))$, and Theorem 4.7 and Proposition 4.10, which require both GAS and LES of $\dot{\xi} = a(\xi, k(\xi))$. In the first case, any stability margin for the ξ-subsystem is also a stability margin for the entire cascade. In the second case, we have to exclude the input uncertainties for which LES of $\xi = 0$ is lost. For a sector margin (ε, ∞), $\varepsilon > 0$, this is not restrictive, because any *static* uncertainty in this sector which preserves GAS of $\xi = 0$ also preserves its LES property. The situation is different for a disk margin because non-LES dynamic IFP uncertainties may destroy LES of $\xi = 0$ and cause instability, despite the fact that ξ converges to 0.

Example 4.16 (*Stability margin with respect to IFP uncertainties*)
Let us consider a cascade without and with a scalar IFP dynamic uncertainty:

$$(C_0)\ \begin{array}{rcl} \dot{z} & = & -\frac{z}{1+z^2} + z\xi \\ \dot{\xi} & = & \xi + u \end{array} \qquad (C_\eta)\ \begin{array}{rcl} \dot{z} & = & -\frac{z}{1+z^2} + z\xi \\ \dot{\xi} & = & \xi + \eta^3 + u \\ \dot{\eta} & = & -\eta^{4k+1} + \eta^2 u \end{array} \tag{4.2.23}$$

The design is performed for the cascade (C_0) which satisfies Assumptions 4.4, 4.5, and 4.6. By Theorem 4.7, $u = -2\xi$ achieves GAS of (C_0). For the subsystem $\dot{\xi} = \xi + u$, $u = -2\xi$ also achieves a disk margin $D(\frac{1}{2})$ because the system $\dot{\xi} = \xi + u_1$, $y_1 = 2\xi$, is OFP$(-\frac{1}{2})$. We now examine the stability of the system (C_η) which consists of the cascade (C_0) perturbed by an IFP(1) uncertainty represented by the η-subsystem. We now distinguish two cases: with $k = 0$ the perturbation is GAS, IFP, and LES, while with $k = 1$ it is GAS, IFP, but not LES. By Theorem 2.34, the feedback interconnection

$$\begin{array}{rcl} \dot{\xi} & = & -\xi + \eta^3 \\ \dot{\eta} & = & -\eta^{4k+1} - 2\eta^2\xi \end{array} \tag{4.2.24}$$

is GAS for both $k = 0$ and $k = 1$. For $k = 0$, the subsystem (4.2.24) is LES and, hence, the nominal control law $u = -2\xi$ achieves GAS of the cascade (C_η), with uncertainty.

The situation is different for $k = 1$, because (4.2.24) is not LES. It can be shown by applying Center Manifold Theorem [16, 56] that $\xi(t)$ converges to 0 as $t^{-\frac{3}{4}}$, which is not fast enough to prevent instability. We show this with a calculation in which, for large z, the function $\frac{z}{z^2+1}$ is approximated by $\frac{1}{z}$. Then, setting $w = \frac{1}{2}z^2$, we have $\dot{w} = -1 + \xi w$ which can be explicitly solved:

$$z^2(t) = e^{\int_0^t \xi(\tau)d\tau}\left(z^2(0) - 2\int_0^t e^{-\int_0^s \xi(\tau)d\tau} ds\right)$$

Now $\int_0^t \xi(\tau)d\tau$, with $\xi(\tau) = \mathcal{O}(\tau^{-\frac{3}{4}})$, diverges as $t \to \infty$, while $\int_0^t e^{-\int_0^s \xi(\tau)d\tau} ds$ remains bounded with a bound which is independent of z. Hence, $z(t)$ grows unbounded if

$$z^2(0) > \int_0^\infty e^{-\int_0^s \xi(\tau)d\tau} ds$$

This illustrates a situation in which the loss of local exponential stability results in the loss of stability. □

When the cascade is partially linear, $\dot{\xi} = A\xi + Bu$, then any LQR-design achieves a disk margin $D(\frac{1}{2})$. When $\dot{\xi} = a(\xi) + b(\xi)u$, stability margins can be guaranteed if a CLF $U(\xi)$ is known for the ξ-subsystem. Then, by Proposition 3.44, the control law given by the Sontag's formula

$$u_s(\xi) = \begin{cases} -\left(c_0 + \frac{L_aU+\sqrt{(L_aU)^2+\|L_bU\|^4}}{\|L_bU\|^2}\right) L_bU & , \quad L_bU(\xi) \neq 0 \\ 0 & , \quad L_bU(\xi) = 0 \end{cases} \tag{4.2.25}$$

minimizes a cost of the form (4.2.22) and guarantees a sector margin $(\frac{1}{2}, \infty)$ for the ξ subsystem and for the entire cascade. The same control law may serve as a starting point for a domination redesign (Proposition 3.35) which, with an increased control effort, achieves a disk margin $D(\frac{1}{2})$. When LES of $\dot{\xi} = a(\xi) + b(\xi)k(\xi)$ is also needed for stabilization of the cascade, then a further restriction is that the CLF U satisfies $\frac{\partial^2 U}{\partial \xi^2}(0) > 0$.

In the above stability margin analysis, a tacit assumption has been made that the cascade form was achieved *without cancellations.* However, this is not so if $\dot{\xi} = A\xi + Bu$ was obtained from the original ξ-subsystem

$$\dot{\xi} = A\xi + B(\alpha(z, \xi) + \beta(z, \xi)v)$$

via the feedback transformation

$$u = \alpha(z, \xi) + \beta(z, \xi)v$$

which involves cancellation of the nonlinear terms.

Example 4.17 *(Loss of stability margins because of cancellations)*
Consider the system

$$\begin{array}{rcl} \dot{z} & = & -z^3 + z^2\xi \\ \dot{\xi} & = & \alpha(z, \xi) + v \end{array} \tag{4.2.26}$$

in which the subsystem $\dot{z} = -z^3 + z^2\xi$ is ISS (see Example 4.15). Using $v = u - \alpha(z, \xi)$ to cancel $\alpha(z, \xi)$, the ξ-subsystem becomes an integrator $\dot{\xi} = u$ which is passive, so that the control law $u = -k\xi$ has a disk margin $D(0)$. However, because of the cancellation, no stability margin is guaranteed for the complete control law

$$v = -\alpha(z, \xi) - k\xi \tag{4.2.27}$$

When $|\alpha(z, \xi)|$ is bounded by a class $\mathcal{K}$ function of ξ, $|\alpha(z, \xi)| \leq \gamma(|\xi|)\xi$, the stability margin can be recovered by domination. The control law

$$u = -(1 + \gamma^2(\xi))\xi$$

guarantees a disk margin $D(\frac{1}{2})$ for the whole cascade. But if $\alpha(z, \xi)$ does not vanish when $\xi = 0$, domination is not possible with partial-state feedback $u(\xi)$.

□

4.3 Feedback Passivation of Cascades

In passivation designs we identify two passive subsystems of a cascade, and use the control to form their feedback interconnection. One path of the feedback interconnection will be created by the control law, while the other path is the interconnection term $\psi(z,\xi)$ which now actively contributes to the task of feedback stabilization. The main tools for passivation designs are Theorem 2.10 on passivity of feedback interconnections, and Theorem 2.28 on stability of passive systems.

The passivation approach, which employs full-state feedback, removes the growth restrictions introduced in Section 4.2.2. It also replaces the GAS assumption for the subsystem $\dot{z} = f(z)$ by a weaker GS assumption.

Assumption 4.18 *(Global stability of the z-subsystem)*
The equilibrium $z = 0$ of $\dot{z} = f(z)$ is globally stable and a C^2 radially unbounded positive definite function $W(z)$ is known such that $L_f W \leq 0$. □

We begin with a passivation design for the partially linear cascade

$$\begin{aligned} \dot{z} &= f(z) + \psi(z,\xi) \\ \dot{\xi} &= A\xi + Bu \end{aligned} \tag{4.3.1}$$

To identify two passive systems H_1 and H_2, we factor the interconnection as

$$\psi(z,\xi) = \tilde{\psi}(z,\xi)C\xi \tag{4.3.2}$$

We have thus created the linear block H_1 with the transfer function

$$H_1(s) = C(sI - A)^{-1}B$$

For this block to be passive, the choice of the output must render $H_1(s)$ a *positive real transfer function.* The block H_2 is the nonlinear system

$$\dot{z} = f(z) + \tilde{\psi}(z,\xi)u_2$$

with the input $u_2 = y_1$ and the output y_2 yet to be defined. We are free to select the output $y_2 = h_2(z,\xi)$ and guarantee passivity via Theorem 2.10 and Proposition 2.11. Using $W(z)$ as a positive definite storage function for H_2, we require that

$$\dot{W} = \frac{\partial W}{\partial z}(f(z) + \tilde{\psi}(z,\xi)y_1) \leq y_2^T u_2 \tag{4.3.3}$$

Knowing that $L_f W(z) \leq 0$, we satisfy (4.3.3) by selecting

$$y_2 = h_2(z,\xi) := (L_{\tilde{\psi}}W)^T(z,\xi) = \tilde{\psi}^T(\frac{\partial W}{\partial z})^T \tag{4.3.4}$$

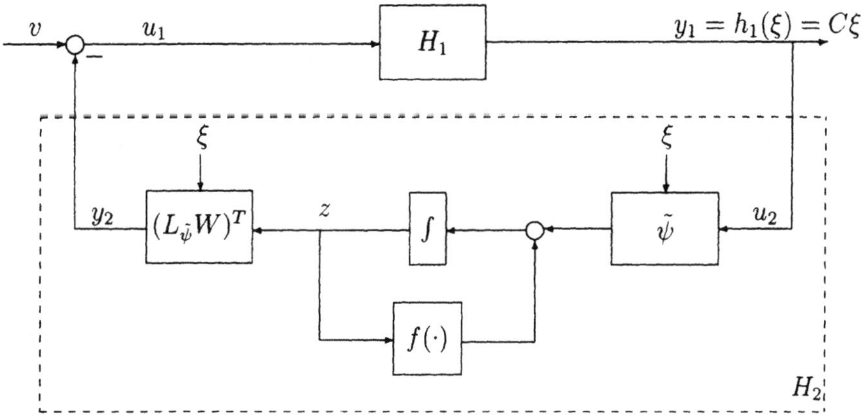

Figure 4.3: Rendering the cascade (4.3.1) passive from v to y_1.

The so defined block H_2 is passive. Next, with the feedback transformation $u = -h_2(z, \xi) + v$ we create the feedback interconnection in Figure 4.3 which, by Theorem 2.10, is passive from v to y_1. By Theorem 2.28, global stability is achieved with the control $v = -y_1$.

Applying an analogous construction to the cascade with a nonlinear ξ-subsystem, we obtain the following result.

Theorem 4.19 *(Feedback passivation design)*
Suppose that for the cascade

$$\begin{aligned} \dot{z} &= f(z) + \psi(z, \xi) \\ \dot{\xi} &= a(\xi) + b(\xi)u \end{aligned} \tag{4.3.5}$$

Assumption 4.18 is satisfied and there exists an output $y = h(\xi)$ such that
(i) the interconnection $\psi(z, \xi)$ can be factored as $\psi(z, \xi) = \tilde{\psi}(z, \xi)y$;
(ii) the subsystem

$$\begin{aligned} \dot{\xi} &= a(\xi) + b(\xi)u \\ y &= h(\xi) \end{aligned} \tag{4.3.6}$$

is passive with a C^1 positive definite, radially unbounded, storage function $U(\xi)$.

Then the entire cascade (4.3.5) is rendered passive with the feedback transformation

$$u = -(L_{\tilde{\psi}}W)^T(z, \xi) + v \tag{4.3.7}$$

and $V(z,\xi) = W(z) + U(\xi)$ is its storage function. If, with the new input v and the output y, the cascade is ZSD, then $v = -ky$, $k > 0$, achieves GAS of the equilibrium $(z,\xi) = (0,0)$. The full control law $u = -(L_{\tilde{\psi}}W)^T - ky$ possesses a $(0,\infty)$ gain margin. □

We have thus overcome the two major limitations of the partial-state feedback designs. First, we have replaced the GAS assumption for the z-subsystem by GS and the detectability condition ZSD. Second, we have achieved global stabilization without a linear growth assumption on $\psi(z,\xi)$.

Example 4.20 (*Global stabilization without growth condition*)
We have shown in Example 4.2 that the system

$$\begin{aligned} \dot{z} &= -z + z^2\xi \\ \dot{\xi} &= u \end{aligned} \tag{4.3.8}$$

is not globally stabilizable by partial-state feedback because of the nonlinear growth of the interconnection term ξz^2. With a passivation design employing full-state feedback we now achieve global stabilization. Using $y_1 = \xi$ we first create a linear passive system H_1. Then, selecting $W(z) = \frac{1}{2}z^2$ as a storage function, we establish that the first equation in (4.3.8) defines a passive system H_2 with $u_2 = \xi$ as the input and $y_2 = z^3$ as the output. Hence, with the feedback transformation

$$u = -y_2 + v = -z^3 + v$$

the cascade (4.3.8) becomes a feedback interconnection of two passive systems. The ZSD property is also satisfied because in the set $y_1 = \xi = 0$ the system reduces to $\dot{z} = -z$. Therefore, a linear feedback control $v = -ky_1$, $k > 0$, makes the whole cascade GAS. □

When the subsystem (4.3.6) is *feedback passive* rather than passive, Theorem 4.19 applies after a passivating feedback transformation. In particular, when the ξ-subsystem is linear as in (4.3.1), we can use Proposition 2.42 which states that the system (A,B,C) is feedback passive if and only if it is weakly minimum phase and has relative degree one. After a linear change of coordinates, the system (A,B,C) can be represented as

$$\begin{aligned} \dot{\xi}_0 &= Q_{11}\xi_0 + Q_{12}y \\ \dot{y} &= Q_{21}\xi_0 + Q_{22}y + CBu \end{aligned} \tag{4.3.9}$$

Then the feedback transformation

$$u = (CB)^{-1}(-2Q_{12}^T P_{11}\xi_0 - Q_{21}\xi_0 - Q_{22}y + v) =: F\xi + Gv \tag{4.3.10}$$

renders the system passive with the storage function $U = \xi_0^T P_{11}\xi_0 + \frac{1}{2}y^T y$.

Proposition 4.21 *(Passivation of partially linear cascades)*
Suppose that for the cascade

$$\begin{aligned} \dot{z} &= f(z) + \psi(z,\xi) \\ \dot{\xi} &= A\xi + Bu \end{aligned} \tag{4.3.11}$$

Assumption 4.18 is satisfied and there exists an output $y = C\xi$ such that
(i) the interconnection $\psi(z,\xi)$ can be factored as $\psi(z,\xi) = \tilde{\psi}(z,\xi)y$;
(ii) the system (A, B, C) has relative degree one and is weakly minimum phase.

Then the entire cascade (4.3.11) with $y = C\xi$ as the output is feedback passive. Its passivity from v to y is achieved with the feedback transformation

$$u = F\xi - G(L_{\tilde{\psi}}W)^T(z,\xi) + Gv \tag{4.3.12}$$

where F and G are defined in (4.3.10). The feedback control $v = -ky$, $k > 0$, guarantees GAS of $(z,\xi) = (0,0)$ if either one of the following two conditions is satisfied:
(iii) $\dot{z} = f(z)$ is GAS and (A, B) is stabilizable, or
(iv) the cascade with output y and input v is ZSD.

The control law $u = F\xi - G(L_{\tilde{\psi}}W)^T(z,\xi) - kGy$, with $k \geq 1 + \|Q_{22}\|^2$, possesses a $(\frac{1}{2}, \infty)$ gain margin provided that
(v) matrix Q_{21} in (4.3.9) is equal to 0. □

Example 4.22 *(Feedback passivation of a partially linear cascade)*
In the cascade

$$\begin{aligned} \dot{z} &= -qz^3 + (c\xi_1 + \xi_2)z^3 \\ \dot{\xi}_1 &= \xi_2 \\ \dot{\xi}_2 &= u \end{aligned} \tag{4.3.13}$$

the z-subsystem $\dot{z} = -qz^3$ is GAS when $q > 0$ and only GS when $q = 0$. With $y_1 = c\xi_1 + \xi_2$, the interconnection term $\psi(z,\xi)$ is factored as $\psi(z,\xi) = y_1 z^3$. The resulting ξ-subsystem is

$$\begin{aligned} \dot{\xi}_1 &= -c\xi_1 + y_1 \\ \dot{y}_1 &= -c^2\xi_1 + cy + u \end{aligned} \tag{4.3.14}$$

It has relative degree one, and its zero-dynamics subsystem is $\dot{\xi}_1 = -c\xi_1$. Hence, the ξ-subsystem is minimum phase if $c > 0$, and nonminimum phase if $c < 0$. For $c \geq 0$, this linear block H_1 is rendered passive by feedback transformation

$$u = -(1 - c^2)\xi_1 - (1 + c)y_1 + v \tag{4.3.15}$$

which achieves $\dot{U} \le vy_1$ with the storage function $U(\xi) = \frac{1}{2}(\xi_1^2 + y_1^2)$. To render the nonlinear block H_2 passive we select $W(z) = \frac{1}{2}z^2$ and let the output y_2 be $y_2 = L_{\tilde{\psi}}W(z) = z^4$. Then, closing the loop with

$$v = -y_2 + w = -z^4 + w \tag{4.3.16}$$

we render the entire system passive from w to y_1. The remaining step is to verify whether the feedback law for $w = -y_1$ achieves GAS. When $q > 0$, GAS is achieved because the property (iii) of Proposition 4.21 holds. However, when $q = 0$, the ZSD property requires $c > 0$, that is, the linear subsystem must be strictly minimum phase: in the set where $y_1 \equiv w \equiv 0$, which implies $\dot{y}_1 = c^2\xi_1 - z^4 \equiv 0$, it is clear that $(z, \xi_1) = (0, 0)$ is the only invariant set of $\dot{z} = 0$, $\dot{\xi}_1 = -c\xi_1$, only if $c > 0$. □

Example 4.23 *(Feedback passivation: nonlinear cascade)*
Theorem 4.19 and Proposition 4.21 do not exhaust all the cases when the passivity of a cascade can be achieved. If the nonlinear cascade

$$\begin{aligned} \dot{z} &= f(z) + \psi(z, \xi) \\ \dot{\xi} &= a(z, \xi) + b(z, \xi)u \end{aligned} \tag{4.3.17}$$

satisfies Assumption 4.18 and $b^{-1}(z, \xi)$ exists for all (z, ξ), then (4.3.17) can be made passive. We choose $y = \xi$ and let $\psi = \tilde{\psi}(z, \xi)\xi$. The feedback transformation

$$u = b^{-1}(z, y)\left(v - a(z, y) - L_{\tilde{\psi}}W(z, y)\right) \tag{4.3.18}$$

renders the entire cascade (4.3.17) passive with the storage function

$$S(z, y) = W(z) + \frac{1}{2}y^T y \tag{4.3.19}$$

Additional flexibility exists when b is a positive definite matrix which depends only on z. Then

$$\tilde{S}(z, y) = W(z) + \frac{1}{2}y^T b^{-1}(z) y \tag{4.3.20}$$

becomes a storage function with the help of the feedback transformation

$$u = v - b^{-1}(z)a(z, y) - (L_{\tilde{\psi}}W)^T(z, y) + \frac{1}{2}b^{-1}(z)\dot{b}b^{-1}(z)y \tag{4.3.21}$$

which is well defined because the entries of the matrix $\dot{b}$ are independent of u:

$$\dot{b}_{ij} = \frac{\partial b_{ij}}{\partial z}(f(z) + \tilde{\psi}(z, y)y)$$

This flexibility of passivation methods will be exploited in Section 4.4 for one of our TORA designs. □

In the feedback passivation designs thus far, global asymptotic stability of the cascade is achieved even when the z-subsystem is only GS, rather than GAS. This means that the stabilization of the z-subsystem is achieved through the action of the state of the ξ-subsystem. We now go one step further in relaxing the stability assumption on the z-subsystem.

Assumption 4.24 *(Global stabilizability of the z-subsystem)*
There exists a C^1 control law $k(z)$ such that the equilibrium $z = 0$ of the system $\dot{z} = f(z) + \tilde{\psi}(z)k(z)$ is globally stable. This is established with a C^2, positive definite, radially unbounded, function $W(z)$ such that

$$\frac{\partial W}{\partial z}(f(z) + \tilde{\psi}(z)k(z)) \leq 0, \qquad \forall z \in I\!R^{n_z} \qquad \Box$$

In the cascade, the control law $k(z)$ is not implementable and its effect must be achieved through the ξ-subsystem. For this task the ξ-subsystem is required to be minimum phase, rather than only weakly minimum phase (compare with Proposition 4.21). The restrictions on the ξ-subsystem and the interconnection are therefore more severe.

Proposition 4.25 *(Stabilization through feedback passivation)*
Suppose that for the cascade

$$\begin{aligned} \dot{z} &= f(z) + \psi(z,\xi) \\ \dot{\xi} &= A\xi + Bu \end{aligned} \tag{4.3.22}$$

there exists an output $y_1 = C\xi$ such that
(i) the interconnection $\psi(z,\xi)$ can be factored as $\psi(z,\xi) = \tilde{\psi}(z)y_1$;
(ii) the system (A, B, C) has relative degree one and is minimum phase.

If Assumption 4.24 is satisfied with the control law $k(z)$, then the entire cascade is feedback passive with respect to the new output

$$y = y_1 - k(z),$$

and its passivity is achieved with the feedback transformation

$$u = F\xi + G\left(\frac{\partial k}{\partial z}(f(z) + \tilde{\psi}(z)(y + k(z)) - L_{\tilde{\psi}}W(z) + v\right) \tag{4.3.23}$$

where F and G are defined in (4.3.10). The feedback control $v = -ky$, $k > 0$, guarantees GAS of the equilibrium $(z,\xi) = (0,0)$ when either one of the following two conditions is satisfied
(iii) the equilibrium $z = 0$ of $\dot{z} = f(z) + \tilde{\psi}(z)k(z)$ is GAS;
(iv) the cascade with output y and input v is ZSD.

Proof: By the minimum phase assumption, Q_{11} in the representation (4.3.9) is Hurwitz. Using $y = y_1 - k(z)$ as a new coordinate, we rewrite (4.3.22) as

$$\begin{aligned} \dot{z} &= f(z) + \tilde{\psi}(z)k(z) + \tilde{\psi}(z)y \\ \dot{\xi}_0 &= Q_{11}\xi_0 + Q_{12}k(z) + Q_{12}y \\ \dot{y} &= Q_{21}\xi_0 + Q_{22}(y + k(z)) + CBu - \tfrac{\partial k}{\partial x}(f(z) + \tilde{\psi}(z)(y + k(z))) \end{aligned} \tag{4.3.24}$$

To show that the feedback transformation (4.3.23) achieves passivity, we use the positive semidefinite storage function

$$V(z, y) = W(z) + \frac{1}{2}y^T y$$

Its time-derivative is $\dot{V} = L_{f+\tilde{\psi}k}W + y^T v$, which, by Assumption 4.24, proves passivity.

With the additional feedback $v = -ky$, $k > 0$ we have $\dot{V} \leq -ky^T y$. Because the closed-loop (z, y)-subsystem is decoupled from the ξ_0-subsystem, this proves global stability of its equilibrium $(z, y) = (0, 0)$ and the convergence of y to zero. With the bounded input $y_1(t) = y(t) + k(z(t))$, the state $\xi_0(t)$ remains bounded because Q_{11} is Hurwitz. Thus, all the states are bounded, the equilibrium $(z, \xi_0, y) = (0, 0, 0)$ is globally stable, and all the solutions converge to the largest invariant set where $y = 0$. If the cascade with the input v and the output y is ZSD, the equilibrium $(z, \xi_0, y) = (0, 0, 0)$ is GAS. ZSD is guaranteed when $\dot{z} = f(z) + \tilde{\psi}(z)k(z)$ is GAS, because then, if $y \equiv 0$, z and y_1 converge to zero and so does ξ_0.

□

In Theorem 4.19 and Proposition 4.21 we were able to avoid cancellations of system nonlinearities and achieve gain margin. This is not the case with the control law in Proposition 4.25 which, in general, does not possess any stability margin. We can recover the margins if our design provides a CLF.

Example 4.26 *(Global stabilization when the z-subsystem is unstable)*
Continuing the theme of Example 4.22, we now let the z-subsystem of the cascade

$$\begin{aligned} \dot{z} &= z^4 + (c\xi_1 + \xi_2)z^3 \\ \dot{\xi}_1 &= \xi_2 \\ \dot{\xi}_2 &= u \end{aligned} \tag{4.3.25}$$

be $\dot{z} = z^4$ which is unstable. We require that for the output $y_1 = c\xi_1 + \xi_2$ the linear subsystem be minimum phase, that is, $c > 0$. Treating y_1 as "virtual" control of the z-subsystem, we stabilize it with $y_1 = -2z$. By Proposition 4.25, the entire cascade with the new output $y = y_1 + 2z$ is made passive by

$$u = -c\xi_2 - 3z^4 - 2(c\xi_1 + \xi_2)z^3 + v \tag{4.3.26}$$

which achieves $\dot{V} \leq yv$ for the storage function $V = \frac{1}{2}(z^2 + y^2)$. Finally, the feedback $v = -y$ achieves GAS of the cascade. □

4.4 Designs for the TORA System

4.4.1 TORA models

In this section we take a respite from the theoretical developments in the preceding two sections and apply them to the TORA system in Figure 4.2.

The TORA system consists of a platform of mass M connected to a fixed frame of reference by a linear spring with spring constant k. The platform can oscillate without friction in the horizontal plane. On the platform, a rotating mass m is actuated by a DC motor. The mass is eccentric with a radius of eccentricity e and can be imagined to be a point mass mounted on a massless rotor. The rotating motion of the mass m creates a force which can be controlled to dampen the translational oscillations of the platform. The motor torque is the control variable.

The design goal is to find a control law to achieve asymptotic stabilization at a desired equilibrium. Our first step toward this goal is to develop TORA models convenient for various designs developed in Sections 4.2 and 4.3. Our initial choice of the state and control is made by physical considerations

x_1 and $x_2 = \dot{x}_1$ – displacement and velocity of the platform

$x_3 = \theta$ and $x_4 = \dot{x}_3$ – angle and angular velocity of the rotor carrying the mass m

u – control torque applied to the rotor.

In these coordinates the state equation of the TORA system is

$$\begin{aligned}
\dot{x}_1 &= x_2 \\
\dot{x}_2 &= \frac{-x_1 + \varepsilon x_4^2 \sin x_3}{1 - \varepsilon^2 \cos^2 x_3} + \frac{-\varepsilon \cos x_3}{1 - \varepsilon^2 \cos^2 x_3} u \\
\dot{x}_3 &= x_4 \\
\dot{x}_4 &= \frac{1}{1 - \varepsilon^2 \cos^2 x_3}[\varepsilon \cos x_3 (x_1 - \varepsilon x_4^2 \sin x_3) + u]
\end{aligned} \tag{4.4.1}$$

All the state variables are in dimensionless units so that the only remaining parameter ε depends on the eccentricity e and the masses M and m. A typical value for ε is 0.1.

In Section 4.1 we have introduced the TORA system as a physical cascade. However, the above state equation (4.4.1) does not exhibit the cascade structure. To exhibit the cascade structure we introduce two new state variables:

$$\begin{aligned} z_1 &= x_1 + \varepsilon \sin x_3 \\ z_2 &= x_2 + \varepsilon x_4 \cos x_3 \end{aligned}$$

With z_1 and z_2 instead of x_1 and x_2, the TORA state equation becomes

$$\begin{aligned} \dot{z}_1 &= z_2 \\ \dot{z}_2 &= -z_1 + \varepsilon \sin x_3 \\ \dot{x}_3 &= x_4 \\ \dot{x}_4 &= \frac{1}{1 - \varepsilon^2 \cos^2 x_3}[\varepsilon \cos x_3 (z_1 - \varepsilon \sin x_3 (1 + x_4^2)) + u] \end{aligned} \tag{4.4.2}$$

This system will be treated as a cascade in two different ways. A physical separation of the translational and rotational dynamics suggests that the subsystems be (z_1, z_2) and (x_3, x_4). This cascade structure will be employed for a partial-state feedback design. We first consider an alternative cascade structure suitable for a passivation design in which the subsystems are (z_1, z_2, x_3) and x_4.

4.4.2 Two preliminary designs

For a better understanding of the TORA system, we start with two designs employing feedback transformations which cancel the nonlinearities in the x_4 equation. We later develop a design which avoids cancellation and guarantees a stability margin. We first force the rotational subsystem into the double integrator form by the feedback transformation

$$v = \frac{1}{1 - \varepsilon^2 \cos^2 x_3}[\varepsilon \cos x_3 (z_1 - \varepsilon \sin x_3 (1 + x_4^2)) + u] \tag{4.4.3}$$

which is well defined because $0 < \varepsilon < 1$.

Example 4.27 *(Preliminary passivation design)*
Using the notation $x_3 = z_3$ and $x_4 = \xi$ and (4.4.3), we rewrite (4.4.2) as

$$\begin{aligned} \dot{z}_1 &= z_2 \\ \dot{z}_2 &= -z_1 + \varepsilon \sin z_3 \\ \dot{z}_3 &= \xi \\ \dot{\xi} &= v \end{aligned} \tag{4.4.4}$$

In this cascade, the z-subsystem is of order three while the ξ-subsystem is a single integrator. The interconnection term is $\psi = [0\ 0\ \xi]^T$. With the output

$y = h(\xi) = \xi$ and the input v, the ξ-subsystem is passive. To apply Theorem 4.19, we need to construct a Lyapunov function $W(z)$ for the z-subsystem

$$\begin{aligned}\dot{z}_1 &= z_2\\ \dot{z}_2 &= -z_1 + \varepsilon \sin z_3\\ \dot{z}_3 &= 0\end{aligned} \tag{4.4.5}$$

Because z_3 is constant, we can treat (4.4.5) as a linear system and select

$$W(z) = \frac{1}{2}(z_1 - \varepsilon \sin z_3)^2 + \frac{1}{2}z_2^2 + \frac{k_1}{2}z_3^2 \tag{4.4.6}$$

where k_1 is a design parameter. The time-derivative of W along the solutions of (4.4.5) is $\dot{W} = 0$. Clearly, (4.4.5) is globally stable, but not asymptotically stable.

Following Theorem 4.19, the feedback transformation

$$v = -L_{\tilde{\psi}}W + w = (z_1 - \varepsilon \sin z_3)\varepsilon \cos z_3 - k_1 z_3 + w \tag{4.4.7}$$

renders the system passive from the new input w to the output $y = \xi$ with respect to the storage function

$$V(z,\xi) = \frac{1}{2}(z_1 - \varepsilon \sin z_3)^2 + \frac{1}{2}z_2^2 + \frac{k_1}{2}z_3^2 + \frac{1}{2}\xi^2 \tag{4.4.8}$$

Indeed, one easily verifies that $\dot{V} = \xi v$.

Next we examine whether the system (4.4.4) with the output $y = \xi$ and the new input w is ZSD. From $y = \xi \equiv 0$ we get $\dot{\xi} \equiv 0$, which, with $w \equiv 0$ gives

$$0 \;\equiv\; \varepsilon \cos z_3(z_1 - \varepsilon \sin z_3) - k_1 z_3 \tag{4.4.9}$$

From (4.4.4), $\xi \equiv 0$ implies that z_3 is constant, and from (4.4.9) z_1 is also a constant so that $\dot{z}_1 = z_2 \equiv 0$. Then $\dot{z}_2 = z_1 - \varepsilon \sin z_3 \equiv 0$ which, together with (4.4.9), shows that $z_3 \equiv 0$. This proves that $y \equiv 0$, $w \equiv 0$ can hold only if $z_1 = z_2 = z_3 = \xi = 0$, that is, the system is ZSD.

Because the system is passive and ZSD, with the positive definite, radially unbounded storage function (4.4.8), we can achieve GAS with $w = -k_2 y = -k_2\xi$. In the coordinates of the model (4.4.2), the so designed passivating control law is

$$\begin{aligned}u &= \beta^{-1}(-\alpha - \tfrac{\partial W}{\partial x_3} - k_2 x_4)\\ &= \varepsilon^2 x_4^2 \,\sin x_3\, \cos x_3 - \varepsilon^3 \cos^2 x_3\,(z_1 - \varepsilon \sin x_3)\\ &\quad -(1 - \varepsilon^2 \cos^2 x_3)(k_1 x_3 + k_2 x_4)\end{aligned} \tag{4.4.10}$$

We remind the reader that this control law includes the terms which cancel some physical nonlinearities. □

Example 4.28 *(Partial-state feedback design)*
Model (4.4.4) cannot be used for a partial feedback design of Section 5.2, because the (z_1, z_2) subsystem is only stable, rather than asymptotically stable. To stabilize it, we imagine that z_3 is a control variable and assign to it a "control law" $z_3 = -\arctan(c_0 z_2)$, $c_0 > 0$, which achieves asymptotic stability. While this "control law" is not implementable, it serves to define a new variable

$$\xi_1 = z_3 + \arctan(c_0 z_2) \tag{4.4.11}$$

which along with $\xi_2 = \dot{\xi}_1$ and one more feedback transformation

$$w = v - \frac{2c_0^3 z_2}{(1 + c_0^2 z_2^2)^2}(-z_1 + \varepsilon \sin z_3)^2 + \frac{c_0}{1 + c_0^2 z_2^2}(-z_2 + \varepsilon\xi \cos z_3) \tag{4.4.12}$$

transforms (4.4.4) into

$$\begin{aligned} \dot{z}_1 &= z_2 \\ \dot{z}_2 &= -z_1 - \varepsilon \sin(\arctan(c_0 z_2)) + \varepsilon\psi(z_2, \xi_1) \\ \dot{\xi}_1 &= \xi_2 \\ \dot{\xi}_2 &= w \end{aligned} \tag{4.4.13}$$

The GAS property of the z-subsystem follows from $W(z) = z_1^2 + z_2^2$ and

$$\dot{W} = -2z_2 \sin(\arctan(c_0 z_2)) \leq 0$$

via the Invariance Principle (Theorem 2.21). The interconnection $\psi(z_2, \xi_1) = \sin(\xi_1 - \arctan(c_0 z_2)) + \sin(\arctan(c_0 z_2))$ is globally Lipschitz and bounded. Hence, a feedback control which renders the ξ-subsystem GAS can be designed disregarding the state z. Such a control is $w = -k_1\xi_1 - k_2\xi_2$. To implement it in the coordinates of the system (4.4.2), we substitute w back into (4.4.12), v back into (4.4.3), ξ_1 into (4.4.11) with $z_3 = x_3$ and ξ_2 evaluated from $\xi_2 = \dot{\xi}_1$ in terms of z_1, z_2, x_3, x_4. Because of these transformations, the final control law employs full-state feedback with undesirable cancellations. We will not give its lengthy expression here. □

4.4.3 Controllers with gain margin

Our goal now is to develop a passivating design which avoids the cancellations performed with the feedback transformation (4.4.3). To this end, we return to the TORA model (4.4.2), and examine the possibility of achieving passivity from the input u to the output $y = x_4$, while avoiding cancellation of nonlinearities. For this we need to modify the storage function (4.4.8). Motivated by Example 4.23, we try the storage function of the form $W(z) + \frac{1}{2}y^T b^{-1} y$. In the notation of the model (4.4.2) this storage function is

$$V(z_1, z_2, x_3, x_4) = \frac{1}{2}(z_1 - \varepsilon \sin x_3)^2 + \frac{1}{2}z_2^2 + \frac{k_1}{2}x_3^2 + \frac{1}{2}x_4^2(1 - \varepsilon^2 \cos^2 x_3) \quad (4.4.14)$$

It is successful because the derivative of $V(z_1, z_2, x_3, x_4)$ along the solutions of (4.4.2) is

$$\dot{V} = -k_1 x_3 x_4 + x_4 u \quad (4.4.15)$$

Hence, $u = -k_1 x_3 + v$ achieves passivity from v to x_4 since $\dot{V}(z, y) = x_4 v$. The ZSD property with respect to the output x_4 is established as before: $x_4 \equiv 0 \Rightarrow z_1 = const. \Rightarrow z_2 \equiv 0 \Rightarrow x_3 \equiv 0 \Rightarrow z_1 \equiv 0$. It follows from Theorem 2.28 that GAS can be achieved with $v = -k_2 x_4$, $k_2 > 0$, that is with

$$u = -k_1 x_3 - k_2 x_4 \quad (4.4.16)$$

The linear controller (4.4.16) is much simpler than either of the two cancellation controllers. It possesses a $(0, \infty)$ gain margin because we can use any positive k_1 in the storage function (4.4.14). Hence, GAS is guaranteed for any positive gains k_1 and k_2.

It is of practical interest to examine if the above linear controller can be modifiedto prevent the control magnitude $|u|$ from exceeding a specific value δ. One possibility is the saturated control law

$$u = -k_1 \frac{x_3}{\sqrt{1 + x_3^2}} - (\delta - k_1), \quad k_1 < \delta \arctan x_4$$

which also achieves GAS as can be verified with the Lyapunov function

$$V_\delta = \frac{1}{2}(z_1 - \varepsilon \sin x_3)^2 + \frac{1}{2}z_2^2 + k_1(\sqrt{1 + x_3^2} - 1) + \frac{1}{2}x_4^2(1 - \varepsilon^2 \cos^2 x_3)$$

This Lyapunov function is positive definite and radially unbounded. Its derivative is $\dot{V}_\delta = -(\delta - k_1)x_4 \arctan x_4 \leq 0$ which proves GAS via the Invariance Principle.

4.4.4 A redesign to improve performance

Typical transient response with the linear passivating controller (4.4.16), henceforth referred to as the P-controller, is shown in Figure 4.4 with the gains $k_1 = 1, k_2 = 0.14$ selected for the fastest convergence. For comparison, the analogous transient response with the cascade controller designed in Example 4.28, referred to as the C-controller, is shown in Figure 4.5.

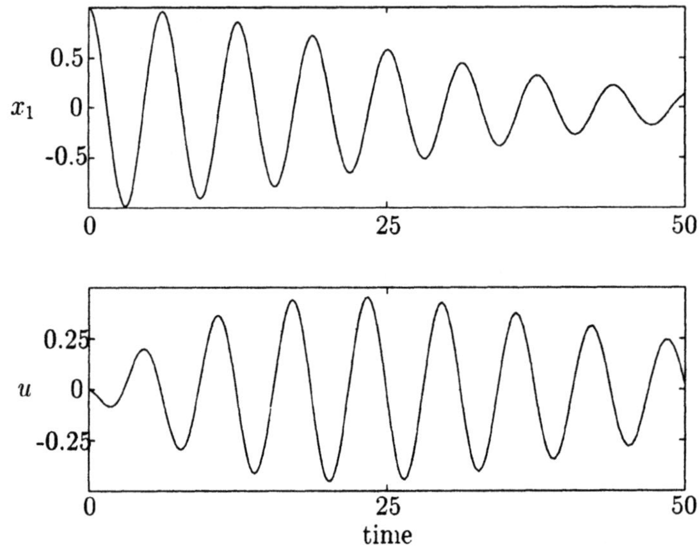

Figure 4.4: Transient response with the P-controller.

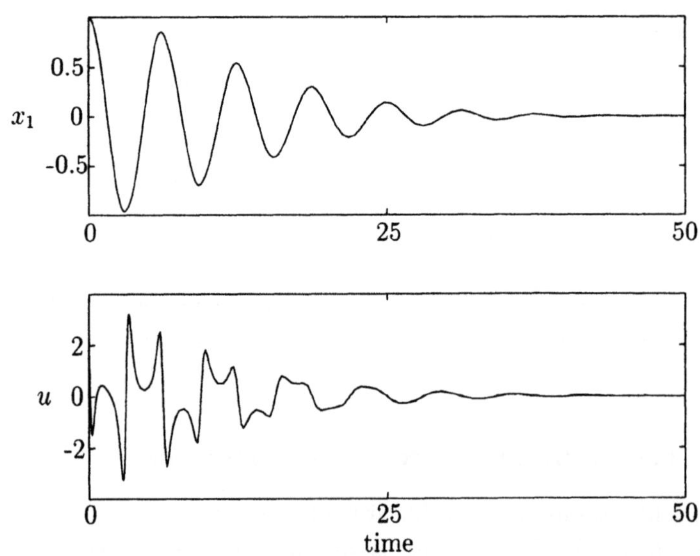

Figure 4.5: Transient response with the C-controller.

The response of the P-controller is considerably slower than that of the C-controller, which is more aggressive, with control magnitudes about seven times larger than with the P-controller.

A drawback of the P-controller is that its response cannot be made faster by adjusting the gains k_1 and k_2. This is explained with the help of a simple linear system

$$\begin{bmatrix} \dot{x}_1 \\ \dot{x}_2 \end{bmatrix} = \begin{bmatrix} 0 & 1 \\ -a & 0 \end{bmatrix} \begin{bmatrix} x_1 \\ x_2 \end{bmatrix} + \begin{bmatrix} 0 \\ 1 \end{bmatrix} u \tag{4.4.17}$$

which is passive from the input u to the output $y = x_2$ with the storage function $V = \frac{a}{2}x_1^2 + \frac{1}{2}x_2^2$. A control law which achieves global asymptotic stability is $u = -ky$. With this control law and $a = 1$, the root locus, as k varies from 0 to ∞, given in Figure 4.6, shows why the exponential decay cannot be made faster than e^{-t} by increasing k.

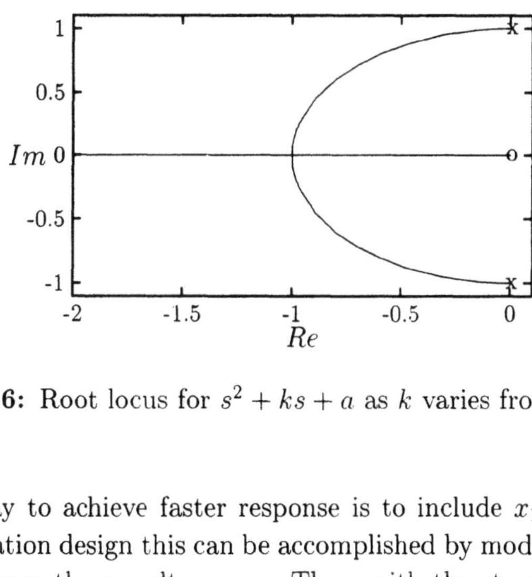

Figure 4.6: Root locus for $s^2 + ks + a$ as k varies from 0 to ∞.

The only way to achieve faster response is to include x_1 in the feedback law. In a passivation design this can be accomplished by modifying the storage function to increase the penalty on x_1. Thus, with the storage function $V = \frac{a+c}{2}x_1^2 + \frac{1}{2}x_2^2$, the resulting control law is $u = -cx_1 - kx_2$ and the response is made as fast as desired by increasing c and k.

Motivated by this linear example we introduce a design parameter k_0 to increase the penalty on the z-variables in the storage function (4.4.14):

$$\begin{aligned} V_R = {} & \frac{k_0+1}{2}[(z_1 - \varepsilon \sin x_3)^2 + z_2^2] + \frac{k_1}{2}x_3^2 \\ & + \frac{1}{2}x_4^2(1 - \varepsilon^2 \cos^2 x_3) \end{aligned}$$

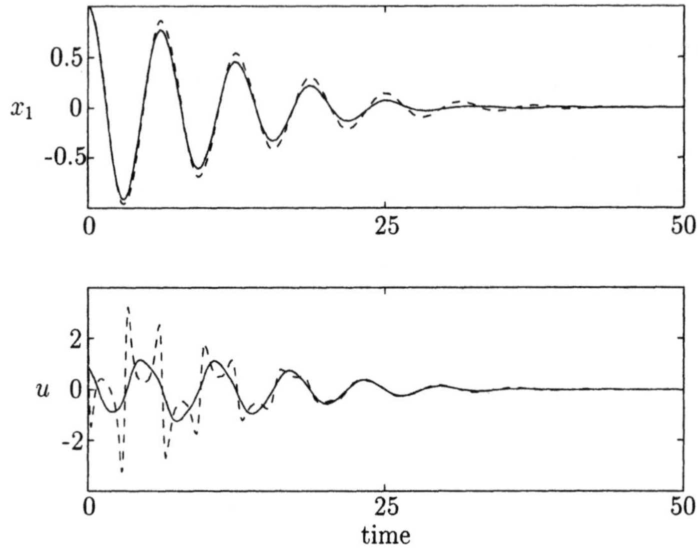

Figure 4.7: Response with P_R-controller (solid) and C-controller (dashed).

The function V_R is made a storage function by the passivating feedback transformation $u = -k_0\varepsilon\cos x_3(-z_1+\varepsilon\sin x_3) - k_1x_3 + v$. Therefore, our redesigned controller, called P_R-controller, is

$$u = -k_0\varepsilon\cos x_3(-z_1 + \varepsilon\sin x_3) - k_1x_3 - k_2x_4$$

It yields $\dot{V}_R = -k_2x_4^2 \leq 0$ and, via the Invariance Principle, guarantees GAS. With $k_0 = 0$, the P_R-controller reduces to the P-controller.

By selecting k_0, the P_R-controller matches the performance of the C-controller as shown in Figure 4.7 where the solid curves represent the P_R-controller and the dashed curves represent the C-controller. With the P_R-controller, the control magnitudes are about half of those with the C-controller. The P_R-controller also has a $(0,\infty)$ gain margin. Using $z_1 + \varepsilon\sin x_3 = x_1$, we can rewrite the P_R controller as

$$u = k_0\varepsilon x_1\cos x_3 - k_1x_3 - k_2x_4$$

Recall that x_1 is scaled displacement of the platform, with the scaling factor depending on the mass of the platform, and x_3 and x_4 are the angle and angular velocity of the rotor. We see that, thanks to its infinite gain margin, the P_R-controller is stabilizing regardless of the values of physical parameters like masses of the platform and the rotor, eccentricity, etc.

4.5 Output Peaking: an Obstacle to Global Stabilization

4.5.1 The peaking phenomenon

We now critically examine the assumptions made in Section 4.2 and 4.3. The two main assumptions impose two structurally different types of restrictions: Assumption 4.4 on the subsystem stability/stabilizability, and Assumption 4.5 on the growth of the interconnection term $\psi(z,\xi)$. The stability properties of the z-subsystem are further characterized by requiring that its Lyapunov function $W(z)$ be bounded by a polynomial (Theorem 4.7 and Proposition 4.8) or that it satisfies an ISS condition (Corollary 4.14). The feedback passivity property, required for passivation designs in Section 4.3, imposes the relative degree one and weak minimum phase constraints. These structural constraints involve the factorization of the interconnection term $\psi(z,\xi) = \tilde{\psi}(z,\xi)h_1(\xi)$ by characterizing the output $y_1 = h_1(\xi)$ of the ξ-subsystem and the function $\tilde{\psi}(z,\xi)$.

We have already suggested, and illustrated by examples, that such restrictions are not introduced deliberately to obtain simpler results. We will now show that most of these restrictions cannot be removed because of the *peaking phenomenon* which is an obstacle to both global and semiglobal stabilizability of nonlinear feedback systems.

In Section 4.3, we have already seen that, using partial-state feedback, global stabilization may be impossible without a *linear* growth restriction on ψ. It was illustrated on the system (4.2.2) that, with an increase in the feedback gain, the region of attraction can be made as large as desired (semiglobal). However, using high-gain feedback to force the state ξ to converge faster will not always make the z-subsystem less perturbed. The reason for this is the *peaking phenomenon* in which the fast stabilization causes large transient "peaks" which increase with faster decay rates.

The controllability of the pair (A, B) in $\dot{\xi} = A\xi + Bu$ is sufficient for a state feedback $u = K\xi$ to place the eigenvalues of the closed-loop system as far to the left of the imaginary axis as desired. This means that any prescribed exponential decay rate $a > 0$ can be achieved with linear feedback $u = K_a\xi$ so that the solutions of $\dot{\xi} = (A + BK_a)\xi$ satisfy

$$\|\xi(t)\| \leq \gamma(a)\|\xi(0)\|e^{-at} \tag{4.5.1}$$

The peaking phenomenon occurs if the growth of γ as a function of a is polynomial. To appreciate this fact consider the simplest case when $\gamma(a) = a$ and let $\|\xi(0)\| = 1$. Then the bound (4.5.1) is maximum at $t = \frac{1}{a}$ and this maximum is ae^{-1}. This is the mildest form of peaking: the peak of ξ grows linearly with a. In general, the transient peak estimated by $\gamma(a)$ grows as a^π where $\pi = 0, 1, 2, \ldots$. This growth is the price paid for achieving the fast decay rate a. The absence of peaking is characterized by $\pi = 0$.

Because a large peak in ξ may force $z(t)$ to escape to infinity in finite time, the peaking phenomenon limits the achievable domain of attraction.

Example 4.29 *(Peaking)*
For the cascade with a cubic nonlinearity

$$\begin{aligned} \dot{z} &= -z^3 + \xi_2 z^3 \\ \dot{\xi}_1 &= \xi_2 \\ \dot{\xi}_2 &= u \end{aligned} \tag{4.5.2}$$

the linear partial-state feedback

$$u(\xi) = -a^2\xi_1 - 2a\xi_2 \tag{4.5.3}$$

places both eigenvalues of the ξ-subsystem at $s = -a$. The state ξ_2 of the ξ-subsystem is a multiplicative disturbance in the z-subsystem. It may seem that if $\xi_2(t)$ converges to zero faster, the interconnection $\xi_2 z^3$ is less destabilizing and that the domain of attraction for the whole system (4.5.2) with (4.5.3) grows as $a \to \infty$.

However, this is false because the explicit solution of the z-subsystem is

$$2z^2(t) = \left(\frac{1}{2z^2(0)} + t - \int_0^t \xi_2(\tau)d\tau\right)^{-1} \tag{4.5.4}$$

The quantity in the parenthesis must remain nonnegative for all $t > 0$ or else $z(t)$ escapes to infinity. But, with the initial condition $\xi_1(0) = 1$, $\xi_2(0) = 0$, the solution $\xi_2(t)$ is

$$\xi_2(t) = -a^2 t e^{-at} \tag{4.5.5}$$

and its peak is ae^{-1} at time $t_p = \frac{1}{a}$. The expression for $z^2(t)$ at $t = t_p$ is

$$z^2(t_p) = \frac{1}{\frac{1}{2z^2(0)} + 2(a+1)(a^{-1} - e^{-1})}$$

For any $a > e$ and $z(0)$ large enough, this implies $z^2(t_p) < 0$ which means that $z(t_p)$ does not exist, that is, $z(t)$ escapes to infinity before $t = t_p$. It is also clear that, along the z-axis, the region of attraction shrinks with an increase in a. □

To proceed with our analysis of peaking, we now characterize the class of linear systems (A, B, C) with the state ξ in which an arbitrarily fast convergence of the output $y = C\xi$ to zero can be achieved without peaking. In our definition of nonpeaking systems, the nonpeaking requirement is imposed on the output only, and some of the states are allowed to peak.

Definition 4.30 *(Nonpeaking systems)*
The system $\dot{\xi} = A\xi + Bu,\ y = C\xi,\ u \in I\!R^m, y \in I\!R^p$ is said to be *nonpeaking* if for each $a > 0$ and $\xi(0)$ there exists a bounded input $u(t)$ such that the state $\xi(t)$ converges to zero and the output $y(t)$ satisfies

$$\|y(t)\| \leq \gamma\|\xi(0)\|(e^{-\sigma at} + \frac{1}{a}) \tag{4.5.6}$$

where the constants γ and σ do not depend on a. In all other cases, (A, B, C) is a *peaking system.* □

For nonpeaking systems we design stabilizing feedback control laws which satisfy the condition (4.5.6). We say that these control laws achieve nonpeaking stabilization of the system (A, B, C) with the understanding that the nonpeaking property applies to the output only.

Example 4.31 *(Nonpeaking design)*
The feedback law (4.5.3) for the system (4.5.2) in Example 4.29 forced both states $\xi_1(t)$ and $\xi_2(t)$ to converge to zero with the same rapid rate a. Because of this, the state ξ_2 reached its peak ae^{-1} which destabilized the z-subsystem.

We will now avoid peaking in ξ_2 by considering it as the output of the nonpeaking system $\dot{\xi}_1 = \xi_2,\ \dot{\xi}_2 = u$. This system is nonpeaking because the fast convergence of $y = \xi_2$ is achieved with the control law

$$u(\xi) = -\xi_1 - a\xi_2 \tag{4.5.7}$$

and the nonpeaking condition (4.5.6) is satisfied. Indeed, for a large, we have

$$\xi_2(t) \approx \xi_2(0)e^{-at} + \mathcal{O}(\frac{|\,\xi_1(0)\,|}{a}) \tag{4.5.8}$$

After a transient, which can be made as short as desired by increasing a, $\xi_2(t)$ is reduced to $\mathcal{O}(\frac{|\xi_1(0)|}{a})$. During the transient, $\xi_2(t)$ does not peak.

Because the output $y = \xi_2$ is nonpeaking, the state z remains bounded for arbitrary large a. The substitution of (4.5.8) into (4.5.4) yields

$$2z^2(t) \leq \left(\frac{1}{2z^2(0)} + (1 - \mathcal{O}(\frac{|\,\xi_1(0)\,|}{a}))t - \frac{|\xi_2(0)|}{a}\right)^{-1} \tag{4.5.9}$$

Given $z(0)$, $\xi_1(0)$ and $\xi_2(0)$, we can always select $a > 0$ large enough to make $1 - \mathcal{O}(\frac{|\xi_1(0)|}{a}) > 0$ and $\frac{|\xi_2(0)|}{a} < \frac{1}{2z^2(0)}$. Then it follows from (4.5.9) that $z(t)$ remains bounded and converges to zero. Hence, the region of attraction grows indefinitely as $a \to \infty$, that is, the stabilization is semiglobal. The price paid for the semiglobal stability is that the convergence of $\xi_1(t)$ to zero is very slow, its rate is approximately $\frac{1}{a}$. This is so, because for a large, one of the eigenvalues of the ξ-subsystem with feedback (4.5.7) is approximately $-\frac{1}{a}$. The other eigenvalue, which determines the decay of ξ_2, is approximately $-a$. □

The clearest insight into the peaking phenomenon is provided by the chain of integrators in which the output of the last integrator is forced to converge to zero with the rapid rate a without peaking. The state of the preceding integrator, being the derivative of the output, must converge with the rate a^2 and peaks as a, that is, with the peaking exponent $\pi = 1$. The states of other intergrators peak with larger exponents.

Proposition 4.32 (*Peaking of output derivatives*)
Assume that for the chain of integrators

$$\begin{aligned} &\dot{\xi}_1 = \xi_2,\ \dot{\xi}_2 = \xi_3,\ \ldots,\ \dot{\xi}_n = u \\ &y = \xi_1 \end{aligned} \tag{4.5.10}$$

a control $u_a(t)$ achieves nonpeaking stabilization of (4.5.10), that is, it forces the output $y = \xi_1$ to satisfy the nonpeaking condition (4.5.6). Then this control also forces each other states ξ_k to peak with the exponent $\pi = k - 1$, that is,

$$\max_{\|\xi(0)\|=1} \max_{t \geq 0} |\xi_k(t)| \geq a^{\pi}\gamma = a^{k-1}\gamma$$

where γ is independent of a.

Given any Hurwitz polynomial $q(s) = s^n + q_{n-1}s^{n-1} + \ldots + q_0$, a nonpeaking feedback control stabilizing the chain of integrators (4.5.10) is

$$u_a = -\sum_{k=1}^{n} a^{n-k+1} q_{k-1}\, \xi_k \tag{4.5.11}$$

Proof: Introducing the magnitude scaling $\bar{\xi}_k = a^{n-k+1}\xi_k$ and the fast time scale $\tau = a\,t$, we rewrite the closed-loop system as

$$\begin{aligned} &\tfrac{d\bar{\xi}_1}{d\tau} = \bar{\xi}_2,\quad \tfrac{d\bar{\xi}_2}{d\tau} = \bar{\xi}_3,\quad \ldots, \\ &\tfrac{d\bar{\xi}_n}{d\tau} = -\textstyle\sum_{k=1}^{n} q_{k-1}\bar{\xi}_k \end{aligned} \tag{4.5.12}$$

This system is asymptotically stable because its characteristic polynomial is Hurwitz. For each $k \in \{1, \ldots, n\}$, we have

$$|\bar{\xi}_k(t)| \le \gamma \|\bar{\xi}(0)\| e^{-\sigma\tau}$$

where the constants γ and σ are independent of a. Returning to the original state ξ and time t, we have

$$|\xi_k(t)| \le c\gamma a^{k-1} \|\bar{\xi}(0)\| e^{-\sigma a t}$$

where c is a constant independent of a. This shows that the output $y = \xi_1$ satisfies the nonpeaking condition (4.5.6).

We now show that peaking in ξ_k cannot be avoided for $k > 1$ if y is to satisfy the nonpeaking condition (4.5.6). We give the proof only for ξ_2 because the proof for the other states follows by induction. Let $\xi_1(0) = 1$ and $\xi_2(0) = 0$, so that $\xi_1(t) \le \frac{1}{2}$ at time $t = \frac{1}{\sigma a} \ln \frac{2\gamma a}{a - 2\gamma} =: \frac{T}{a}$. Then

$$\xi_1(\frac{T}{a}) - \xi_1(0) = \int_0^{\frac{T}{a}} \xi_2(t) dt - 1 \le -\frac{1}{2}$$

implies

$$\int_0^{\frac{T}{a}} |\xi_2(t)| dt \ge \frac{1}{2}$$

This shows that, as $a \to \infty$, the the maximum value of $|\xi_2(t)|$ on the interval $[0, \frac{T}{a}]$ grows linearly with a, that is the peaking exponent of ξ_2 is $\pi = k - 1 = 1$.

□

4.5.2 Nonpeaking linear systems

We will now characterize the structural properties of nonpeaking linear systems (A, B, C) and design control laws which achieve nonpeaking stabilization. As always, we assume that (A, B) is stabilizable. For what follows we recall that when the output $y = C\xi$ is required to track a prescribed function of time, the solution involves the *right inverse* of the system (A, B, C), see Appendix A. Therefore, it is not surprising that every nonpeaking system (A, B, C) is right-invertible. We will first consider the case when $m = p$, and the relative degree is $\{r_1, \ldots, r_m\}$, $r := r_1 + \ldots + r_m \le n$. The non-square case will be discussed at the end of this subsection.

As described in Appendix A, a change of coordinates and a preliminary feedback will put the system (A, B, C) in the normal form

$$\begin{aligned} \dot{\xi}_0 &= A_0 \xi_0 + B_0 y, \qquad \xi_0 \in I\!R^{n-mr} \\ y_i^{(r_i)} &= u_i, \quad i = 1, \ldots, m \end{aligned} \tag{4.5.13}$$

which consists of the zero-dynamics subsystem (A_0, B_0) and $m = p$ separate chains of integrators. The eigenvalues of A_0 are the zeros of the transfer function $H(s) = C(sI-A)^{-1}B$. Because the original pair (A, B) is stabilizable, the pair (A_0, B_0) inherits this property.

When the system (A, B, C) is minimum phase, that is, when A_0 is Hurwitz, the convergence to zero of the output y implies the convergence to zero of ξ_0. From this fact and Proposition 4.32, we deduce that minimum phase systems are nonpeaking.

Proposition 4.33 *(Minimum phase systems are nonpeaking)*
Every square, right-invertible minimum phase system (A, B, C) is nonpeaking. Consider such a system and let $q(s) = s^n + q_{n-1}s^{n-1} + \ldots + q_0$ be any Hurwitz polynomial. Then the linear feedback

$$u_i = -\sum_{k=1}^{n} a^{n-k+1} q_{k-1} y_i^{(k-1)}, \quad i = 1, \ldots, m \tag{4.5.14}$$

with a as large as desired, achieves nonpeaking stabilization of (A, B, C). □

The strict nonminimum phase property and peaking are directly related. If the system (A, B, C) is strictly nonminimum phase, that is if at least one of its zeros have positive real part, then it cannot be stabilized without peaking. This is shown by the following result of Braslavsky and Middleton [78].

Proposition 4.34 *(Peaking in nonminimum phase systems)*
Let (A, B, C) be a SISO system with a zero ν in the open right half-plane. If $y(t)$ is the bounded response to a bounded input $u(t)$ and initial condition $\xi(0)$, then

$$\int_0^{\infty} e^{-\nu\tau} y(\tau) d\tau = C(\nu I - A)^{-1}\xi(0) \tag{4.5.15}$$

which implies that $y(t)$ is peaking. □

To see that this equality prevents nonpeaking stabilization, we show that in $y(t) \leq \gamma\|\xi(0)\|(e^{-\sigma a t} + \frac{1}{a})$, γ increases with a. The substitution into (4.5.15) and integration yield

$$\left(\frac{\gamma}{\sigma a + \nu} + \frac{\gamma}{a\nu}\right)\|\xi(0)\| \geq \|C(\nu I - A)^{-1}\xi(0)\| \tag{4.5.16}$$

The right hand side of this inequality is independent of a. Clearly, the only possibility for the inequality (4.5.16) to hold for an arbitrary $\xi(0)$ and all a is that γ increases with a. This means that $y(t)$ is peaking.

When A_0 is not Hurwitz, the output y must be employed as an input u_0 to stabilize the zero dynamics. This explains the close relationship between peaking in the output and the location of the zeros of (A, B, C). For a nonpeaking stabilization of (A, B, C), we must be able to stabilize the zero-dynamics subsystem

$$\dot{\xi}_0 = A_0\xi_0 + B_0u_0 \tag{4.5.17}$$

with an input which satisfies

$$\|u_0(t)\| \leq \gamma\|\xi(0)\|(e^{-\sigma at} + \frac{1}{a}) \tag{4.5.18}$$

This imposes a constraint on the feedback gains admissible for the stabilization of the zero dynamics which then becomes a constraint on the eigenvalues of A_0, that is, on the zeros of (A, B, C). The unstable eigenvalues of A_0 are constrained to be on the imaginary axis, giving rise to the Jordan block canonical form:

$$A_0 = \begin{bmatrix} A_u & A_J \\ 0 & A_s \end{bmatrix}$$

where A_s is Lyapunov stable.

Theorem 4.35 *(Low-gain stabilization of the zero dynamics)*
If (A_0, B_0) is stabilizable and the eigenvalues of A_0 are in the closed left half plane, then the pair (A_0, B_0) is stabilizable by a low-gain feedback control $u_0 = K_0(a)\xi_0$ which does not peak and, for a large, satisfies

$$\|u_0(t)\| = \|K_0(a)e^{(A_0+B_0K_0(a))t}\xi_0(0)\| \leq \frac{\gamma_1}{a}e^{-\sigma at}\|\xi_0(0)\| \tag{4.5.19}$$

where γ_1 and σ are positive constants independent of a. Moreover, for A_0 in the Jordan block form, the low-gain matrix $K(a)$ can be chosen such that the state ξ_s corresponding to A_s does not peak:

$$\|\xi_s(t)\| \leq \gamma_2\|\xi_0(0)\| \tag{4.5.20}$$

where γ_2 is a positive constant independent of a. □

The proof of the theorem is given in Appendix B.

Starting with $u_0 = K_0(a)\xi_0$, which achieves a low-gain stabilization of the zero-dynamics subsystem (4.5.17), we proceed to the nonpeaking stabilization of the whole system

$$\begin{aligned} \dot{\xi}_0 &= (A_0 + B_0K_0(a))\xi_0 + B_0(y - K_0(a)\xi_0), \qquad && \xi_0 \in \mathbb{R}^{n-r} \\ y_i^{(r_i)} &= u_i, && i = 1, \ldots, m \end{aligned} \tag{4.5.21}$$

Defining $e = y - K_0(a)\xi_0$, this system is rewritten as

$$\begin{aligned} \dot{\xi}_0 &= (A_0 + B_0 K_0(a))\xi_0 + B_0 e, \\ e_i^{(r_i)} &= u_i + \phi_i^T \xi \end{aligned} \tag{4.5.22}$$

where ϕ_i's are known vectors. With e treated as the new output, the system (4.5.22) is minimum phase because $\tilde{A}_0 = A_0 + B_0 K_0(a)$ is Hurwitz for all a. Thus, by Proposition 4.33, a high-gain feedback of $[e_i, \dot{e}_i, \ldots, e^{(r_i-1)}]$, $i = 1, \ldots, m$, achieves a fast stabilization of e without peaking. Returning to the original system, the next proposition shows that the same feedback achieves nonpeaking stabilization of the system (A, B, C).

Proposition 4.36 *(Nonpeaking design)*
Let $q(s) = s^r + q_{r-1}s^{r-1} + \ldots + q_0$ be an arbitrary Hurwitz polynomial. Under the assumptions of Theorem 4.35, the feedback

$$u_i = -\varphi_i^T \xi - \sum_{k=1}^{r_i} a^{r-k+1} q_{k-1} e_i^{(k-1)} \tag{4.5.23}$$

achieves nonpeaking stabilization of the system (A, B, C).

Proof: By Proposition 4.33, the feedback (4.5.23) is stabilizing. When a is large, the convergence to zero of e and its derivatives is *fast*. In particular, we have

$$\|e(t)\| \leq \gamma_2 \|\xi(0)\| e^{-at} \tag{4.5.24}$$

for some constant γ_2 independent of a. Using the explicit solution of (4.5.21), we have

$$u_0(t) = K_0(a)\xi_0(t) = K_0(a)e^{\tilde{A}_0 t}\xi_0(0) + \int_0^t K_0(a)e^{\tilde{A}_0(t-\tau)}B_0 e(\tau)d\tau$$

With (4.5.24) and (4.5.19), this yields the estimate

$$\|u_0(t)\| \leq \frac{\gamma_1}{a}\|\xi_0(0)\| + \int_0^t \frac{\gamma_1}{a}\gamma_2\|\xi(0)\|e^{-at}d\tau \tag{4.5.25}$$

and, hence, the bound

$$\|u_0(t)\| \leq \frac{\gamma_3}{a}\|\xi_0(0)\|$$

where γ_3 is a constant independent of a. The output $y(t) = e(t) + K_0(a)\xi_0(t)$ satisfies

$$\|y(t)\| \leq \|e(t)\| + \|K_0(a)\xi_0(t)\| \leq \gamma_2\|\xi(0)\|e^{-at} + \frac{\gamma_3}{a}\|\xi(0)\| \tag{4.5.26}$$

and the nonpeaking constraint (4.5.6) is satisfied with $\gamma = \max(\gamma_2, \gamma_3)$. □

In the construction of the feedback $u = K(a)\xi$, a *high-gain* feedback stabilization of the output is combined with a *low-gain* feedback stabilization of the zero dynamics. The fast decay of the output y implies that the derivatives of y peak. The small magnitude of y, which remains after its fast decay, is used for low-gain stabilization of the zero dynamics. This results in a slow convergence of $\xi_0 = (\xi_s\ \xi_u)^T$, during which ξ_u peaks. A chain of integrators shows that these limitations are structural, and cannot be altered by design.

Proposition 4.37 *(Peaking states)*
Let the system (A, B, C) be a single chain of integrators in which the output is the i-th state:

$$\begin{array}{ll} \dot{\xi}_1 = \xi_2,\ \dot{\xi}_2 = \xi_3,\ \ldots,\ \dot{\xi}_n = u \\ y = \xi_i, \qquad i \in \{1, \ldots, n\} \end{array} \tag{4.5.27}$$

Then every input $u(t)$ which forces the output to satisfy the nonpeaking condition (4.5.6) causes the peaking of the following states:

(i) for $k \in \{1, \ldots, n-i\}$ the state ξ_{i+k} peaks with an exponent $\pi = k$

(ii) for $k \in \{1, \ldots, i-1\}$ the state ξ_{i-k} peaks with an exponent $\pi = k - 1$

Proof: The peaking in the derivatives of y, as stated in (i), has been shown in Proposition 4.32. We only prove that ξ_{i-2} peaks with exponent $\pi = 1$. The rest of the proof follows by induction. Let $\xi_{i-1}(0) = 1$ and $\xi_{i-2}(0) = 0$. Using the fact that $|\xi_i(\tau)| \le \gamma(e^{-a\tau} + \frac{1}{a})$, we have for all $t \ge 0$

$$\xi_{i-1}(t) = 1 - \int_0^t \xi_i(\tau)d\tau \ge 1 - \frac{\gamma}{a}(1 + t)$$

In particular, this shows that $\xi_{i-1}(t) \ge \frac{1}{2}$ on a time interval $[0, T(a)]$, where $T(a)$ grows linearly in a. This implies that

$$\xi_{i-2}(T(a)) \ge \frac{1}{2}T(a)$$

Hence, ξ_{i-2} peaks with exponent $\pi = 1$. □

Example 4.38 *(Peaking states)*
Consider a chain of four integrators in which the output is $y = \xi_3$, that is $n = 4$ and $i = 3$. Then by Proposition 4.37 the nonpeaking states are ξ_3 and ξ_2, while both ξ_1 and ξ_4 peak with the exponent $\pi = 1$. The peaking of ξ_4 is fast and of ξ_1 is slow. The state ξ_2 is nonpeaking because it represents the ξ_s part of the zero-dynamics subsystem $\dot{\xi}_1 = \xi_2$, $\dot{\xi}_2 = 0$. □

Proposition 4.36 provides us with a design methodology for nonpeaking stabilization of the square right-invertible systems. As shown by Saberi, Kokotović, and Sussmann [92], this methodology can be extended to non-square right-invertible systems as follows: if the system (A, B, C) is stabilizable and right-invertible, then there exists a linear dynamic feedback transformation such that the new system is stabilizable and square-invertible. In addition, the zeros introduced by the dynamic transformation are freely assignable [93]. Thus, the problem of nonpeaking stabilization of non-square right-invertible systems is reduced to the same problem for the square right-invertible systems.

The right-invertibility condition is necessary to prevent peaking. If a system is not right-invertible, then there exist at least two components of the output which cannot be controlled by two independent components of the input. This is the case when, in the chain of integrators (4.5.27), two different states ξ_i and ξ_j are the components of a two-dimensional output. If $i < j$, then $\xi_j(t)$ necessarily peaks during a fast stabilization of ξ_i. Hence, a system (A, B, C) which is not right-invertible is necessarily peaking.

Example 4.39 *(Lack of right-invertibility implies peaking)*
For the two-input system

$$\begin{aligned}\dot{\xi}_1 &= u_1\\ \dot{\xi}_2 &= -\xi_2 + \xi_3\\ \dot{\xi}_3 &= u_2\end{aligned}$$

consider the three choices of the output pair (y_1, y_2):

$$(\xi_1, \xi_2), \quad (\xi_1, \xi_3), \quad (\xi_2, \xi_3)$$

The systems with the first two choices are right-invertible. The first system is without zeros and the second system has a zero at -1. Hence these two systems are nonpeaking. The third system with the output (ξ_2, ξ_3) is not right-invertible because $\xi_2(t)$ and $\xi_3(t)$ cannot be specified independently from each other. The output $y_1 = \xi_2$ is controlled by the output $y_2 = \xi_3$ and for y_1 to be fast, y_2 must peak. Hence, the output $y^T = [y_1, y_2]$ cannot satisfy the nonpeaking condition (4.5.6). □

We summarize our characterization of nonpeaking systems in the following theorem.

Theorem 4.40 *(Nonpeaking systems)*
The system (A, B, C) is nonpeaking if and only if it is stabilizable, right-invertible, and has no zeros in the open right-half plane. Every such system can be stabilized without peaking using linear state feedback. □

4.5.3 Peaking and semiglobal stabilization of cascades

We now analyze the peaking phenomenon as an obstacle to semiglobal stabilization of the partially linear cascade

$$\begin{aligned} \dot{z} &= f(z) + \tilde{\psi}(z,\xi)y \\ \dot{\xi} &= A\xi + Bu \\ y &= C\xi \end{aligned} \tag{4.5.28}$$

For semiglobal stabilization with partial-state feedback $u = K\xi$ the assumption that (A, B, C) is a nonpeaking system is not sufficient. From the decomposition $\xi_0 = (\xi_u, \xi_s)$ of Theorem 4.35 and Proposition 4.37, we know that a fast decay of the output y induces peaking in the derivatives of y and in ξ_u. For semiglobal stabilization these states are not allowed to enter the interconnection $\tilde{\psi}(z, \xi)$.

Theorem 4.41 *(Nonpeaking cascade)*
Suppose that Assumption 4.4 holds. If (A, B, C) is a nonpeaking system, and the state ξ enters the interconnection $\tilde{\psi}$ only with its nonpeaking components y and ξ_s, that is $\tilde{\psi} = \tilde{\psi}(z, y, \xi_s)$, then semiglobal asymptotic stability of the cascade (4.5.28) can be achieved with partial-state feedback, that is, (4.5.28) is a *nonpeaking cascade.*

Proof: Let Ω be the desired compact region of attraction of $(z, \xi) = (0, 0)$ and let constants R_z and R_ξ be such that

$$\forall (z,\xi) \in \Omega: \ \|z\| \leq R_z, \quad \|\xi\| \leq R_\xi$$

If (A, B, C) is nonpeaking, we know from Theorem 4.40 that a partial-state feedback stabilizes the ξ-subsystem with the additional property that the output y decays fast without peaking,

$$\|y(t)\| \leq \gamma \|\xi(0)\| (e^{-at} + \frac{1}{a})$$

and that the state ξ_s defined in Theorem 4.35 does not peak,

$$\xi_s(t) \leq \tilde{\gamma} \|\xi(0)\|$$

where the constants γ and $\tilde{\gamma}$ are independent of a. We will show that a can be chosen such that, for any initial condition in Ω, the solution $z(t)$ is bounded. By Proposition 4.1, this will imply that the set Ω is included in the region of attraction of $(z, \xi) = (0, 0)$.

To establish the boundedness of $z(t)$, we first augment the system (4.5.28) with

$$\dot{\chi} = -a\chi + \gamma R_\xi, \quad \chi(0) = \gamma R_\xi(1 + \frac{1}{a}), \qquad \chi \in I\!R, \tag{4.5.29}$$

noting that $0 < \chi(t) \leq \chi(0) =: \chi_{max}$ for all $t \geq 0$. Then, for all initial conditions $(z(0), \xi(0)) \in \Omega$, we have

$$\|y(t)\| \leq |\chi(t)| \ \leq \chi_{max} \tag{4.5.30}$$

Because the system $\dot{z} = f(z)$ is GAS, there exists a smooth, radially unbounded, positive definite function $W(z)$ such that for all $z \neq 0$,

$$\frac{\partial W}{\partial z}(z)f(z) < 0 \tag{4.5.31}$$

We pick a level set W_c such that $\|z\| \leq R_z \Rightarrow W(z) \leq W_c$ and, for the positive definite function

$$V(z, \chi) = W(z) + \chi^2$$

we pick the level set $V_c = W_c + (\chi_{max})^2$. By definition of $\chi_{\max}$,

$$W(z(t)) \leq W_c \Rightarrow V(z(t), \chi(t)) \leq V_c$$

so for each initial condition $(z(0), \xi(0)) \in \Omega$, we have $V(z(0), \chi(0)) \leq V_c$. If the solution $z(t)$ grows unbounded, so does $V(z(t), \chi(t))$ and the solution $(z(t), \chi(t))$ eventually leaves the compact region $V(z, \chi) \leq V_c$. Then, because $V(z(0), \chi(0)) \leq V_c$, there exists a finite time $T \geq 0$ such that

$$V(z(T), \chi(T)) = V_c \ \text{ and } \ \dot{V}(z(T), \chi(T)) > 0 \tag{4.5.32}$$

By definition, we have $0 < \chi(T) \leq \chi_{max}$. This implies $W_c \leq W(z(T)) < V_c$. Hence, there exist two positive constants z_m and z_M such that $\|z(T)\| \in [z_m, z_M]$.

The time-derivative of V is

$$\dot{V} = \frac{\partial W}{\partial z}(z)f(z) + \frac{\partial W}{\partial z}\tilde{\psi}(z, y, \xi_s)y - 2a\chi^2 + 2\gamma R_\xi \chi$$

Using (4.5.31), we can define constants $\alpha_1 > 0$, $\alpha_2 > 0$, such that

$$\| z \| \in [z_m, z_M] \ \Rightarrow \ \frac{\partial W}{\partial z}(z)f(z) \leq -\alpha_1 \|z\|^2$$

$$\| z \| \in [z_m, z_M] \Rightarrow \|\frac{\partial W}{\partial z}(z)\tilde{\psi}(z, y, \xi_s)\| \leq \alpha_2 \|z\|$$

Because ξ_s are the nonpeaking components of ξ_0 and $0 \leq \|y\| \leq \chi_{max}$, α_1 and α_2 can be chosen independent of a. Using these two inequalities and (4.5.30) we obtain

$$\dot{V}(z(T), \chi(T)) \leq \alpha_1\|z(T)\|^2 + \alpha_2\|z(T)\|\ |\chi(T)| - 2a\chi^2(T) + 2\gamma R_\xi|\chi(T)| \tag{4.5.33}$$

In view of $-a\chi^2(T) + 2\gamma R_\xi|\chi(T)| \leq \frac{\gamma R_\xi}{a}$ for all $\chi(T)$, we obtain

$$\dot{V}(z(T), \chi(T)) \leq -\frac{\alpha_1}{2}\|z(T)\|^2 + \frac{\gamma R_\xi}{a} \leq -\frac{\alpha_1}{2}z_m^2 + \frac{\gamma R_\xi}{a}$$

for all $a > \frac{\alpha_2}{2\alpha_1}$. Because α_1, z_m, γ, and R_ξ are independent of a, the right-hand side can be made strictly negative if a is chosen sufficiently large. This shows that (4.5.32) cannot be satisfied if a is large enough. Therefore $z(t)$ is bounded, and Ω is included in the region of attraction of $(z, \xi) = (0, 0)$. □

Example 4.42 *(Semiglobal stabilization of a nonpeaking cascade)*
The partially linear cascade

$$\begin{aligned} \dot{z} &= -\delta z + \xi_3 z^2, \quad \delta > 0 \\ \dot{\xi}_1 &= \xi_2 \\ \dot{\xi}_2 &= \xi_3 \\ \dot{\xi}_3 &= u \end{aligned} \tag{4.5.34}$$

is nonpeaking because the output $y = \xi_3$ can be factored out of the interconnection $\psi = \xi_3 z^2 = z^2 y$ and the chain of integrators

$$\dot{\xi}_1 = \xi_2, \dot{\xi}_2 = \xi_3, \dot{\xi}_3 = u, y = \xi_3 \tag{4.5.35}$$

is a nonpeaking system. Its nonpeaking stabilization is achieved with the linear high-low gain feedback

$$u = -a\xi_3 - \xi_2 - \frac{1}{a}\xi_1 \tag{4.5.36}$$

By Theorem 4.41, this control law ensures asymptotic stability of $(z, \xi) = (0, 0)$ with a prescribed compact region of attraction if the constant a is large enough. This constant must increase to enlarge the domain of attraction. From the explicit solution

$$z(t) = e^{-\delta t}\left(\frac{1}{z(0)} - \int_0^t e^{-\delta\tau}\xi_3(\tau)d\tau\right)^{-1}$$

we obtain that, to avoid a finite time escape of $z(t)$, it is necessary that

$$\frac{a\delta^2\xi_3(0) - (a\delta\xi_2(0) + \delta\xi_1(0) + \xi_2(0))}{(a\delta^3 + a^2\delta^2 + a\delta + 1)} < \frac{1}{z(0)} \tag{4.5.37}$$

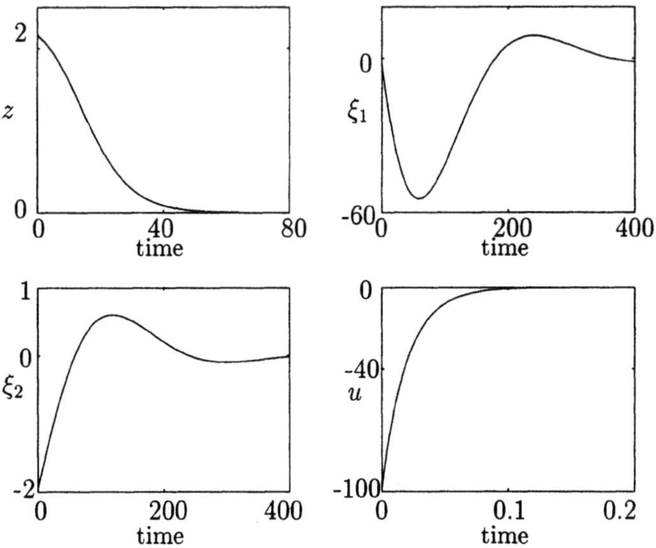

Figure 4.8: Linear high-low gain design (4.5.36): fast peaking of u and slow peaking of ξ_1.

Thus, for the initial condition $(2, -2, -2, 2)$ to be in the region of attraction, when $\delta = 0.1$, we must use $a > 42$. The large value of a in the control law (4.5.36) results in a high-gain for ξ_3 and a low-gain for ξ_1.

As shown in Proposition 4.37, the fast stabilization without peaking of the output ξ_3 causes the fast peaking of its derivative u in the fast time scale $\mathcal{O}(\frac{1}{a})$, and the slow peaking of the state ξ_1 in the slow time scale $\mathcal{O}(a)$. The $\mathcal{O}(a)$ large transients of u and ξ_1 in different time scales are illustrated in Figure 4.8. The figure also shows that the convergence of z is governed by the z-subsystem $\dot{z} = -\delta z$. The slow convergence for $\delta = 0.1$ is not improved by the partial-state feedback design (4.5.36) because the state z is not used for feedback.

In Chapter 6, we will return to the cascade (4.5.34) and obtain a considerable improvement of the design (4.5.36) by using a full-state feedback forwarding design.

□

The nonpeaking property is necessary for semiglobal stabilization (and, a fortiori, for global stabilization) if no further assumptions are made on the z-subsystem. This was shown by Saberi, Kokotović, and Sussmann [92] for *global* stabilization, and more recently by Braslavsky and Middelton [8] for *semiglobal* stabilization. This result, which shows that peaking is a structural obstacle to achieving an arbitrarily large region of attraction, applies to full-state feedback as well.

Theorem 4.43 *(Lack of semiglobal stabilizability)*
If (A, B, C) is a peaking system, then there exists $f(z)$ and $\tilde{\psi}(z, \xi)y$ such that Assumption 4.4 holds, but the cascade (4.5.28) is not semiglobally stabilizable.

Proof: By Theorem 4.40, the peaking system (A, B, C) is either not stabilizable, or not right-invertible, or has at least one unstable zero. For the case when (A, B, C) is not stabilizable the statement is obvious. We prove the remaining two cases by counter examples.

(A, B, C) is not right-invertible. For the cascade

$$\begin{aligned} \dot{z} &= (-1 + |\xi_1| + |\xi_2|)z^3 \\ \dot{\xi}_1 &= \xi_2 \\ \dot{\xi}_2 &= u \end{aligned} \tag{4.5.38}$$

we select two outputs $y_1 = \xi_1$, $y_2 = \xi_2$ so that the ξ-subsystem is a peaking system because it is not right-invertible.

That (4.5.38) cannot be semiglobally stabilized is clear from its solution:

$$z(t) = \frac{z(0)}{\sqrt{1 + 2z(0)(t - \int_0^t (|y_1| + |y_2|)\, ds)}} \tag{4.5.39}$$

If $z(0) = \frac{1}{2}$ the denominator does not vanish for $t \leq 1$ only if $\int_0^1 (|y_1| + |y_2|)\, dt < 2$ which implies that $\int_0^1 |y_1|\, dt < 2$ and $\int_0^1 |y_2|\, dt < 2$. The latter inequality provides

$$|y_1(t)| \geq |y_1(0)| - \int_0^t |y_2|\, ds > |y_1(0)| - 2$$

for all $t \in [0, 1]$. Hence, if $\int_0^1 |y_2|\, dt < 2$ and $|y_1(0)| > 4$, we have

$$\int_0^1 |y_1|\, dt > |y_1(0)| - 2 > 2$$

so that , with $z(0) = \frac{1}{2}$ and $y_1(0) > 4$, the denominator of (4.5.39) vanishes at some $t_f < 1$ and $z(t)$ escapes to infinity in finite time.

(A, B, C) is strictly nonminimum phase. For any (A, B, C) with a zero in the open right half plane, say at $s = \nu$, it has been shown in [8] that in the cascade (4.5.28) with the scalar z-subsystem

$$\dot{z} = -\alpha z + \beta z^{q+1} y^{2s} \tag{4.5.40}$$

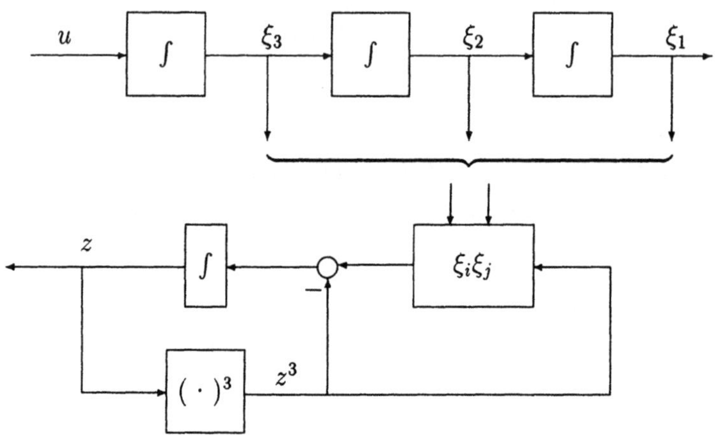

Figure 4.9: A system exhibiting several peaking situations.

one can find positive integers q, s, and positive real numbers α, β such that, there exists an initial condition $(z(0), \xi(0))$ for which $z(t)$ escapes to infinity in finite time, regardless of the control input $u(t)$. Because of the unstable zero, the output must act as a stabilizing control for the zero dynamics and therefore, its "energy" cannot be arbitrarily reduced. At the same time, this "energy" of the output perturbs the z-subsystem and causes a finite escape time of $z(t)$. □

In view of Theorem 4.41 and 4.43, the restriction of the interconnection term to the form $\psi = \tilde{\psi}(z, y, \xi_s)y$ is a key condition for semiglobal stabilization of a cascade system. Our final example will illustrate how the choice of a particular factorization of ψ is dictated by the input-output properties of the system (A, B, C).

Example 4.44 *(Factorization of ψ and the I/O structure of (A, B, C))*
In the cascade in Figure 4.9, a scalar nonlinear system is connected with a chain of three integrators through the product $\xi_i\xi_j$ of any two $(i, j = 1, 2, 3)$ integrator states

$$\begin{aligned} \dot{z} &= (-1 + \xi_i\xi_j)z^3 \\ \dot{\xi}_1 &= \xi_2 \\ \dot{\xi}_2 &= \xi_3 \\ \dot{\xi}_3 &= u \end{aligned} \tag{4.5.41}$$

We now present an analysis of several peaking situations in this cascade. This analysis remains unchanged if instead of ξ_j we have ξ_j^{2k+1}, $k = 1, 2, 3$, etc. Only in the case ξ_j^{2k} and $|i - j| = 2$, a more intricate analysis is needed to establish whether the effect of slow peaking is destabilizing. It can be shown that, using partial-state feedback, semiglobal stabilization is achievable when $\psi = \xi_1 \xi_3^{2k} z^3$ and it is not achievable when $\psi = \xi_3 \xi_1^{2k} z^3$.

Depending on the integrator states which enter the interconnection $\psi = \xi_i \xi_j z^3$, the two cases which lead to different peaking situations are: first, when ξ_i and ξ_j are the same $(i = j)$ or separated by one integrator $(j = i + 1)$; second, when ξ_i and ξ_j are separated by two integrators $(j = i + 2)$.

Case one: $|i - j| \leq 1$. By Proposition 4.36, there exist control laws $u = -k_1\xi_1 - k_2\xi_2 - k_3\xi_3$, which stabilize the chain of integrators and force the output $y = \xi_j$ to rapidly decay to zero *without peaking* of ξ_j and $\xi_i = j - 1$. Then Theorem 4.41 establishes that any such control law achieves semiglobal stabilization of the cascade (4.5.41). The same result applies to the interconnection $\psi = \xi_i \xi_j^k z^3$, where the exponent k is any positive integer.

Case two: $\psi = \xi_1 \xi_3 z^3$. The assumptions of Theorem 4.41 are not satisfied with either of the obvious choices $y = \xi_1$ or $y = \xi_3$, because, by Proposition 4.37, in each case the interconnection ψ contains a peaking state. With the choice $y = \xi_1$, the state ξ_3 peaks in a fast time scale with the exponent two. while with $y = \xi_3$, the state ξ_1 peaks in a slow time scale with the exponent one.

This peaking situation motivates us to search for a less obvious choice of output. Rewriting $\xi_1\xi_3$ as

$$\xi_1 \xi_3 z^3 = -\xi_1^2 z^3 + \xi_1(\xi_1 + \xi_3) z^3$$

we examine the possibility of using $y = \xi_1 + \xi_3$ as the output. To treat $\xi_1 z^3 y$ as the interconnection requires that the z-subsystem be augmented by the term $-\xi_1^2 z^3$ to $\dot{z} = -(1 + \xi_1^2) z^3$, which is acceptable because the GAS property is preserved uniformly in ξ_1. Using $y = \xi_1 + \xi_3$ as a change of variables, we rewrite the cascade (4.5.41) as

$$\begin{aligned} \dot{z} &= -(1 + \xi_1^2) z^3 + (\xi_1 z^3) y \\ \dot{\xi}_1 &= \xi_2 \\ \dot{\xi}_2 &= -\xi_1 + y \\ \dot{y} &= \xi_2 + u \end{aligned} \tag{4.5.42}$$

Now, the ξ-subsystem is weakly minimum phase because its zero-dynamics subsystem is a harmonic oscillator. A partial-state feedback control law which stabilizes the linear subsystem for any $a > 0$ is

$$u = -ay - 2\xi_2$$

With a large, $y(t)$ rapidly decays to zero without peaking. Hence, for the whole cascade the equilibrium at the origin is asymptotically stable with a region of attraction which can be made as large as desired by increasing a.

Because the linear subsystem in (4.5.42) is weakly minimum phase and has relative degree one, global stabilization is also achievable without using high gain to make y fast. The feedback passivation design of Proposition 4.21 is directly applicable. A particular control law of the form (4.3.12) is

$$u = -\xi_1 z^4 - 2\xi_2 - y$$

With this control law the cascade is GAS.

Backstepping and Forwarding: This example also serves as a good motivation for the recursive designs to be developed in Chapter 6. In the case $\psi = \xi_1\xi_2$, feedback passivation is not applicable because the relative degree of the output $y = \xi_2$ is two. This higher relative degree obstacle will be removed by backstepping. In the case $\psi = \xi_2\xi_3$, feedback passivation is not applicable because, with the output $y = \xi_3$, the system is not weakly minimum phase. This nonminimum phase obstacle will be removed by forwarding. □

4.6 Summary

We have analyzed the key structural properties of cascade systems which motivate several feedback stabilization designs and determine limits to their applicability. The simplest cascades are those in which the linear ξ-subsystem is controllable and the z-subsystem is GAS. Even in these cascades, the *peaking phenomenon* in the ξ-subsystem can destabilize the z-subsystem.

Our new characterization of output peaking shows that in a chain of integrators, only two consecutive states can be nonpeaking. All other states exhibit peaking which is fast for the "upstream" states and slow for the "downstream" ones. Every nonminimum phase system is peaking: its output cannot be rapidly regulated to zero without first reaching a high peak which is determined by the unstable modes of the zero dynamics.

Peaking is a structural obstacle to global and semiglobal stabilization in both partial- and full-state feedback designs. It may appear in both, fast and slow time scales. Although it is not an obstacle to local stabilization, peaking causes the region of attraction to shrink as the feedback gain increases.

To avoid the destabilizing effect of peaking, we have required that either the peaking states be excluded from the interconnection term, or the growth with respect to z be linear. We have shown that global stabilization can be achieved with partial-state feedback if the stability properties of the z-subsystem are guaranteed by either polynomial or ISS-type Lyapunov functions. With the help of such characterizations, we have determined when partial-state feedback designs can achieve desired stability margins.

Our full-state feedback designs employ passivation and remove the linear growth restriction which was imposed by partial-state feedback designs. The GAS assumption on the z-subsystem is replaced, first, by a GS assumption, and then, by a stabilizability assumption. In the latter case, the output of the ξ-subsystem plays the part of the stabilizing control for the z-subsystem.

All our designs have the potential to guarantee desired gain or phase margins, provided they avoid cancellations. Alternatively, if a design constructs a control Lyapunov function (CLF), desired margins can be guaranteed by employing Sontag's formula or with a domination redesign. Using the TORA system, we have compared performance of several designs and illustrated their abilities to improve transient performance and robustness.

4.7 Notes and References

Stabilization studies of the cascade nonlinear system have been stimulated by both, physical configuration of system components, such as in the large scale systems literature [77, 96, 116], and by system structural properties uncovered by input-output linearization [14, 43]. In the latter case the ξ-subsystem is linear, while the z-subsystem represents the nonlinear part of the zero-dynamics subsystem. Conditions for local stabilization via partial-state feedback were formulated by Sontag [100]. Peaking phenomenon was analyzed by Sussmann and Kokotović [105], who gave sufficient conditions for semiglobal stabilization, termed potentially global stabilization by Bacciotti [5]. For the normal form such conditions were given by Byrnes and Isidori [14]. Sussmann and Kokotović [105] and Saberi, Kokotović and Sussmann [92] prevent the destabilizing effect of peaking by imposing the linear growth constraint on the interconnection and the GES property of the nonlinear subsystem. The latter condition was relaxed by Lin [65] who replaced the GES requirement by a quadratic-like property of the Lyapunov function. In an alternative approach initiated by Sontag [103], the stability of the cascade is guaranteed by imposing the ISS property on the z-subsystem.

Various passivation ideas appeared earlier in adaptive control. They were introduced the nonlinear stabilization by Kokotović and Sussmann [59] and Saberi, Kokotović, and Sussmann [92], as an extension of a result by Byrnes and Isidori [13]. A nonlinear version of feedback passivation was given by Ortega [85], while structural conditions for feedback passivity were derived by Byrnes, Isidori, and Willems [15]. Further extensions are due to Lozano, Brogliato, and Landau [69], Kanellakopoulos [55], and Krstić, Kanellakopoulos, and Kokotović [61].

Our treatment of output peaking extends the results of Mita [79], Francis and Glover [20], and Kokotović and Sussmann [59], and, in addition, stresses the importance of not only fast, but also slow peaking. A systematic high- and low-gain design of linear systems which addresses these phenomena was developed by Lin and Saberi [67]. The awareness of the destabilizing effects of peaking has led to saturation designs by Teel [110] and to observer-based feedback by Esfandiary and Khalil [19].

Chapter 5

Construction of Lyapunov functions

Several designs in the preceding chapters require the knowledge of Lyapunov functions which need to be constructed during the design. This construction is a crucial part of the design and is the main topic of this chapter.

For a general nonlinear system $\dot{x} = f(x)$, the construction of a Lyapunov function is an intractable problem. There are globally stable time-invariant systems for which no time-invariant Lyapunov function exists [33]. However, structural properties of practically important classes of nonlinear systems can make the construction of Lyapunov functions a feasible task. This is the case with the basic cascade structures in this chapter.

For a stable (z, ξ)-cascade, the construction of a Lyapunov function assumes that the subsystem Lyapunov functions $W(z)$ and $U(\xi)$ are known. When one of the subsystems is only stable, then $c_1 W + c_2 U$ usually fails, and a composite Lyapunov function with the "nonlinear weights" $l(W) + \rho(U)$ proposed by Mazenc and Praly [75] is a better choice. This construction, presented in Section 5.1, requires a preliminary change of coordinates restricted by a "nonresonance condition".

A more general construction with a cross-term, presented in Section 5.2, is the main tool for the forwarding design of Chapter 6. This construction, which in most situations requires numerical integration, is based on the recent work by the authors [46]. *Relaxed* constructions in Section 5.3 avoid numerical integrations.

Designs based on the Lyapunov constructions are presented in Section 5.4. Adaptive controllers for systems with unknown parameters are designed in Section 5.5.

5.1 Composite Lyapunov functions for cascade systems

5.1.1 Benchmark system

The construction of the two main types of Lyapunov functions, composite and with cross-term, will be introduced with the help of the benchmark system

$$\begin{aligned}\dot{x}_1 &= x_2 + x_3^2\\ \dot{x}_2 &= x_3\\ \dot{x}_3 &= u\end{aligned} \tag{5.1.1}$$

As a simple representative of nonlinear systems which are not feedback linearizable, this system will be used throughout this chapter in a series of illustrative examples.

For a passivation design, the benchmark system can be treated as one of the two cascade structures: first, the (x_1, x_2)-subsystem cascaded with the x_3-integrator, and, second, the x_1-subsystem cascaded with the double integrator (x_2, x_3).

Since the uncontrolled $(u = 0)$ benchmark system is unstable, each of the two cascades contains an unstable subsystem. Prior to a Lyapunov construction, they must be converted into cascades with one stable and one asymptotically stable subsystem.

First cascade: $(1, 2) + (3)$. The feedback passivation design of Section 5.4. directs us to select an output for which the relative degree is one and the system is weakly minimum phase. In the benchmark system (5.1.1), the relative degree requirement is met with the output $y = x_2 + x_3$. Using y instead of x_3, we rewrite (5.1.1) as

$$\begin{aligned}\dot{x}_1 &= x_2 + x_2^2 + (2x_2 + y)y\\ \dot{x}_2 &= -x_2 + y\\ \dot{y} &= -x_2 + y + u\end{aligned} \tag{5.1.2}$$

To show that this system satisfies the weak minimum phase requirement we prove global stability of its zero-dynamics subsystem

$$\begin{aligned}\dot{x}_1 &= x_2 + x_2^2\\ \dot{x}_2 &= -x_2\end{aligned} \tag{5.1.3}$$

The proof is an explicit calculation in which $\tilde{x}(s)$ denotes the solution at time $s \geq 0$, for the initial condition $\tilde{x}(0) = x$. From the solution of (5.1.3)

$$\tilde{x}_2(s) = x_2 e^{-s}, \quad \tilde{x}_1(s) = x_1 + x_2(1 - e^{-s}) + \frac{x_2^2}{2}(1 - e^{-2s}) \tag{5.1.4}$$

we see that the equilibrium $(\tilde{x}_1, \tilde{x}_2) = (0, 0)$ of (5.1.3) is globally stable because $\tilde{x}_2(s)$ decays exponentially while $|\tilde{x}_1(s)|$ is bounded by $|x_1| + |x_2| + \frac{x_2^2}{2}$.

We have thus arrived at a cascade structure to which a feedback passivation design could be applied if a Lyapunov function $V(x_1, x_2)$ for the zero-dynamics subsystem (5.1.3) were available. Using V, the control law

$$u = x_2 - y - \frac{\partial V}{\partial x_1}(2x_2 + y) - \frac{\partial V}{\partial x_2} + v \tag{5.1.5}$$

would achieve passivity from the new input v to the output y. Upon verification that the ZSD condition is satisfied, the GAS would be guaranteed with the feedback $v = -y$.

Therefore, the remaining major task in this design is the construction of a Lyapunov function $V(x_1, x_2)$ for the zero-dynamics (5.1.3). We treat (5.1.3) as the cascade of the exponentially stable subsystem $\dot{x}_2 = -x_2$ with the stable subsystem $\dot{x}_1 = 0$. The attempt to use the simplest composite Lyapunov function $c_1x_1^2 + c_2x_2^2$ fails because its derivative is sign-indefinite due to the interconnection $x_2 + x_2^2$. We show in Example 5.5 that a composite Lyapunov function with "nonlinear weights" $l(x_1^2) + \rho(x_2^2)$ also fails because of the linear term x_2 in the interconnection $x_2 + x_2^2$. However, in Example 5.7 we succeed with a construction which employs the change of coordinates $\zeta = x_1 + x_2$ to eliminate the linear interconnection term. As it will be explained later, this construction, which requires a preliminary change of coordinates, is restricted by a nonresonance condition, which is satisfied in the benchmark system (5.1.3).

A more general construction, to which we devote most of this chapter, is for Lyapunov functions with a cross-term Ψ. Using the expressions (5.1.4), in Example 5.9 we explicitly evaluate

$$\Psi(x_1, x_2) = \int_0^\infty \tilde{x}_1(s)(\tilde{x}_2(s) + \tilde{x}_2^2(s))ds = \frac{1}{2}(x_1 + x_2 + \frac{x_2^2}{2})^2 - \frac{1}{2}x_1^2 \tag{5.1.6}$$

which is the cross-term in the Lyapunov function $\frac{1}{2}x_1^2 + \Psi(x_1, x_2) + x_2^2$ constructed for the zero-dynamics subsystem (5.1.3). Both Lyapunov constructions, composite and with cross-term, have taken advantage of the "nested cascade structure", treating the zero-dynamics subsystem (5.1.3) as a cascade within the larger cascade (5.1.2).

Second cascade: $(1) + (2, 3)$. In this cascade structure we first stabilize the unstable linear subsystem $\dot{x}_2 = x_3$, $\dot{x}_3 = u$ using a linear feedback transformation such as

$$u = -x_2 - 2x_3 + v$$

With the new input v set to zero, the system

$$\begin{aligned} \dot{x}_1 &= x_2 + x_3^2 \\ \dot{x}_2 &= x_3 \\ \dot{x}_3 &= -x_2 - 2x_3 + v \end{aligned} \tag{5.1.7}$$

is globally stable, as shown by its solution

$$\tilde{x}_2(s) = x_2 e^{-s} + (x_2 + x_3)se^{-s}, \quad \tilde{x}_3(s) = x_3 e^{-s} - (x_2 + x_3)se^{-s},$$
$$\tilde{x}_1(s) = x_1 + \int_0^s d(-2\tilde{x}_2 - \tilde{x}_3) - \int_0^s d(\frac{\tilde{x}_2^2 + \tilde{x}_3^2}{4}) \tag{5.1.8}$$

Here $\tilde{x}_2(s)$ and $\tilde{x}_3(s)$ decay exponentially while $|\tilde{x}_1(s)|$ is bounded by $|x_1| + |2x_2| + |x_3| + \frac{x_2^2+x_3^2}{4}$. We have thus satisfied the requirements for a passivation design. If a Lyapunov function $V(x)$ were available for the whole system (5.1.7), and if this system with output $y = \frac{\partial V}{\partial x_3}$ were ZSD, then the damping control

$$v = -\frac{\partial V}{\partial x_3} \tag{5.1.9}$$

would achieve GAS, as shown in Section 3.5.2. Again, the remaining design task is the construction of $V(x)$.

The construction of a "nonlinearly weighted" composite Lyapunov function in Example 5.7 employs a change of coordinates $\zeta = x_1 + 2x_2 + x_3$. The construction with cross-term in Example 5.9 uses the subsystem Lyapunov functions $W = \frac{1}{2}x_1^2$ and $U = \frac{1}{2}x_2^2 + \frac{1}{2}x_3^2$ with the cross-term explicitly evaluated from (5.1.8):

$$\Psi(x_1, x_2, x_3) = \int_0^\infty \tilde{x}_1(s)(\tilde{x}_2(s) + \tilde{x}_3^2(s))ds = \frac{1}{2}(x_1 + 2x_2 + x_3 + \frac{x_1^2 + x_2^2}{4})^2 - \frac{1}{2}x_1^2$$

and the Lyapunov function for (5.1.7) is $V = W + \Psi + U$.

With either of the two constructed Lyapunov functions, the final design step achieves GAS with damping control (5.1.9).

5.1.2 Cascade structure

Our basic Lyapunov construction is for the cascade structure

$$(\Sigma_0) \left\{ \begin{aligned} \dot{z} &= f(z) + \psi(z, \xi) \\ \dot{\xi} &= a(\xi) \end{aligned} \right.$$

where $\dot{z} = f(z)$ is globally stable, and $\dot{\xi} = a(\xi)$ is GAS and LES. This construction will be generalized to various augmentations (Σ_0).

The cascade structure (Σ_0) is easily recognized in the zero-dynamics subsystem (5.1.3) where the ξ-subsystem is $\dot{x}_2 = -x_2$, the z-subsystem is $\dot{x}_1 = 0$, while the interconnection term is $\psi = x_2 + x_2^2$. Similarly, for the cascade (5.1.7), the ξ-subsystem is $\dot{x}_2 = x_3$, $\dot{x}_3 = -x_2 - x_3$, the z-subsystem is $\dot{x}_1 = 0$, and the interconnection is $\psi = x_2 + x_3^2$. In each case, the interconnection trivially satisfies the linear growth condition of Assumption 4.5 because it is independent of x_1. The Lyapunov function $W(z) = z^2$ for the z-subsystem satisfies the polynomial growth assumption of Theorem 4.7. These two assumptions are repeated here for convenience:

Assumption 5.1 *(Linear growth)*
The function $\psi(z, \xi)$ satisfies a *linear growth assumption*, that is, there exist two class-$\mathcal{K}$ functions $\gamma_1(.)$ and $\gamma_2(.)$, differentiable at $\xi = 0$, such that

$$\| \psi(z, \xi) \| \leq \gamma_1(\| \xi \|) \| z \| + \gamma_2(\| \xi \|)$$

□

Assumption 5.2 *(Growth of the Lyapunov function W(z))*
The positive definite function $W(z)$ is C^2, radially unbounded, and satisfies $L_f W(z) \leq 0$ for all z. In addition, there exist constants c and M such that, for $\| z \| > M$,

$$\| \frac{\partial W}{\partial z} \| \| z \| \leq c\, W(z)$$

□

We have seen in Section 4.2 that, even when $\dot{z} = f(z)$ is GAS, boundedness of the solutions of (Σ_0) cannot be guaranteed in the absence of either one of these two assumptions, which, taken together, are sufficient for global stability.

Proposition 5.3 *(Global stability)*
If Assumptions 5.1 and 5.2 are satisfied, then the equilibrium $(z, \xi) = (0, 0)$ of (Σ_0) is globally stable.

Proof: Global boundedness of the solutions has been established in Theorem 4.7 and all we need to prove is *local* stability of $(z, \xi) = (0, 0)$. The Jacobian linearization of (Σ_0) is triangular and hence, its eigenvalues are the union of the eigenvalues of $A = \frac{\partial a}{\partial \xi}(0)$ and $F = \frac{\partial f}{\partial z}(0)$. The LES of $\dot{\xi} = a(\xi)$ implies that A is Hurwitz and the system (Σ_0) has a center manifold [16, 56] which is a submanifold of the hyperplane $\xi = 0$. The system (Σ_0) reduced to the invariant hyperplane $\xi = 0$ is $\dot{z} = f(z)$, which, by the Center Manifold Theorem [16, 56], proves stability of the equilibrium $(z, \xi) = (0, 0)$. □

5.1.3 Composite Lyapunov functions

In the literature dealing with stability of interconnected systems, it is usually assumed that each isolated subsystem is GAS. Then a *composite* Lyapunov function for the entire system is formed as a weighted sum of the subsystem Lyapunov functions. For the cascade (Σ_0) such a composite Lyapunov function would be $V(z,\xi) = c_1 W(z) + c_2 U(\xi)$. Its time-derivative contains the negative definite terms $c_1 L_f W(z)$ and $c_2 L_a U(\xi)$. However, it also contains an indefinite cross-term $c_1 L_\psi W(z,\xi)$ due to the interconnection ψ. For this construction to succeed, we must be able to choose the weights c_1 and c_2 so that the indefinite cross-term $c_1 L_\psi W(z,\xi)$ is dominated by the negative definite terms. This is not an easy task and severe restrictions must be imposed [56].

The construction of a composite Lyapunov function is even more challenging when one of the two subsystems, in our case the z-subsystem in the cascade (Σ_0), is only stable rather than asymptotically stable. In this case, the term $L_f W$ is only semidefinite and, in general, will not dominate the indefinite cross-term $L_\psi W$. This has led Mazenc and Praly [75] to replace the constants c_1 and c_2 by nonlinear "weights" $l(.)$ and $\rho(.)$ and construct

$$V(z,\xi) = l(W(z)) + \rho(U(\xi)) \tag{5.1.10}$$

as a composite Lyapunov function for (Σ_0). Henceforth, the term composite Lyapunov function will refer to this type of function. For this construction, one more assumption is needed which implies the LES property of $\dot{\xi} = a(\xi)$.

Assumption 5.4 (*Negativity of $L_a U(\xi)$*)
A C^2, positive definite, radially unbounded, function $U(\xi)$ is known such that $L_a U(\xi)$ is negative definite and locally quadratic, that is $\frac{\partial^2 L_a U}{\partial \xi^2}(0) < 0$.

□

We note that this assumption is not necessarily satisfied if $\dot{\xi} = a(\xi)$ is LES.

To construct $l(W)$ and $\rho(V)$ we examine the inequality

$$\begin{aligned} \dot{V} &= l'(W)[L_f W + L_\psi W] + \rho'(U) L_a U \\ &\leq l'(W) L_\psi W + \rho'(U) L_a U \end{aligned} \tag{5.1.11}$$

where l' and ρ' are the derivatives of l and ρ with respect to W and U, respectively. The term $\rho'(U)L_a U$ depends on ξ only, and is negative definite if $\rho'(U) > 0$. The term $l'(W)L_\psi W$ depends on both z and ξ and is indefinite. For the negative term to dominate, the indefinite term must be bounded for each fixed ξ, independently of z. Under Assumptions 5.1 and 5.2 we have

$$\begin{aligned}\|L_\psi W\| &\leq \|\frac{\partial W}{\partial z}\|(\gamma_1(\|\xi\|)\|z\| + \gamma_2(\|\xi\|)) \\ &\leq cW\gamma_1 + \|\frac{\partial W}{\partial z}\|\gamma_2\end{aligned}$$

Returning to (5.1.11) this means that both $l'(W)W$ and $l'(W)\|\frac{\partial W}{\partial z}\|$ must be bounded uniformly in z. In view of Assumption 5.2, both of these requirements are satisfied by $l(W) = \ln(W+1)$ because

$$l'(W)W = \frac{W}{W+1} < 1 \text{ and } \frac{1}{W+1}\|\frac{\partial W}{\partial z}\| \leq \frac{\alpha_1 W + \alpha_2}{W+1} \leq \tilde{\alpha}$$

for some constant $\tilde{\alpha}$. When ψ is uniformly bounded as a function of z, then $l(W) = \sqrt{W+1} - 1$ is also a good choice.

A more difficult requirement is that near $\xi = 0$ both $\gamma_1(\|\xi\|)$ and $\gamma_2(\|\xi\|)$ be quadratic or higher-order in ξ. If this is not the case, the negative definite term will not be able to dominate the indefinite term because the C^1 property of ρ, a, and U implies that $\|\rho'(U)L_aU\| \leq k\|\xi\|^2$ near $\xi = 0$ for some $k > 0$. This is the case with the zero-dynamics subsystem of the benchmark example (5.1.2).

Example 5.5 (*Linear interconnection terms*)
Let us reconsider the cascade (5.1.2) with its zero-dynamics subsystem (5.1.3) rewritten in the (z, ξ)-notation as

$$\begin{aligned}\dot{z} &= \xi + \xi^2 = \psi(\xi) \\ \dot{\xi} &= -\xi\end{aligned} \qquad (5.1.12)$$

With the subsystem Lyapunov functions $W(z) = z^2$ and $U(\xi) = \xi^2$, we examine whether

$$V = \ln(z^2 + 1) + \rho(\xi^2)$$

qualifies as a composite Lyapunov function. For this we need to find a C^1 function ρ to make

$$\dot{V} = 2\frac{z}{z^2+1}(\xi + \xi^2) - 2\rho'(\xi^2)\xi^2 \qquad (5.1.13)$$

nonpositive for all (z, ξ). We pick any z, say $z = 1$, and check if the negative term $-2\rho'(\xi^2)\xi^2$ dominates the indefinite term $\xi + \xi^2 = \psi(\xi)$ near $\xi = 0$. In this attempt we fail because, whatever C^1 function ρ we choose, the term $-2\rho'(\xi^2)\xi^2$ is quadratic near $\xi = 0$ and cannot dominate a term which is linear in ξ. If this linear term were absent, then the choice $\rho(\xi^2) = \xi^2$ would guarantee that $\dot{V} \leq 0$ for all (z, ξ). We also see the role of Assumption 5.4. If we had chosen $U(\xi) = \xi^4$ as a Lyapunov function for the ξ-subsystem, the domination would be impossible even if the linear term is removed from the interconnection. □

We have thus made a key observation: a composite Lyapunov function (5.1.10) is successful if the interconnection ψ does not contain a term linear in ξ. This observation holds in general.

Theorem 5.6 (*Composite Lyapunov functions*)
Suppose that (Σ_0) satisfies Assumptions 5.1, 5.2, and 5.4. If the interconnection $\psi(z, \xi)$ satisfies the condition

$$\frac{\partial \psi}{\partial \xi}(z, 0) \equiv 0 \tag{5.1.14}$$

then a continuous positive function $\gamma(.)$ can be found such that the radially unbounded positive definite function

$$V(z, \xi) = \ln(W(z) + 1) + \int_0^{U(\xi)} \gamma(s) ds \tag{5.1.15}$$

is nonincreasing along the solutions of (Σ_0).

Proof: By inspection, $V(z, \xi)$ in (5.1.15) is positive definite and radially unbounded. Its time-derivative along the solutions of (Σ_0) is

$$\begin{aligned} \dot{V} &= \frac{1}{W(z)+1}(L_f W(z) + L_\psi W(z, \xi)) + \gamma(U(\xi)) L_a U(\xi) \\ &\leq \frac{1}{W(z)+1} L_\psi W(z, \xi) + \gamma(U(\xi)) L_a U(\xi) \end{aligned}$$

By Assumption 5.1, we have

$$\left|\frac{1}{W+1} L_\psi W\right| \leq \frac{1}{W+1} \left\| \frac{\partial W}{\partial z} \right\| (\gamma_1(\| \xi \|) \| z \| + \gamma_2(\| \xi \|))$$

and, by Assumption 5.2, this implies

$$\dot{V} \leq \gamma_3(\|\xi\|) + \gamma(U(\xi)) L_a U(\xi)$$

for some function $\gamma_3 \in \mathcal{K}$. From (5.1.14) we know that γ_1 and γ_2 can be chosen such that $\gamma_1'(0) = \gamma_2'(0) = 0$ and, therefore, $\gamma_3'(0) = 0$. Thus $\gamma_3(\|\xi\|) = \gamma_4(\|\xi\|)\|\xi\|^2$ for some continuous function γ_4.

By Assumption 5.4 there is a constant $\alpha > 0$ such that $L_aU(\xi) \leq -\alpha||\xi||^2$ in a neighborhood of $\xi = 0$. Therefore, there exists a function

$$\gamma(U(\xi)) > \frac{\gamma_4(||\xi||)||\xi||^2}{|L_aU(\xi)|}$$

which achieves $\dot{V} \leq 0$.

□

An important issue to be resolved in this chapter is whether the requirement that the interconnection ψ does not contain a term linear in ξ is a severe structural constraint. Can such a term be removed by a preliminary change of coordinates? Let's examine this issue on the zero-dynamics subsystem (5.1.3).

Example 5.7 *(Change of coordinates for a composite Lyapunov function)* Returning to the zero-dynamics system (5.1.12) our goal now is to find a change of coordinates which will remove the linear term ξ from the interconnection $\psi(\xi) = \xi + \xi^2$. After a quick examination we notice that, with the change of coordinates $\zeta = z + \xi$, the system (5.1.12) becomes

$$\begin{aligned} \dot{\zeta} &= \xi^2 \\ \dot{\xi} &= -\xi \end{aligned} \tag{5.1.16}$$

so that the interconnection is now only ξ^2. The composite Lyapunov function $V(\zeta, \xi) = \ln(\zeta^2+1)+\xi^2$ has the time-derivative which is negative semidefinite:

$$\dot{V} = 2(\frac{\zeta}{\zeta^2+1} - 1)\xi^2 \leq 0$$

Using this Lyapunov function in the (x, y) coordinates of (5.1.2), the passivating transformation (5.1.5) is

$$u = -x_2 - y - 2\frac{x_1 + x_2}{(x_1 + x_2)^2 + 1}(2x_2 + y + 1) + v \tag{5.1.17}$$

It is easy to verify that the system (5.1.2) with input v and output y is ZSD. The design is completed with feedback $v = -y$ which achieves GAS.

Let us repeat the same construction for the second cascade (5.1.7). The linear change of coordinates $(\zeta, \xi_1, \xi_2) = (x_1 + 2x_2 + x_3, x_2, x_3)$ transforms it into

$$\begin{aligned} \dot{\zeta} &= \xi_2^2 \\ \dot{\xi}_1 &= \xi_2 \\ \dot{\xi}_2 &= -2\xi_2 - \xi_1 \end{aligned} \tag{5.1.18}$$

The composite Lyapunov function

$$V(\zeta,\xi)=\frac{1}{2}\ln(\zeta^2+1)+\frac{\xi_1^2+\xi_2^2}{2}$$

has a nonpositive time-derivative:

$$\dot{V}=(\frac{\zeta}{\zeta^2+1}-1)\xi_2^2\leq 0$$

The damping control

$$v=\frac{\partial V}{\partial \xi_2}=-\frac{\zeta}{\zeta^2+1}-\xi_2=-\frac{x_1+x_2+x_3}{(x_1+x_2+x_3)^2+1}-x_3 \qquad (5.1.19)$$

achieves GAS of (5.1.7). In this case $\dot{\zeta}=\xi_2^2$ is independent of ξ_1 and we were able to dominate the indefinite term in $\dot{V}$ even though $\dot{U}=-\xi_2^2$ is only negative semidefinite. □

For our future reference it is important to note that both control laws (5.1.17) and (5.1.19) do not grow unbounded in $|x_1|$ with fixed x_2 and y. Instead, they saturate and even tend to 0 as $|x_1|\to\infty$, which is a consequence of the nonlinear weighting $\ln(W+1)$ in the composite Lyapunov function (5.1.10).

Is it always possible to find a change of coordinates to remove from the interconnection $\psi(z,\xi)$ the terms which are linear in ξ? The answer to this question is negative even for the linear cascade

$$\begin{aligned}\dot{z}&=Fz+M\xi\\ \dot{\xi}&=A\xi\end{aligned} \qquad (5.1.20)$$

where M is a constant matrix. For the existence of a decoupling change of coordinates $\zeta=z+N\xi$ it is necessary and sufficient that N be the solution of the Sylvester equation

$$FN-NA=M$$

It is well-known that N exists if and only if the "nonresonance" condition $\lambda_i(A)\neq\lambda_j(F),\quad i=1,\ldots,n_\xi,\quad j=1,\ldots,n_z$ is satisfied by the eigenvalues of A and F. An example violating this condition is $F=-1$, $A=-1$, and $M\neq 0$. Then the matrix of the whole system (5.1.20) is a single Jordan block which cannot be diagonalized.

When the Jacobian linearization cannot be diagonalized, a composite function (5.1.10), in general, fails to be a Lyapunov function for the cascade. To overcome this difficulty, and to reach a larger class of (z,ξ)-cascades, we now proceed to the construction of a Lyapunov function with a cross-term.

5.2 Lyapunov Construction with a Cross-Term

5.2.1 The construction of the cross-term

Instead of restricting ourselves to a combination of nonlinearly weighted $W(z)$ and $U(\xi)$ or searching for a decoupling change of coordinates which may not exist, we will now proceed to construct a Lyapunov function with a cross-term $\Psi(z,\xi)$:

$$V_0(z,\xi) = W(z) + \Psi(z,\xi) + U(\xi) \tag{5.2.1}$$

The cross-term must guarantee that V_0 is nonincreasing along the solutions of (Σ_0). The time-derivative of V_0 is

$$\dot{V}_0 = L_f W + L_\psi W + \dot{\Psi} + L_a U \tag{5.2.2}$$

The terms $L_f W$ and $L_a U$ are nonpositive. Therefore, to ensure the negativity of $\dot{V}_0$, the cross-term $\Psi(z,\xi)$ is chosen as

$$\dot{\Psi} = -L_\psi W = -\frac{\partial W}{\partial z}\psi$$

This means that Ψ is the line-integral of $\frac{\partial W}{\partial z}\psi$ along the solution of (Σ_0) which starts at (z,ξ):

$$\Psi(z,\xi) = \int_0^\infty L_\psi W(\tilde{z}(s,z,\xi),\tilde{\xi}(s,\xi))\, ds \tag{5.2.3}$$

The following theorem shows that the integral is well defined and that the resulting V_0 is a Lyapunov function for (Σ_0).

Theorem 5.8 *(Lyapunov function with a cross-term)*
If Assumptions 5.1 and 5.2 are satisfied then the following holds:

(i) $\Psi(z,\xi)$ exists and is continuous in $\mathbb{R}^{n_z} \times \mathbb{R}^{n_\xi}$;

(ii) $V_0(z,\xi)$ is positive definite;

(iii) $V_0(z,\xi)$ is radially unbounded;

Proof: (i) We first prove the existence of the function $\Psi(z,\xi)$. Arguing as in the proof of Theorem 4.7, we have that for each $\tau \geq 0$

$$\left|\frac{\partial W}{\partial z}(\tilde{z}(\tau))\ \psi(\tilde{z}(\tau),\tilde{\xi}(\tau))\right| \leq \| \frac{\partial W}{\partial z} \| \left(\gamma(\|\xi\|)e^{-\alpha\tau} + \gamma(\|\xi\|)e^{-\alpha\tau} \| \tilde{z}(\tau) \|\right) \tag{5.2.4}$$

Because $W(z)$ is radially unbounded, Theorem 4.7 implies that $\| \tilde{z}(\tau) \|$ and $\| \frac{\partial W}{\partial z}(\tilde{z}(\tau)) \|$ are bounded on $[0, +\infty)$. From (5.2.4) there exists $\gamma_1 \in \mathcal{K}$ such that

$$\left| \frac{\partial W}{\partial z}(\tilde{z}(\tau)) \; \psi(\tilde{z}(\tau), \tilde{\xi}(\tau)) \right| \leq \gamma_1(\| (z, \xi) \|) e^{-\alpha \tau} \tag{5.2.5}$$

We conclude that, as a time function, $\frac{\partial W}{\partial z}(\tilde{z})\psi(\tilde{z}, \tilde{\xi})$ is integrable on $[0, \infty)$, and hence, the integral (5.2.3) exists and is bounded for all bounded (z, ξ).

Next we prove continuity of Ψ at any fixed $(\bar{z}, \bar{\xi})$. Denote by $B(\bar{z}, \delta)$ the ball around $\bar{z}$ with radius δ. Let $(z, \xi) \in U_\delta := B(\bar{z}, \delta) \times B(\bar{\xi}, \delta)$. We will show that

$$| \Psi(z, \xi) - \Psi(\bar{z}, \bar{\xi}) | \leq \epsilon$$

for δ sufficiently small.

Without loss of generality, we can choose $\delta < 1$. Using (5.2.5) we can find a finite time $T > 0$ such that for all $(z, \xi) \in U_1$

$$\int_T^\infty \left| \frac{\partial W}{\partial z}(\tilde{z}(s)) \; \psi(\tilde{z}(s), \tilde{\xi}(s)) \right| \, ds < \frac{\epsilon}{4}$$

Denote by $(\bar{z}(\tau), \bar{\xi}(\tau))$ the solution $(\tilde{z}(\tau; \bar{z}, \bar{\xi}), \tilde{\xi}(\tau; \bar{\xi}))$. It remains to show that

$$\left| \int_0^T \left(\frac{\partial W}{\partial z}(\tilde{z}) \; \psi(\tilde{z}, \tilde{\xi}) - \frac{\partial W}{\partial z}(\bar{z}) \; \psi(\bar{z}, \bar{\xi}) \right) \, ds \right| < \frac{\epsilon}{2} \tag{5.2.6}$$

for $\|z - \bar{z}\| + \|\xi - \bar{\xi}\|$ sufficiently small.

The solutions of (Σ_0) are continuous with respect to initial conditions over the finite time interval $[0, T]$ and belong to a compact set for all initial conditions in U_1. It follows that the integrand on the left-hand side of (5.2.6) uniformly converges to zero when δ tends to zero. Inequality (5.2.6) is, therefore, satisfied for δ sufficiently small, which establishes continuity.

(ii) The function $W(\tilde{z}(\tau))$, along the solution of (Σ_0) for an initial condition (z, ξ), satisfies

$$W(\tilde{z}(\tau)) = W(z) + \int_0^\tau \dot{W}(\tilde{z}(s), \tilde{\xi}(s)) ds$$

Evaluating $\dot{W}$ yields

$$W(\tilde{z}(\tau)) - \int_0^\tau \frac{\partial W}{\partial z}(\tilde{z}(s)) \; \psi(\tilde{z}(s), \tilde{\xi}(s)) ds = W(z) + \int_0^\tau \frac{\partial W}{\partial z}(\tilde{z}(s)) \; f(\tilde{z}(s)) ds \tag{5.2.7}$$

The proof of (i) shows that the second term on the left-hand side converges as $\tau \to \infty$ and, because $W(\tilde{z}(\tau)) \geq 0$, the whole left hand side is bounded from below for all $\tau \geq 0$. Since the right-hand side of (5.2.7) is nonincreasing as a function of τ, we conclude that as $\tau \to \infty$ the limits on both sides exist:

$$\lim_{\tau\to\infty} W(\tilde{z}(\tau)) - \int_0^\infty \frac{\partial W}{\partial z}(\tilde{z}(s))\, \psi(\tilde{z}(s), \tilde{\xi}(s))\, ds = W(z) + \int_0^\infty \frac{\partial W}{\partial z}(\tilde{z}(s))\, f(\tilde{z}(s)) ds$$

The second term on the left hand side is $\Psi(z,\xi)$, so, as $\tau \to \infty$, the function $W(\tilde{z}(\tau))$ converges to some finite nonnegative value

$$W_\infty(z,\xi) = W(z) + \Psi(z,\xi) + \int_0^\infty \frac{\partial W}{\partial z}(\tilde{z})\, f(\tilde{z})\, ds \tag{5.2.8}$$

Substituting (5.2.8) into (5.2.1) we obtain V_0 as the sum of three nonnegative terms:

$$V_0(z,\xi) = W_\infty(z,\xi) - \int_0^\infty \frac{\partial W}{\partial z}(\tilde{z})\, f(\tilde{z})\, ds + U(\xi) \geq 0 \tag{5.2.9}$$

It follows that $V_0(z,\xi) = 0$ implies $\xi = 0$. By construction, $V_0(z,0) = W(z)$, so we conclude that

$$V_0(z,\xi) = 0 \Rightarrow (z,\xi) = (0,0) \tag{5.2.10}$$

Equalities (5.2.9) and (5.2.10) imply that V_0 is positive definite.

(iii) It follows immediately from (5.2.9) that V_0 tends to infinity when $\| \xi \|$ tends to infinity. It is therefore sufficient to prove that for all $\xi \in I\!R^m$

$$\lim_{\|z\|\to\infty} \left(W_\infty(z,\xi) - \int_0^{+\infty} \frac{\partial W}{\partial z}(\tilde{z}(\tau)) f(\tilde{z}(\tau)) d\tau \right) = +\infty \tag{5.2.11}$$

Fix $\xi \in I\!R^m$ so that the class $\mathcal{K}$ function γ used in the inequality (5.2.4) becomes a constant C. We then write for each $\tau \geq 0$

$$\begin{aligned} \dot{W} - L_f W &= L_\psi W \geq - | L_\psi W | \\ &\geq - \| \frac{\partial W}{\partial z} \| (Ce^{-\alpha\tau} + Ce^{-\alpha\tau} \| \tilde{z} \|) \\ &\geq - \| \frac{\partial W}{\partial z} \| \| \tilde{z} \| Ce^{-\alpha\tau} - (1 - \| \tilde{z} \|) \| \frac{\partial W}{\partial z} \| Ce^{-\alpha\tau} \end{aligned}$$

Now we examine the second term on the right hand side. If $(1 - \| \tilde{z} \|) \leq 0$ this term can be dropped without affecting the inequality. When $(1 - \| \tilde{z} \|) > 0$ we have to keep this term, but now $\|\tilde{z}\| < 1$ so the term is bounded by $K_2 e^{-\alpha\tau}$. Therefore, we can write

$$\dot{W} - L_f W \geq - \| \frac{\partial W}{\partial z} \| \| \tilde{z} \| Ce^{-\alpha\tau} - K_2 e^{-\alpha\tau} \tag{5.2.12}$$

Using Assumption 5.2 we obtain

$$\begin{aligned} \dot{W} &\geq -Ke^{-\alpha\tau}W - K_2e^{-\alpha\tau} + L_fW \qquad &\text{when } \|z\| > \kappa \\ \dot{W} &\geq -K_1e^{-\alpha\tau} - K_2e^{-\alpha\tau} + L_fW \qquad &\text{when } \|z\| \leq \kappa \end{aligned} \tag{5.2.13}$$

for some positive κ, K and K_1 which may depend only on ξ.

Inequalities (5.2.13) yield the following lower bounds on $W(\tilde{z}(\tau))$:

$$\begin{aligned} &\|\tilde{z}(t)\| > \kappa \text{ for } t \in [0,\tau) \Rightarrow \\ &\quad \Rightarrow W(\tilde{z}(\tau)) \geq \phi(\tau,0)W(z) + \int_0^\tau \phi(\tau,s)(-K_2e^{-\alpha_1 s} + L_fW)ds \\ &|\tilde{z}(t)\| \leq \kappa \text{ for } t \in [0,\tau) \Rightarrow \\ &\quad \Rightarrow W(\tilde{z}(\tau)) \geq W(z) + \int_0^\tau (-K_1e^{-\alpha s} - K_2e^{-\alpha s} + L_fW)\, ds \end{aligned}$$

where $\phi(\tau,s) := e^{-\frac{K}{\alpha}(e^{-\alpha s} - e^{-\alpha\tau})}$. Noting that $1 \geq \phi(\tau,s) \geq e^{-\frac{K}{\alpha}}$ for all $\tau \geq s \geq 0$, we can combine the two bounds on W to obtain that $\forall\tau \geq 0$

$$W(\tilde{z}(\tau)) \geq \phi(\tau,0)W(z) + \int_0^\tau (-K_1e^{-\alpha s} - K_2e^{-\alpha s} + L_fW)\, ds \tag{5.2.14}$$

Hence for all $\tau \geq 0$

$$W(\tilde{z}(\tau)) \geq e^{-\frac{K}{\alpha}}W(z) + \int_0^\tau L_fW\, ds + \kappa(\tau) \tag{5.2.15}$$

where $\kappa(\tau) := -\int_0^\tau (K_1e^{-\alpha s} + K_2e^{-\alpha s})ds$ exists and is bounded over $[0,+\infty)$. Subtracting from both sides of (5.2.15) the term $\int_0^\tau L_fWds$ and taking the limit when τ tends to infinity, we obtain

$$W_\infty(z,\xi) - \int_0^\infty L_fWds \;\geq\; K_3W(z) + \kappa^\star \tag{5.2.16}$$

with $\kappa^\star$ finite. It is clear form the construction that $\kappa^\star$ and K_3 may depend on $\|\xi\|$ but are independent of $\|z\|$. When $\|\, z\, \|$ tends to infinity, the right-hand side of (5.2.16) tends to infinity which proves (5.2.11). □

Let us illustrate the construction of V_0 with the benchmark system of Section 6.1.1.

Example 5.9 (*Cross-term construction for the benchmark system*)
We now construct a Lyapunov function $V_0(z,\xi)$ for the zero-dynamics subsystem of (5.1.2) rewritten as

$$\begin{aligned} \dot{z} &= \xi + \xi^2 \\ \dot{\xi} &= -\xi \end{aligned} \tag{5.2.17}$$

Let $W(z) = \frac{1}{2}z^2$ and $U(\xi) = \frac{1}{2}\xi^2$ be the Lyapunov functions for the isolated subsystems of (5.2.17). Then the cross-term is

$$\Psi(z,\xi) = \int_0^\infty \tilde{z}(s)(\tilde{\xi}(s) + \tilde{\xi}^2(s))\, ds = \int_0^\infty d(\frac{\tilde{z}^2}{2})$$

Substituting the solution (5.1.4) and integrating, we obtain

$$\Psi(z,\xi) = \frac{1}{2}(z + \xi + \frac{\xi^2}{2})^2 - \frac{1}{2}z^2$$

Hence, in the original x-coordinates, the Lyapunov function is

$$V_0(x_1, x_2) = \frac{1}{2}(x_1 + x_2 + \frac{x_2^2}{2})^2 + \frac{1}{2}x_2^2$$

With this Lyapunov function, the passivating transformation (5.1.5) for the whole cascade (5.1.2) is

$$u = -y - (x_1 + x_2 + \frac{x_2^2}{2})(3x_2 + y + 1) + v \tag{5.2.18}$$

It is easy to verify that the system (5.1.2) with input v and output y is ZSD. Hence the feedback $v = -y$ achieves GAS.

Let us now apply the same construction with the cross-term to the alternative cascade (5.1.7) rewritten here as

$$\begin{aligned} \dot{z} &= \xi_1 + \xi_2^2 \\ \dot{\xi}_1 &= \xi_2 \\ \dot{\xi}_2 &= -2\xi_2 - \xi_1 \end{aligned} \tag{5.2.19}$$

Using the z-subsystem Lyapunov function $W(z) = \frac{1}{2}z^2$, the cross-term $\Psi(z,\xi)$ is

$$\Psi(z,\xi) = \int_0^\infty \tilde{z}(s)(\tilde{\xi}_1(s) + \tilde{\xi}_2^2(s))\, ds = \int_0^\infty d(\frac{\tilde{z}^2}{2})$$

and, from the solution (5.1.8), we obtain

$$\Psi(z,\xi) = \frac{1}{2}(z + 2\xi_1 + \xi_2 + \frac{\xi_1^2 + \xi_2^2}{4})^2 - \frac{1}{2}z^2$$

Hence, in the original coordinates, the Lyapunov function is

$$V_0(x_1, x_2, x_3) = \frac{1}{2}(x_1 + 2x_2 + x_3 + \frac{x_2^2 + x_3^2}{4})^2 + \frac{1}{2}(x_2^2 + x_3^2)$$

The damping control

$$v = -\frac{\partial V}{\partial x_3} = -(x_1 + 2x_2 + x_3 + \frac{x_2^2 + x_3^2}{4})(1 + \frac{1}{2}x_3) - x_3 \tag{5.2.20}$$

achieves GAS. □

5.2.2 Differentiability of the function Ψ

Because the control laws based on the Lyapunov function V_0 will use its partial derivatives, it is important to establish the differentiability properties of the cross-term $\Psi(z,\xi)$. If the system (Σ_0) is C^∞ we prove that the function Ψ is C^∞ provided that the following assumption is satisfied.

Assumption 5.10 *(Restriction on the z-subsystem – smoothness of Ψ)*
The vector field $f(z)$ in (Σ_0) has the form

$$f(z) = \begin{pmatrix} f_1(z_1) \\ F_2 z_2 + f_2(z_1, z_2) \end{pmatrix}, \quad z = \begin{pmatrix} z_1 \\ z_2 \end{pmatrix} \tag{5.2.21}$$

Furthermore, $f_2(0, z_2) = 0$, the equilibrium $z_1 = 0$ of $\dot{z}_1 = f_1(z_1)$ is GAS, and the system $\dot{z}_2 = F_2 z_2$ is Lyapunov stable. □

We first show that Ψ is C^1.

Theorem 5.11 *(Continuous differentiability of the cross-term)*
Under Assumptions 5.1, 5.2, and 5.10, the function Ψ defined by (5.2.3) is C^1 in $\mathbb{R}^{n_z} \times \mathbb{R}^{n_\xi}$.

Proof: By standard results for ordinary differential equations (see [56] or Theorem 2, p.302 in [39]), the partial derivatives of $\tilde{z}(\tau; z, \xi)$ and $\tilde{\xi}(\tau; \xi)$ with respect to z and ξ exist for each z, ξ, and $\tau \geq 0$. The time behavior of these partial derivatives is governed by the *variational equation* of (Σ_0). It is well known that the variational equation of a stable nonlinear system is not necessarily stable. Below we show that, under Assumption 5.10, its solutions cannot grow exponentially.

For an arbitrary constant $a \geq 0$, the time-varying matrix

$$\chi(\tau) := \frac{\partial \tilde{z}(\tau)}{\partial z} e^{-a\tau}$$

satisfies the linear time-varying differential equation

$$\frac{d\chi}{d\tau} = -a\chi + \left(\frac{\partial f}{\partial z} + \frac{\partial \psi}{\partial z}\right)\bigg|_{(\tilde{z}(\tau),\tilde{\xi}(\tau))} \chi \tag{5.2.22}$$

with the initial condition $\chi(0) = I$. For $a = 0$, this is the variational equation of $\frac{\partial \tilde{z}(\tau)}{\partial z}$ along the solution $(\tilde{z}(s), \tilde{\xi}(s))$. We will show that the solution of (5.2.22) converges to zero for any $a > 0$.

Assumption 5.10 provides the decomposition

$$\frac{\partial f}{\partial z} = \begin{pmatrix} \frac{\partial f_1}{\partial z_1} & 0 \\ \frac{\partial f_2}{\partial z_1} & F_2 + \frac{\partial f_2}{\partial z_2} \end{pmatrix} \tag{5.2.23}$$

with the asymptotic property (due to asymptotic stability of $z_1 = 0$ in $\dot{z}_1 = f_1(z_1)$)

$$\lim_{\tau\to\infty} \frac{\partial f_1}{\partial z_1}(\tau) = \frac{\partial f_1}{\partial z_1} |_{z_1=0} := F_1, \ \lim_{\tau\to\infty} \frac{\partial f_2}{\partial z_2} = 0$$

Therefore we rewrite (5.2.22) as

$$\frac{d\chi}{d\tau} = \begin{bmatrix} -aI + F_1 & 0 \\ \frac{\partial f_2}{\partial z_1}(\tau) & -aI + F_2 \end{bmatrix} \chi + \mathcal{B}(\tau)\chi \tag{5.2.24}$$

where $\mathcal{B}(\tau)$ converges to zero as $\tau \to \infty$. Because the constant matrices F_1 and F_2 cannot have eigenvalues with positive real parts and because $\frac{\partial f_2}{\partial z_1}(\tau)$ remains bounded on $(0, \infty)$, we conclude that the system (5.2.24) is asymptotically stable for any strictly positive constant a. Hence, $\chi(\tau)$ is bounded on $[0, +\infty)$ and, moreover, converges to zero as $\tau \to \infty$.

With a similar argument we establish boundedness and convergence of the time-varying matrices

$$\nu(\tau) := \frac{\partial \tilde{z}(\tau)}{\partial \xi} e^{-a\tau}$$
$$\eta(\tau) := \frac{\partial \tilde{\xi}(\tau)}{\partial \xi}$$

which satisfy

$$\begin{aligned} \frac{d\nu}{d\tau} &= -a\nu + \left(\frac{\partial f}{\partial z} + \frac{\partial \psi}{\partial z}\right)\bigg|_{(\tilde{z}(\tau),\tilde{\xi}(\tau))} \nu + \frac{\partial \psi}{\partial \xi}\bigg|_{(\tilde{z}(\tau),\tilde{\xi}(\tau))} \eta e^{-a\tau} \\ \frac{d\eta}{d\tau} &= \frac{\partial a}{\partial \xi}\bigg|_{\tilde{\xi}(\tau)} \eta \end{aligned} \tag{5.2.25}$$

for the initial condition $\nu(0) = 0$, $\eta(0) = I$.

Next we prove the differentiability of Ψ. Using the chain rule we obtain

$$\frac{\partial \Psi}{\partial z}(z, \xi) = \int_0^\infty d_{\tilde{z}}(\tau) \frac{\partial \tilde{z}(\tau)}{\partial z} d\tau \tag{5.2.26}$$

$$\frac{\partial \Psi}{\partial \xi}(z, \xi) = \int_0^\infty \left(d_{\tilde{z}}(\tau) \frac{\partial \tilde{z}(\tau)}{\partial \xi} + d_{\tilde{\xi}}(\tau) \frac{\partial \tilde{\xi}(\tau)}{\partial \xi}\right) d\tau \tag{5.2.27}$$

where

$$d_{\tilde{z}}(\tau) := \left(\psi^T \frac{\partial^2 W}{\partial z^2} + \frac{\partial W}{\partial z} \frac{\partial \psi}{\partial z}\right)\bigg|_{(\tilde{z}(\tau),\tilde{\xi}(\tau))} \tag{5.2.28}$$

$$d_{\tilde{\xi}}(\tau) := \frac{\partial W}{\partial z} \frac{\partial \psi}{\partial \xi}\bigg|_{(\tilde{z}(\tau),\tilde{\xi}(\tau))} \tag{5.2.29}$$

Since $\dot{\xi} = a(\xi)$ is GAS and LES, there exists a constant $\alpha > 0$ and a class $\mathcal{K}$ function κ such that

$$\|\tilde{\xi}(s,\xi)\| \leq \kappa(\|\xi\|)e^{-\alpha s} \tag{5.2.30}$$

Because ψ and $\frac{\partial\psi}{\partial z}$ vanish when $\xi = 0$ we have

$$\begin{aligned} \|\psi(\tilde{z}(\tau),\tilde{\xi}(\tau))\| &\leq \gamma_5(\|(z,\xi)\|)\ e^{-\alpha\tau} \\ \|\frac{\partial\psi}{\partial z}(\tilde{z}(\tau),\tilde{\xi}(\tau))\| &\leq \gamma_6(\|(z,\xi)\|)\ e^{-\alpha\tau} \end{aligned} \tag{5.2.31}$$

with functions $\gamma_5, \gamma_6 \in \mathcal{K}_\infty$. This yields the estimates

$$\begin{aligned} \|d_{\tilde{z}}(\tau)\| &\leq \gamma_7(\|(z,\xi)\|)\ e^{-\alpha\tau} \\ \|d_{\tilde{\xi}}(\tau)\| &\leq \gamma_8(\|(z,\xi)\|) \end{aligned} \tag{5.2.32}$$

for some $\gamma_7, \gamma_8 \in \mathcal{K}_\infty$. Using the definition of χ, ν, and the fact that $\|\eta(\tau)\| \leq \gamma_9(\|\xi\|)\ e^{-\alpha\tau}$ for some $\gamma_9 \in \mathcal{K}_\infty$ we finally obtain

$$\begin{aligned} \left\|\frac{\partial\Psi}{\partial z}(z,\xi)\right\| &\leq \gamma_7(\|(z,\xi)\|)\int_0^\infty \|\chi(\tau)\|\ e^{-(\alpha-a)\tau}d\tau \\ \left\|\frac{\partial\Psi}{\partial \xi}(z,\xi)\right\| &\leq \gamma_7(\|(z,\xi)\|)\int_0^\infty \|\nu(\tau)\|\ e^{-(\alpha-a)\ \tau}\ d\tau + \gamma_{10}(\|(z,\xi)\|) \end{aligned}$$

for some $\gamma_{10} \in \mathcal{K}$. Since we can choose $a < \alpha$, the integrals exist, which proves the existence of the partial derivatives of Ψ. The continuity of the partial derivatives can be proven along the same lines as the continuity of Ψ. □

We now verify that, under Assumption 5.10, the function Ψ can be differentiated as many times as f and W.

Corollary 5.12 *(Smoothness of the cross-term)*
Under Assumptions 5.1, 5.2, and 5.10, the function Ψ defined by (5.2.3) is C^∞ in $\mathbb{R}^n \times \mathbb{R}^m$.

Proof: As in the proof of Theorem 5.11 we show the existence and continuity of

$$\frac{\partial^2\Psi}{\partial z_i \partial z_j},\ 1 \leq i \leq n,\ 1 \leq j \leq n$$

Existence and continuity of partial derivatives of any order then follows by induction.

First recall that, if f is smooth, the partial derivatives of any order of $\tilde{z}(\tau;z,\xi)$ and $\tilde{\xi}(\tau;\xi)$ exist and are continuous for any $\tau \geq 0$ and any $(z,\xi) \in$

$I\!R^n \times I\!R^m$. Similarly, smoothness of W implies that the partial derivatives of any order of W exist and are bounded along the solutions of (Σ_0).

Using the chain rule, from (5.2.26) we have

$$\frac{\partial^2 \Psi}{\partial z_i \partial z_j} = \int_0^\infty \left(\frac{\partial \tilde{z}}{\partial z_i}(\tau)\right)^T \frac{\partial d_{\tilde{z}}^T}{\partial \tilde{z}}(\tau) \frac{\partial \tilde{z}}{\partial z_j}(\tau)\, d\tau + \int_0^\infty d_{\tilde{z}}(\tau) \frac{\partial^2 \tilde{z}}{\partial z_j \partial z_i}(\tau)\, d\tau \quad (5.2.33)$$

Recall from the proof of Theorem 5.11 that

$$\|\frac{\partial d_{\tilde{z}}}{\partial \tilde{z}}(\tau)\| \leq \gamma_{11}(\|(z,\xi)\|) e^{-\alpha\tau} \quad (5.2.34)$$

for some function $\gamma_{11} \in \mathcal{K}_\infty$. From Theorem 5.11 and (5.2.34) we conclude that the first integral on the right hand side of (5.2.33) exists. It is therefore sufficient to prove the existence of the integral

$$\int_0^\infty d_{\tilde{z}}(\tau) \frac{\partial^2 \tilde{z}}{\partial z_j \partial z_i}(\tau)\, d\tau \quad (5.2.35)$$

or, using the estimate (5.2.32), to prove the boundedness on $(0, \infty)$ of the time function

$$\mu(\tau) := \frac{\partial^2 \tilde{z}}{\partial z_j \partial z_i}(\tau)\, e^{-a\tau} \quad (5.2.36)$$

for $0 < a < \alpha$.

Proceeding as in the proof of Theorem 5.11, we note that $\mu(\tau)$ satisfies the time-varying differential equation

$$\frac{d\mu}{d\tau} = -a\mu + \left(\frac{\partial f}{\partial z} + \frac{\partial \psi}{\partial z}\right)\Big|_{(\tilde{z}(\tau), \tilde{\xi}(\tau))} \mu + R(\tau) \quad (5.2.37)$$

with initial condition $\mu(0) = 0$; denoting by $\mathcal{F}_k$ the k-th column of the matrix $(\frac{\partial f}{\partial z} + \frac{\partial \psi}{\partial z})\Big|_{(\tilde{z}(\tau), \tilde{\xi}(\tau))}$, the k-th component of the vector $R(\tau)$ given by

$$R_k(\tau) := \left(e^{-\frac{1}{2}a\tau} \frac{\partial \tilde{z}}{\partial z_i}(\tau)\right)^T \frac{\partial \mathcal{F}_k}{\partial z}\Big|_{(\tilde{z}(\tau), \tilde{\xi}(\tau))} \left(e^{-\frac{1}{2}a\tau} \frac{\partial \tilde{z}}{\partial z_j}(\tau)\right) \quad (5.2.38)$$

By Theorem 5.11, $(e^{-\frac{1}{2}a\tau} \frac{\partial \tilde{z}}{\partial z_j})$ converges to 0 and hence, $R(\tau)$ converges to zero. As a consequence, the differential equation (5.2.37) for μ has the same structure as the differential equation (5.2.22) for χ. The rest of the proof of Theorem 5.11 can be used to conclude that $\mu(\tau)$ converges to zero as $s \to 0$. □

Examining the variational equations in the proof of Theorem 5.11, we observe that their asymptotic behavior occurs in the neighborhood of the limit sets of $\dot{z} = f(z)$. The differentiability properties of $\Psi(z, \xi)$ are determined by this asymptotic behavior. When the limit sets of $\dot{z} = f(z)$ are equilibria we give a condition under which $\Psi(z, \xi)$ is a C^r function.

Assumption 5.13 *(Restriction on limit sets for the z-subsystem)*
The limits sets of $\dot{z} = f(z)$ consist of equilibria only, and at each equilibrium z_e the eigenvalues of the Jacobian linearization of $f(z)$ have real parts strictly smaller than $\frac{1}{r}\alpha$, where $r \in \{1, 2, \ldots\}$ and α is defined in (5.2.30). □

We note that Assumption 5.13 includes the possibility of unstable equilibria away from the origin, which does not contradict global stability of the equilibrium at the origin.

Theorem 5.14 *(C^r differentiability of the cross-term)*
Under Assumptions 5.1, 5.2, and 5.13, the function Ψ defined by (5.2.3) is C^r in $\mathbb{R}^n \times \mathbb{R}^m$.

Proof: We first prove the theorem for the case $r = 1$. For an arbitrary initial condition (z, ξ), the Assumption 5.13 on the limit sets implies

$$\left.\frac{\partial f}{\partial z}\right|_{\tilde{z}(\tau)} \to F, \text{ as } \tau \to \infty$$

with F a constant matrix with eigenvalues with real parts strictly smaller than α. Now the constant a has to be chosen such that $\max\{Re(\lambda_i(F)),\ i = 1, \ldots, n\} < a < \alpha$. Assumption 5.13 guarantees that such a constant exists.

Then the differential equations for χ and ν, defined in the proof of Theorem 5.11, are of the form

$$\dot{\chi} = (F - aI)\chi + \mathcal{B}_1\chi$$
$$\dot{\nu} = (F - aI)\nu + \mathcal{B}_2\nu + \beta$$

Because the matrix $F - aI$ is Hurwitz and $\mathcal{B}_i$ and β converge to 0 we conclude that χ and ν converge to 0. The rest of the proof for the case $r = 1$ is identical to the proof of Theorem 5.11.

To prove that Ψ is twice continuously differentiable when $r = 2$ we consider again $\mu(\tau)$ defined by (5.2.36) and rewrite its dynamics as

$$\frac{\partial \mu}{\partial \tau} = (F - aI)\mu + \mathcal{B}(\tau) + R(\tau)$$

where $\mathcal{B}$ converges to 0 as $\tau \to \infty$. The vector $R(\tau)$, given by (5.2.38), converges to 0 provided that a can be chosen such that $0 < a < \alpha$ and $e^{-\frac{1}{2}a\tau}\frac{\partial \tilde{z}}{\partial z}$ is bounded. The latter will be satisfied if a can be found such that $\frac{1}{2}a > \max\{Re(\lambda_i(F))\}$. That such an a exists is guaranteed by Assumption 5.13, since for $r = 2$, $\frac{1}{2}\alpha > \max\{Re(\lambda_i(F))\}$. Thus, μ is bounded and converges to 0; so, the existence of the second partial derivatives of Ψ can be concluded as in the proof of Corollary 5.12. The existence of partial derivatives of order higher than 2 when $r > 2$ can be shown in the same way. □

Assumption 5.13 restricts $\dot{z} = f(z)$ to have special limit sets, that is, equilibria. For more complex limit sets, such as limit cycles, analogous differentiability property can be expected to hold as we later illustrate by Example 5.18. However, in the absence of Assumption 5.10 or 5.13, the cross-term Ψ may fail to be continuously differentiable.

Example 5.15 *(Lack of continuous differentiability)*
Consider the system

$$\begin{aligned} \dot{z} &= -z(z-1)(z-2) + \xi \\ \dot{\xi} &= -\tfrac{1}{2}\xi \end{aligned} \tag{5.2.39}$$

The subsystem $\dot{z} = -z(z-1)(z-2) := f(z)$ has three equilibria, $0, 1$, and 2, where the first and third are locally asymptotically stable and the second is unstable. Nevertheless, the equilibrium at 0 is globally stable. A C^1 polynomial Lyapunov function $W(z)$ for $\dot{z} = f(z)$ is given by

$$W(z) = \begin{cases} z^2 & z \le \frac{1}{2} \\ \frac{1}{2} - (z-1)^2 & \frac{1}{2} < z \le \frac{5}{4} \\ \frac{1}{4} + \frac{1}{3}(z-2)^2 & z > \frac{5}{4} \end{cases} \tag{5.2.40}$$

and it can easily be smoothened in the neighborhood of $z = \frac{1}{2}$ and $z = \frac{5}{4}$ to be C^r for any $r > 1$. Assumptions 5.1 and 5.2 are satisfied and, by Theorem 5.8, the cross-term $\Psi(z, \xi)$ exists and is continuous. We will now show that it is not differentiable.

The three equilibria of the z-subsystem yield three different equilibria for the cascade (5.2.39): $x_{e1} = (0,0), x_{e2} = (1,0)$, and $x_{e3} = (2,0)$ where we used the notation $x = (z, \xi)$. The Jacobian linearization of (5.2.39) at x_{e2} is

$$\dot{x} = \begin{bmatrix} 1 & 1 \\ 0 & -\frac{1}{2} \end{bmatrix} x \;:= A_l x$$

Hence the equilibrium x_{e2} is hyperbolic and has a smooth stable invariant manifold [32, 56]. This stable manifold is not tangent to the x_1 axis. This means that $\xi \neq 0$ in this manifold except at x_{e2}. If the partial derivative $\frac{\partial \Psi}{\partial z}$ exists, it must satisfy

$$\frac{\partial \Psi}{\partial z} = \int_0^\infty \frac{\partial^2 W}{\partial z^2} \frac{\partial \tilde{z}}{\partial z} \tilde{\xi}\, d\tau \tag{5.2.41}$$

With $\chi := \frac{\partial \tilde{z}}{\partial z}$, the variational equation of $\dot{z} = f(z)$ is

$$\frac{d\chi}{d\tau} = \frac{\partial f}{\partial z}\chi, \qquad \chi(0) = 1 \tag{5.2.42}$$

With the initial condition (z, ξ) chosen in the stable manifold of x_{e2} with $\xi \neq 0$, the solution of (5.2.39) converges to x_{e2}. Because $W(z)$ has a local maximum at $z = 1$, $\frac{\partial^2 W}{\partial z^2} \to c < 0$, and because $\frac{\partial f}{\partial z} \to 1$ we can write $\chi(\tau) \geq \gamma_{12}(\|(z,\xi)\|)e^{(1-\delta)\tau}$ for any fixed $\delta > 0$, and for some continuous strictly positive function γ_{12}. With $\delta = \frac{1}{4}$ we obtain

$$\|\frac{\partial \Psi}{\partial z}\| \geq \gamma_{13}(\|(z,\xi)\|) \int_0^\infty e^{\frac{1}{4}\tau}\, d\tau$$

and, because the integral diverges, we conclude that Ψ is not C^1. □

5.2.3 Computing the cross-term

In general, the function Ψ is a solution of the following partial differential equation

$$\frac{\partial \Psi}{\partial z}(f(z) + \psi(z,\xi)) + \frac{\partial \Psi}{\partial \xi}a(\xi) = -\frac{\partial W}{\partial z}\psi(z,\xi) \tag{5.2.43}$$

with the boundary condition $\Psi(z, 0) = 0$. This PDE is obtained by taking the time-derivatives of both sides of (5.2.3). Various numerical methods can be used to approximate the solution Ψ and its partial derivatives. Their values at a point (x, ξ) can be obtained by integration of a set of ordinary differential equation. Let us first present a number of cases in which analytical expressions can be obtained.

To obtain a closed-form solution for the line integral which defines the function Ψ

$$\Psi(z,\xi) = \int_0^\infty \frac{\partial W}{\partial z}(\tilde{z})\psi(\tilde{z},\tilde{\xi})\, ds$$

we need closed-form solutions $(\tilde{z}(s), \tilde{\xi}(s))$ of (Σ_0). An expression for $\tilde{z}(s)$ and $\tilde{\xi}(s)$ can be obtained for a number of particular cases when (Σ_0) is in the form

$$\begin{aligned} \dot{z} &= (F + H(\xi))z + \psi(\xi) \\ \dot{\xi} &= A\xi \end{aligned} \tag{5.2.44}$$

After the substitution of the solution $\tilde{\xi}(s) = e^{As}\xi$, the z-subsystem becomes a time-varying linear differential equation. Its solution can be substituted in the line integral (5.2.3) as illustrated by the following examples.

Example 5.16 *(Second-order systems)*
For the second order system

$$\begin{aligned} \dot{z} &= \psi_1(\xi)z + \psi_2(\xi) \\ \dot{\xi} &= -a\xi \end{aligned} \tag{5.2.45}$$

we select $W(z) = z^2$ which yields the cross-term

$$\Psi(z,\xi) = \int_0^\infty 2\tilde{z}(s)(\psi_1(\tilde{\xi}(s))\tilde{z}(s) + \psi_2(\tilde{\xi}(s))\, ds \tag{5.2.46}$$

Substituting the solutions of (5.2.45)

$$\begin{aligned}\tilde{z}(s) &= e^{\int_0^s \psi_1(\tilde{\xi}(\mu))d\mu} z + \int_0^s e^{\int_\tau^s \psi_1(\tilde{\xi}(\mu))d\mu} \psi_2(\tilde{\xi}(\tau))d\tau \\ \tilde{\xi}(s) &= e^{-as}\xi\end{aligned}$$

in the integral (5.2.46), the expression for Ψ can be written as

$$\begin{aligned}\Psi(z,\xi) &= z^2 \int_0^\infty \frac{d}{ds}\left\{e^{2\int_0^s \psi_1(\tilde{\xi}(\mu))d\mu}\right\} ds \\ &\quad +2z \int_0^\infty \frac{d}{ds}\left\{e^{\int_0^s \psi_1(\tilde{\xi}(\mu))d\mu} \int_0^s e^{\int_\tau^s \psi_1(\tilde{\xi}(\mu))d\mu} \psi_2(\tilde{\xi}(\tau))d\tau\right\} ds \\ &\quad + \int_0^\infty \frac{d}{ds}\left\{\int_0^s e^{\int_\tau^s \psi_1(\tilde{\xi}(\mu))d\mu} \psi_2(\tilde{\xi}(\tau))d\tau\right\}^2 ds \\ &= -z^2 + \left(z e^{\int_0^\infty \psi_1(\tilde{\xi}(\mu))d\mu} + \int_0^\infty e^{\int_\tau^\infty \psi_1(\tilde{\xi}(\mu))d\mu} \psi_2(\tilde{\xi}(\tau))d\tau\right)^2\end{aligned}$$

Because the function $\psi(z,\xi) := \psi_1(\xi)z + \psi_2(\xi)$ vanishes at $\xi = 0$ we can write $\psi_1(\xi) = \bar{\psi}_1(\xi)\xi$ and $\psi_2(\xi) = \bar{\psi}_2(\xi)\xi$. Using these expressions and the change of variables $\sigma = \tilde{\xi}(\mu) = \xi e^{-a\mu}$ and $u = \tilde{\xi}(\tau) = \xi e^{-a\tau}$ we obtain

$$\Psi(z,\xi) = -z^2 + \left(z e^{\frac{1}{a}\int_0^\xi \bar{\psi}_1(\sigma)d\sigma} + \frac{1}{a}\int_0^\xi e^{\frac{1}{a}\int_0^u \bar{\psi}_1(\sigma)d\sigma} \bar{\psi}_2(u)du\right)^2 \tag{5.2.47}$$

Finally a Lyapunov function for the system (5.2.45) is given by

$$V_1(z,\xi) = W(z)+\Psi(z,\xi)+\xi^2 = \left(z e^{\frac{1}{a}\int_0^\xi \bar{\psi}_1(\sigma)d\sigma} + \frac{1}{a}\int_0^\xi e^{\frac{1}{a}\int_0^u \bar{\psi}_1(\sigma)d\sigma} \bar{\psi}_2(u)du\right)^2 + \xi^2 \tag{5.2.48}$$

The above integrals can be explicitly solved for certain functions ψ_1 and ψ_2 or else they can be approximated. □

Example 5.17 *(Polynomial interconnection)*
When in the system

$$\begin{aligned}\dot{z} &= Fz + p(\xi) \\ \dot{\xi} &= A\xi\end{aligned} \tag{5.2.49}$$

the interconnection term $p(\xi)$ is a polynomial, Ψ is also a polynomial. In particular, if p is a linear vector function of ξ, then Ψ is a quadratic form.

For the sake of illustration, when z and ξ are scalars and p is a second order polynomial, the cross-term is

$$\Psi(z,\xi) = a_1 z\xi + a_2 z\xi^2 + a_3\xi^2 + a_4\xi^3 + a_5\xi^4$$

where the coefficients are independent of z and ξ. □

If (Σ_0) is not in the form (5.2.44) then it is usually not possible to obtain a closed-form solution for $\tilde{z}(s)$ and in turn for Ψ. Nevertheless, the next example illustrates a situation where a closed-form solution for Ψ does not require the solution of the differential equation $\dot{z} = f(z) + \psi(z,\xi)$.

Example 5.18 *(Skew-symmetric z-subsystem)*
Consider the system

$$\begin{aligned} \dot{z} &= F(z)z + \psi(\xi)z \\ \dot{\xi} &= A\xi \end{aligned} \tag{5.2.50}$$

where ψ is a scalar function and the matrix $F(z)$ satisfies $F^T(z)P + P^T F(z) \equiv 0$ for some positive definite matrix P. The quadratic Lyapunov function $W(z) = z^T P z$ satisfies $\dot{W}(z) = \psi(\xi)W(z)$ and, therefore,

$$W(\tilde{z}(\tau)) = W(z)e^{\int_0^\tau \psi(\tilde{\xi}(s))ds}. \tag{5.2.51}$$

On the other hand, we have

$$\Psi(z,\xi) = \int_0^\infty 2\tilde{z}^T P\psi(\tilde{\xi})\,\tilde{z}\,d\tau = \int_0^\infty W(\tilde{z})\psi(\tilde{\xi})\,d\tau \tag{5.2.52}$$

Substituting (5.2.51) in (5.2.52) we obtain the expression

$$\Psi(z,\xi) = W(z)\int_0^\infty e^{\int_0^\tau \psi(\tilde{\xi})ds}\psi(\tilde{z})d\tau = W(z)\left(e^{\int_0^\infty \psi(\tilde{\xi}(s))ds} - 1\right)$$

We remark that Ψ is smooth although Assumptions 5.10 and 5.13 may not be satisfied. □

The control laws of the next section will employ the partial derivatives of $\Psi(z,\xi)$. For on-line computation of these control laws, when z and ξ are known at time t, we need to evaluate $\frac{\partial\Psi}{\partial z}$ and $\frac{\partial\Psi}{\partial\xi}$ with desired accuracy.

Denote by $\Psi^\star(z,\xi,\tau)$ the line integral evaluated up to the time τ:

$$\Psi^\star(z,\xi,\tau) \triangleq \int_0^\tau \frac{\partial W}{\partial z}(\tilde{z})\,\psi(\tilde{z},\tilde{\xi})ds \tag{5.2.53}$$

We write $\Psi^\star$ as a function of τ only, but we keep in mind that it also depends on z and ξ. $\Psi^\star$ is the solution of the differential equation

$$(\Psi^\star)'(\tau) = \frac{\partial W}{\partial z}\psi\bigg|_{(\tilde{z}(\tau),\tilde{\xi}(\tau))}, \qquad \Psi^\star(0) = 0 \tag{5.2.54}$$

where the notation $(\Psi^\star)'$ stands for $\frac{d\Psi^\star}{d\tau}$. By taking the partial derivatives with respect to z and ξ, we obtain the following differential equations (in the notation defined in (5.2.28) and (5.2.29))

$$\left(\frac{\partial\Psi^\star}{\partial z}\right)'(\tau) = \psi_{\tilde{z}}(\tau)\,\chi(\tau)\,e^{a\tau} \tag{5.2.55}$$

$$\left(\frac{\partial\Psi^\star}{\partial \xi}\right)'(\tau) = \psi_{\tilde{z}}(\tau)\,\nu(\tau)\,e^{a\tau} + \psi_{\tilde{\xi}}(\tau)\eta(\tau) \tag{5.2.56}$$

with the initial conditions $\dfrac{\partial\Psi^\star}{\partial z}(0) = 0$ and $\dfrac{\partial\Psi^\star}{\partial \xi}(0) = 0$. The proof of Theorem 5.11 provides the bound

$$\|\frac{\partial\Psi}{\partial z} - \frac{\partial\Psi^\star}{\partial z}(T)\| \leq M(\|(z,\xi)\|)\int_T^\infty e^{-(\alpha-a)s}ds = \frac{1}{\alpha-a}M(\|(z,\xi)\|)e^{-(\alpha-a)T}$$

for some $M \in \mathcal{K}_\infty$. The same bound can be established for the difference $\|\dfrac{\partial\Psi}{\partial \xi} - \dfrac{\partial\Psi^\star}{\partial \xi}(T)\|$. We summarize this as follows.

Proposition 5.19 *(Finite time integration)*
For any given $\varepsilon > 0$ and a compact set $\Omega \subset R^{n_z+n_\xi}$, there exists a constant $T > 0$ such that

$$\|\frac{\partial\Psi}{\partial z} - \frac{\partial\Psi^\star}{\partial z}(\tau)\| < \varepsilon \tag{5.2.57}$$

$$\|\frac{\partial\Psi}{\partial \xi} - \frac{\partial\Psi^\star}{\partial \xi}(\tau)\| < \varepsilon \tag{5.2.58}$$

for every $\tau > T$ whenever $(z,\xi) \in \Omega$. □

In other words, to obtain the partial derivatives with the desired accuracy we have to integrate the set of equations (5.2.22), (5.2.25), (5.2.55), and (5.2.56) on an interval of sufficient length. In general, to achieve the accuracy as in Proposition 5.19 the interval of integration has to increase with the size of the compact set Ω.

5.3 Relaxed Constructions

5.3.1 Geometric interpretation of the cross-term

In the preceding two sections we presented two different constructions of Lyapunov functions for cascade systems: composite Lyapunov functions and Lyapunov functions with a cross-term. When, with a change of coordinates $\zeta = \zeta(z, \xi)$, a cascade system can be decoupled into two separate subsystems, then in the new coordinates (ζ, ξ), a composite Lyapunov function is the sum of the subsystem Lyapunov functions. Because a Lyapunov function with the cross-term $\Psi(z, \xi)$ can be calculated for the same cascade in the original coordinates (z, ξ), the link between the two Lyapunov functions gives a geometric interpretation to the cross-term. We show this for the special cascade

$$\begin{aligned} \dot{z} &= Fz + \psi(\xi), \quad F + F^T = 0 \\ \dot{\xi} &= a(\xi) \end{aligned} \tag{5.3.1}$$

where all the eigenvalues of F are on the imaginary axis. Using $W(z) = z^T z$ and the fact that $z^T F z = 0$ we calculate the cross-term

$$\begin{aligned} \Psi(z, \xi) &= 2\int_0^\infty \tilde{z}^T(s)(\dot{\tilde{z}}(s) - F\tilde{z}(s))ds = \int_0^\infty d(\tilde{z}^T(s)\tilde{z}(s)) \\ &= (\tilde{z}^T(s)\tilde{z}(s))_\infty - (\tilde{z}^T(s)\tilde{z}(s))_0 = (\tilde{z}^T(s)\tilde{z}(s))_\infty - W(z) \end{aligned}$$

Thus, the Lyapunov function $V_0(z, \xi)$ for the cascade (5.3.1) is

$$V_0(z, \xi) = W(z) + \Psi(z, \xi) + U(\xi) = (\tilde{z}^T(s)\tilde{z}(s))_\infty + U(\xi) \tag{5.3.2}$$

We observe that $\|\tilde{z}(s)\|^2$ has a limit as $s \to \infty$, although the solution

$$\tilde{z}(s) = e^{Fs}z + e^{Fs}\int_0^s e^{-F\tau}\psi(\tilde{\xi}(\tau))d\tau$$

itself does not have a limit, except when $F \equiv 0$.

We now proceed to find a change of coordinates needed to construct a composite Lyapunov function. Because $A = \frac{\partial a}{\partial \xi}(0)$ is Hurwitz and F has all its eigenvalues on the imaginary axis, the cascade (5.3.1) has the stable manifold in which the behavior of (5.3.1) is described by $\dot{\xi} = a(\xi)$. The change of coordinates which exhibits the stable manifold is

$$\zeta = z + \int_t^\infty e^{-F(\tau - t)}\psi(\xi(\tau + t; t; \xi))d\tau \tag{5.3.3}$$

where $\xi(\tau + t; t; \xi) = \xi(\tau; 0; \xi) = \tilde{\xi}(\tau)$, because $\dot{\xi} = a(\xi)$ is time-invariant. It is easy to check by differentiating with respect to t, that (5.3.3) decouples (5.3.1) into two systems

$$\begin{aligned} \dot{\zeta} &= F\zeta \\ \dot{\xi} &= a(\xi) \end{aligned} \tag{5.3.4}$$

This decoupled form identifies two invariant manifolds of the cascade: the stable manifold $\zeta = 0$ and the center manifold $\xi = 0$.

A composite Lyapunov function is the sum of the subsystem Lyapunov functions:

$$V(\zeta, \xi) = \zeta^T \zeta + U(\xi) \tag{5.3.5}$$

To link $V_0(z, \xi)$ with $V(\zeta, \xi)$ we evaluate $\zeta^T\zeta$ in the coordinates (z, ξ). Noting that $e^{(F+F^T)s} = I$ for any s, we obtain for $t = 0$ and all s

$$\begin{aligned}
\zeta^T\zeta &= \left(z + \int_0^\infty e^{-F\tau}\psi(\tilde{\xi}(\tau))d\tau\right)^T (e^{Fs})^T e^{Fs} \left(z + \int_0^\infty e^{-F\tau}\psi(\tilde{\xi}(\tau))d\tau\right) \\
&= \left(e^{Fs}z + \int_0^\infty e^{F(s-\tau)}\psi(\tilde{\xi}(\tau))d\tau\right)^T \left(e^{Fs}z + \int_0^\infty e^{F(s-\tau)}\psi(\tilde{\xi}(\tau))d\tau\right) \\
&= (\tilde{z}(s) + \int_s^\infty e^{F(s-\tau)}\psi(\tilde{\xi}(\tau))d\tau)^T (\tilde{z}(s) + \int_s^\infty e^{F(s-\tau)}\psi(\tilde{\xi}(\tau))d\tau)
\end{aligned}$$

Because the integrals converge to 0 as $s \to \infty$, we obtain

$$\zeta^T\zeta = \lim_{s\to\infty} (\tilde{z}^T(s)\tilde{z}(s))$$

Thus, in the original coordinates (z, ξ), the two Lyapunov functions are identical:

$$V(\zeta(z, \xi), \xi)) = V_0(z, \xi)$$

Properties of $V_0(z, \xi)$. The construction of the Lyapunov function with the cross-term eliminates the intermediate task of finding a decoupling change of coordinates. Moreover, $V_0(z, \xi)$ can be constructed even when a decoupling change of coordinates does not exist, that is when the cascade is not reducible to the decoupled form (5.3.4).

Another property that does not require the existence of a decoupling change of coordinates is that the sum $W(z) + \Psi(z, \xi)$ in $V_0(z, \xi)$ equals the limit of $W(\tilde{z}(s))$ as $s \to \infty$. We show below that this holds whenever $L_f W \equiv 0$ because then $\dot{W}$ reduces to

$$\dot{W} = \frac{\partial W}{\partial z}\psi(z, \xi)$$

Integrating along the solution $(\tilde{z}(s), \tilde{\xi}(s))$ of the cascade (Σ_0) we obtain

$$\begin{aligned}
\lim_{s\to\infty} \int_0^s \dot{W}(\tilde{z}(\tau))\, d\tau &= \lim_{s\to\infty} W(\tilde{z}(s)) - W(z) = \\
&= \int_0^\infty L_\psi W(\tilde{z}(\tau), \tilde{\xi}(\tau))\, d\tau = \Psi(z, \xi)
\end{aligned}$$

and hence

$$W(z) + \Psi(z,\xi) = W(z) + \lim_{s\to\infty} W(\tilde{z}(s)) - W(z) =: W_\infty(z,\xi) \tag{5.3.6}$$

If $f(z) \equiv 0$, then the limit of $\tilde{z}(s)$ is

$$z_\infty(z,\xi) := \lim_{s\to\infty} \tilde{z}(s) = z + \int_0^\infty \psi(\tilde{z}(s),\tilde{\xi}(s))\, ds \tag{5.3.7}$$

and $W_\infty(z,\xi) = W(z_\infty(z,\xi))$.

The mapping $(z,\xi) \to (z_\infty,\xi)$ defines a local change of coordinates because

$$\frac{\partial z_\infty}{\partial z} = I + \int_0^\infty \frac{\partial \psi}{\partial z}(\tilde{z}(s),\tilde{\xi}(s))\, ds \tag{5.3.8}$$

and the integral vanishes at $\xi = 0$. It is clear from (5.3.8) that when ψ does not depend on z, this change of coordinates is defined globally and decouples (Σ_0) into the two subsystems $\dot{\zeta} = 0$, $\dot{\xi} = a(\xi)$.

The following example illustrates a situation when a global change of coordinates exists even though the interconnection depends on z.

Example 5.20 (*Cross-term as a global decoupling change of coordinates*)
The system

$$\begin{aligned} \dot{z} &= \frac{z^2}{1+z^2}\xi = \psi(z,\xi) \\ \dot{\xi} &= -\xi \end{aligned} \tag{5.3.9}$$

satisfies Assumption 5.1 because for all z

$$|\psi(z,\xi)| = \left|\frac{z^2}{1+z^2}\xi\right| \le |\xi|$$

Using $W(z) = z^2$ we obtain from (5.3.6)

$$V_0(z,\xi) = W(z) + \Psi(z,\xi) + \xi^2 = z_\infty^2 + \xi^2 \tag{5.3.10}$$

The explicit solution of (5.3.9) is $\tilde{z}(s) \equiv 0$, if $z = 0$, and

$$\tilde{z}(s) = \frac{z^2 - 1 + z\xi(1-e^{-s}) + \sqrt{(z^2 - 1 + z\xi(1-e^{-s}))^2 + 4z^2}}{2z} \quad \text{if } z \neq 0$$

The limit as $s \to \infty$ is

$$\tilde{z}(s) \to z_\infty = \frac{z^2 - 1 + z\xi + \sqrt{(z^2 - 1 + z\xi)^2 + 4z^2}}{2z} \quad \text{if } z \neq 0$$

The change of coordinates $(z,\xi) \to (z_\infty,\xi)$ is globally defined because the matrix

$$\begin{bmatrix} \frac{\partial z_\infty}{\partial z} & \frac{\partial z_\infty}{\partial \xi} \\ 0 & 1 \end{bmatrix} = \begin{bmatrix} \frac{z^2+1}{\sqrt{(z^2-1+z\xi)^2+4z^2}}\frac{z_\infty}{z} & z_\infty z \\ 0 & 1 \end{bmatrix}$$

is nonsingular and $\|(z_\infty,\xi)\| \to \infty$ as $\|(z,\xi)\| \to \infty$.

We thus obtain the Lyapunov function $V_0(z,\xi) = z_\infty^2 + \xi^2$ for the system (5.3.9) as

$$V_0(z,\xi) = \begin{cases} \dfrac{\left(z^2-1+z\xi+\sqrt{(z^2-1+z\xi)^2+4z^2}\right)^2}{4z^2} + \xi^2 & \text{if} \quad z \neq 0 \\ \xi^2 & \text{if} \quad z = 0 \end{cases} \tag{5.3.11}$$

which, by Theorem 5.11 and Corollary 5.12, is C^∞. □

5.3.2 Relaxed change of coordinates

A decoupling change of coordinates was found for (5.3.1) because the system has a global stable manifold $\zeta = 0$ given in the integral form (5.3.3). Using the graph $z = \eta(\xi)$ of this manifold the decoupling change of coordinates is $\zeta = z - \eta(\xi)$ and the PDE defining $\eta(\xi)$ is

$$\frac{\partial \eta}{\partial \xi} a(\xi) = F\eta + \psi(\xi), \quad \eta(0) = 0$$

This PDE is obtained by differentiating $z = \eta(\xi)$ with respect to time and substituting (5.3.1) in $\dot{z} = \frac{\partial \eta}{\partial \xi}\dot{\xi}$.

We proceed to investigate the existence of a stable manifold for the cascade

$$(\Sigma_0) \begin{cases} \dot{z} = f(z) + \psi(z,\xi) \\ \dot{\xi} = a(\xi) \end{cases}$$

In this case we let the manifold expression be implicit, $\zeta(z,\xi) = 0$. If the manifold exists, the decoupling transformation $\zeta = \zeta(z,\xi)$ satisfies $\dot{\zeta} = f(\zeta)$, and the PDE defining $\zeta(z,\xi)$ is

$$\frac{\partial \zeta}{\partial z}(f(z) + \psi(z,\xi)) + \frac{\partial \zeta}{\partial \xi} a(\xi) = f(\zeta), \quad \zeta(z,0) = z \tag{5.3.12}$$

This equation is impractical for computation and we use it only to define a *relaxed change of coordinates.* We recall from Section 5.1.3 that the presence in $\psi(z,\xi)$ of terms linear in ξ prevented the construction of composite Lyapunov functions in Theorem 5.6. This motivates us to seek a relaxed change of coordinates which removes only these terms linear in ξ. This can be accomplished by finding a function $\bar{\zeta}(z,\xi)$ which satisfies

$$\frac{\partial \bar{\zeta}}{\partial z}(f(z)+\psi(z,\xi))+\frac{\partial \bar{\zeta}}{\partial \xi}a(\xi)=f(\bar{\zeta})+\mathcal{R}(\bar{\zeta},\xi), \quad \bar{\zeta}(z,0)=z \tag{5.3.13}$$

where $\mathcal{R}(\bar{\zeta},\xi)$ contains only quadratic and higher-order terms in ξ.

Proposition 5.21 *(Relaxed manifold PDE)*
Let Assumptions 5.1, 5.2, and 5.4 be satisfied and suppose that $\bar{\zeta}(z,\xi)$ is a solution of the relaxed manifold PDE (5.3.13) where $\mathcal{R}(\bar{\zeta},\xi)$ is quadratic or higher-order in ξ and satisfies the linear growth assumption in $\bar{\zeta}$ (Assumption 5.1). Then if $\zeta=\bar{\zeta}(z,\xi)$ qualifies for a global change of coordinates, it transforms (Σ_0) into a cascade in which the interconnection ψ does not contain terms linear in ξ. In the new coordinates (ζ,ξ), a Lyapunov function for (Σ_0) is given by

$$V(\zeta,\xi)=\ln(W(\zeta)+1)+\int_0^{U(\xi)}\gamma(s)ds$$

where the function $\gamma(.)$ is constructed as in Theorem 5.6. □

Requiring that the decoupling be achieved only up to the quadratic terms in ξ has the advantage that such a relaxed change of coordinates exists and is explicit when the z-subsystem is linear. This follows from the results of Mazenc and Praly [75].

Proposition 5.22 *(Relaxed change of coordinates)*
Suppose that in addition to Assumption 5.1, the cascade (Σ_0) satisfies:

(i) $\dot{z}=Fz$, and $\psi(z,\xi)=M\xi+\sum_{l=1}^{n_\xi}\xi_l M_l z+r(z,\xi)$, where $r(z,\xi)$ is quadratic or higher-order in ξ.

(ii) $\lambda_i(A)\neq\lambda_j(F)$ and $\lambda_i(A)+\lambda_j(F)\neq\lambda_k(F)$, $i=1,\ldots,n_\xi$, $j,k=1,\ldots,n_z$

Then a constant $\nu>0$ and matrices N, N_l, $l=1,\ldots,n_\xi$ exist such that the global change of coordinates

$$\zeta=\bar{\zeta}(z,\xi)=(I+\frac{\sum_{l=1}^{n_\xi}N_l\xi_l}{1+\nu\|\xi\|^2})z+N\xi \tag{5.3.14}$$

transforms (Σ_0) into the partially decoupled form

$$\begin{aligned}\dot{\zeta} &= F\zeta + \bar{\psi}(\zeta,\xi)\\ \dot{\xi} &= a(\xi)\end{aligned}$$

where $\bar{\psi}$ does not contain terms linear in ξ, that is $\frac{\partial\bar{\psi}}{\partial\xi}(\zeta,0) \equiv 0$. Matrices N, N_l, $l = 1,\ldots,n_\xi$ can be obtained by solving a set of linear algebraic equations.

Proof: Below we denote by $\mathcal{O}(\|\xi\|^2)(\|z\|+1)$ any term which is quadratic or higher-order in ξ and which, for a fixed ξ, is bounded by $k(\|z\|+1)$. The change of coordinates (5.3.14) yields

$$\begin{aligned}\frac{\partial\zeta}{\partial z} &= I + \sum_{l=1}^{n_\xi} N_l\xi_l + \mathcal{O}(\|\xi\|^2)(\|z\|+1)\\ \frac{\partial\zeta}{\partial\xi} &= N + [N_1 z \ldots N_{n_\xi} z] + \mathcal{O}(\|\xi\|^2)(\|z\|+1)\end{aligned}$$

Substituting these expressions in the PDE (5.3.13) we obtain

$$\begin{aligned}(I + \sum_{l=1}^{n_\xi} N_l\xi_l)(Fz + M\xi + \sum_{l=1}^{n_\xi}\xi_l M_l z) + (N + [N_1 z \ldots N_{n_\xi} z])A\xi =\\ = F[(I + \sum_{l=1}^{n_\xi} N_l\xi_l)z + N\xi] + \mathcal{O}(\|\xi\|^2)(\|z\|+1)\end{aligned} \quad (5.3.15)$$

Equating the linear terms yields $NA - FN = -M$ and N exists and is unique because $\lambda_i(A) \neq \lambda_j(F)$.

Equating the second-order terms yields

$$N_l F - FN_l + \Pi_l(N_1,\ldots,N_{n_\xi}) = M_l, \quad l = 1,\ldots,n_\xi \quad (5.3.16)$$

where the i-th column of Π_l is $\Pi_l^{(i)} = \sum_{k=1}^{n_\xi} N_k^{(i)} a_{kl}$. From [11] it is known that (5.3.16) has a unique solution if $\lambda_i(A) + \lambda_j(F) \neq \lambda_k(F)$.

Finally, given the matrices N_l one can always find a constant $\nu > 0$ such that $0.5 < |I + \frac{\sum_{l=1}^{n_\xi} N_l\xi_l}{1+\nu\|\xi\|^2}| < 1.5$, which guarantees that (5.3.14) is a globally invertible change of coordinates $\zeta \leftrightarrow z$. □

5.3.3 Lyapunov functions with relaxed cross-term

We have seen that the construction of Lyapunov functions with cross-term remains applicable even when the decoupling change of coordinates does not exist. However, the cross-term $\Psi(z,\xi)$ has to be calculated either by integration

$$\Psi(z,\xi) = \int_0^\infty \frac{\partial W}{\partial z}(\tilde{z}(s))\psi(\tilde{z}(s),\tilde{\xi}(s))ds$$

or by solving the cross-term PDE

$$\frac{\partial \Psi}{\partial z}(f(z)+\psi(z,\xi)) + \frac{\partial \Psi}{\partial \xi}a(\xi) = -\frac{\partial W}{\partial z}\psi(z,\xi), \quad \Psi(z,0)=0 \tag{5.3.17}$$

as shown in Section 5.2.3. To avoid the burden of computation, we again employ relaxation and obtain $\bar{\Psi}$ from the PDE

$$\frac{\partial \bar{\Psi}}{\partial z}(f(z)+\psi(z,\xi)) + \frac{\partial \bar{\Psi}}{\partial \xi}a(\xi) = -\frac{\partial W}{\partial z}\psi(z,\xi) + \mathcal{R}(z,\xi), \quad \bar{\Psi}(z,0)=0 \tag{5.3.18}$$

where $\mathcal{R}(z,\xi)$ is quadratic or higher-order in ξ near $\xi=0$. We note that

$$\bar{\Psi} = \int_0^\infty \frac{\partial W}{\partial z}\psi(\tilde{z},\tilde{\xi}) - \mathcal{R}(\tilde{z},\tilde{\xi})\, ds$$

satsifes (5.3.18). Its existence is proven in the same way as that of Ψ.

Proposition 5.23 *(Relaxed cross-term for composite Lyapunov functions)* Let Assumptions 5.1, 5.2, and 5.4 be satisfied and let $\bar{\Psi}(z,\xi)$ be a solution of (5.3.18), where $\mathcal{R}(z,\xi)$ satisfies

(i) $\|\mathcal{R}(z,\xi)\|$ is quadratic or higher-order in $\|\xi\|$ near $\xi=0$,

(ii) For $\|\xi\|$ fixed and $\|z\|$ large, $\|\mathcal{R}(z,\xi)\|$ is bounded by $cW(z)$.

If, for some $c_2>0$ and $\gamma_1(.)\in\mathcal{K}$, the function $\bar{V}(z,\xi)$ satisfies

$$\bar{V}(z,\xi) = W(z) + \bar{\Psi}(z,\xi) + \gamma_1(U(\xi)) \tag{5.3.19}$$

then a composite Lyapunov function for (Σ_0) is

$$V(z,\xi) = \ln(\bar{V}(z,\xi)+1) + \int_0^{U(\xi)} \gamma(s)ds$$

where $\gamma(.)$ is constructed as in Theorem 5.6.

Proof: The time-derivative of $V(z,\xi)$ is

$$\dot{V} = \frac{\dot{\bar{V}}}{\bar{V}+1} + \gamma(U(\xi))L_aU(\xi)$$

and the time-derivative of $\bar{V}$ satisfies

$$
\begin{aligned}
\dot{\bar{V}} &= \frac{\partial W}{\partial z}(f(z) + \psi(z,\xi)) + \dot{\bar{\Psi}}(z,\xi) + \gamma_1'(U(\xi))L_aU(\xi) \\
&\leq \frac{\partial W}{\partial z}\psi(z,\zeta) + \dot{\bar{\Psi}}(z,\zeta) \leq \mathcal{R}(z,\xi)
\end{aligned} \tag{5.3.20}
$$

Returning to $\dot{V}$, and using (i) and (ii), we conclude

$$
\dot{V} \leq \frac{\mathcal{R}(z,\xi)}{\bar{V}+1} + \gamma(U(\xi))L_aU(\xi) \leq \gamma_3(\|\xi\|) + \gamma(U(\xi))L_aU(\xi)
$$

where $\gamma_3(.)$ is quadratic or higher-order in ξ. The function γ which achieves $\dot{V} \leq 0$ can then be constructed as in Theorem 5.6. □

An advantage in relaxing the cross-term Ψ to $\bar{\Psi}$ is that the construction of $\bar{\Psi}$ is explicit when the z-subsystem is linear.

Proposition 5.24 *(Construction of the relaxed cross-term)*
Suppose that, in addition to Assumptions 5.1 and 5.4,

$$
\dot{z} = Fz, \quad \text{and} \quad \psi(z,\xi) = M\xi + \sum_{i=1}^{n_\xi} \xi_l M_l z + r(z,\xi) \tag{5.3.21}
$$

where $r(z,\xi)$ is quadratic or higher-order in ξ.
Then a relaxed cross-term which satisfies Proposition 5.23 is

$$
\bar{\Psi}(z,\xi) = \xi^T\Psi_0\xi + \xi^T\Psi_1 z + z^T\left(\sum_{i=1}^{n_\xi} \frac{\xi_i}{1+\nu\|\xi\|^2}\Psi_{2i}\right)z \tag{5.3.22}
$$

where Ψ_0, Ψ_1, and Ψ_{2i}, $i = 1,\ldots,n_\xi$ are constant matrices which can be obtained by solving a set of linear algebraic equations.

Proof: Let $W(z) = \frac{1}{2}z^T\bar{W}z$ where $\bar{W} > 0$ satisfies $\bar{W}F + F^T\bar{W} \leq 0$. The partial derivatives of the relaxed cross-term (5.3.22) are

$$
\begin{aligned}
\frac{\partial \bar{\Psi}}{\partial z} &= \xi^T\Psi_1 + 2z^T\left(\sum_{l=1}^{n_\xi} \xi_l\Psi_{2l}\right) + \mathcal{O}(\|\xi\|^2)(\|z\|^2+1) \\
\frac{\partial \bar{\Psi}}{\partial \xi} &= 2\xi^T\Psi_0 + (\Psi_1 z)^T + z^T[\Psi_{21}\ldots\Psi_{2n_\xi}]z + \mathcal{O}(\|\xi\|^2)(\|z\|^2+1)
\end{aligned}
$$

Substituting these expressions in (5.3.18)

$$
\frac{\partial \bar{\Psi}}{\partial z}(Fz+M\xi+\sum_{l=1}^{n_\xi}\xi_l M_l z)+\frac{\partial \bar{\Psi}}{\partial \xi}A\xi = -z^T\bar{W}(M\xi+\sum_{l=1}^{n_\xi}\xi_l M_l z)+\mathcal{O}(\|\xi\|^2(\|z\|^2+1) \tag{5.3.23}
$$

and equating the quadratic terms, we obtain

$$\begin{aligned} F^T\Psi_1^T + \Psi_1^T A &= -\bar{W}M \\ A^T\Psi_0 + \Psi_0 A &= -\tfrac{1}{2}(M^T\Psi_1^T + \Psi_1 M) \end{aligned} \tag{5.3.24}$$

The unique solutions Ψ_0 and Ψ_1 exist because A is Hurwitz and F is Lyapunov stable, so that $\lambda_i(A) \neq -\lambda_j(F)$.

Equating the terms of the form $z^T(\sum_{l=1}^{n_\xi}(.)\xi_l)z$ we obtain

$$\Psi_{2l}F + F^T\Psi_{2l} + \Pi_l(\Psi_{21}, \dots, \Psi_{2n_\xi}) = -\bar{W}M_l, \quad l = 1, \dots, n_\xi \tag{5.3.25}$$

where the i-th column of Π_l is $\Pi_l^{(i)} = \sum_{k=1}^{n_\xi} \Psi_{2k}^{(i)} a_{kl}$. It is known from [11] that (5.3.25) has a unique solution if $\lambda_i(A) + \lambda_j(F) \neq -\lambda_k(F)$, which is satisfied because $\lambda_i(A) + \lambda_j(F) < 0$ and $-\lambda_k(F) \geq 0$. This completes the calculation of $\bar{\Psi}$ in (5.3.22).

By Assumption 5.4 the Lyapunov function $U(\xi)$ is locally quadratic. Hence, for each $\alpha_2 > 0$, a function $\gamma_1 \in \mathcal{K}$ can be found such that $\gamma_1(U) + \xi^T\Psi_0\xi \geq \alpha_2\xi^T\xi$ and we obtain

$$\begin{aligned} W(z) + \bar{\Psi}(z,\xi) + \gamma_1 U(\xi) &\geq \\ \geq z^T\Big(\frac{1}{2}\bar{W} + \sum_{i=1}^{n_\xi} \frac{\xi_i}{1+\nu\|\xi\|^2}\Psi_{2i}\Big)z &+ \xi^T\Psi_1 z + \xi^T\Psi_0\xi + \gamma_1 U(\xi) \\ &\geq \alpha_1(\nu)\|z\|^2 + \xi^T\Psi_1 z + \alpha_2\|\xi\|^2 \end{aligned}$$

where $\alpha_1(\nu) = \frac{1}{2}\lambda_{min}(\bar{W}) + \mathcal{O}(\frac{1}{\nu})$. In view of

$$|\xi^T\Psi_1 z| \leq k_1\|\xi\|\|z\| \leq \frac{k_1}{2\mu}\|\xi\|^2 + \frac{k_1\mu}{2}\|z\|^2$$

which is true for any $\mu > 0$, there exist positive constants ν, α_2, μ, and $c > 0$ such that

$$\alpha_1(\nu)\|z\|^2 + \xi^T\Psi_1 z + \alpha_2\|\xi\|^2 \geq (\alpha_1(\nu) - \frac{k_1\mu}{2})\|z\|^2 + (\alpha_2 - \frac{k_1}{2\mu})\|\xi\|^2 \geq \frac{c}{2}z^T\bar{W}z$$

which proves (5.3.19). □

We stress that the above explicit construction of $\bar{\Psi}(z,\xi)$ is valid even when an invariant manifold $\zeta(z,\xi) = 0$ and the corresponding decoupling change of coordinates do not exist. In other words, $\bar{\Psi}(z,\xi)$ can be constructed even when the nonresonance conditions of Proposition 5.22 are violated.

Example 5.25 *(Construction of a relaxed cross-term)*
The system

$$\begin{aligned} \dot{z} &= -z + \xi + z\xi \\ \dot{\xi} &= a(\xi), \ \tfrac{\partial a}{\partial \xi}(0) = -1 \end{aligned} \tag{5.3.26}$$

does not have an invariant manifold $z = \eta(\xi)$ because its Jacobian linearization is the Jordan block

$$\begin{pmatrix} -1 & 1 \\ 0 & -1 \end{pmatrix}$$

Hence, a change of coordinates which removes the terms ξ and $z\xi$ does not exist. We choose $W(z) = \frac{1}{2}z^2$, and, using Proposition 5.24, construct the cross-term $\bar{\Psi}$ as

$$\bar{\Psi}(z,\xi) = \frac{1}{4}\xi^2 + \frac{1}{2}\xi z + \frac{1}{3}z^2\frac{\xi}{1+\nu\xi^2}$$

By selecting the constants $\gamma_1 = 1$ and $\nu = 4$ we obtain

$$\bar{V}(z,\xi) = \frac{1}{2}z^2 + \bar{\Psi}(z,\xi) + \gamma_1\xi^2 = z^2(\frac{1}{2} + \frac{1}{3}\frac{\xi}{1+\nu\xi^2}) + \frac{1}{2}\xi z + (\gamma_1 + \frac{1}{4})\xi^2 \geq \frac{1}{4}z^2$$

A Lyapunov function for (5.3.26) is thus $V(z,\xi) = \ln(\bar{V}(z,\xi)+1) + \int_0^{U(\xi)}\gamma(s)ds$ where the function $\gamma(.)$ can be constructed as in Theorem 5.6. □

When the interconnection $\psi(z,\xi)$ contains no bilinear terms $z_i\xi_j$, then the quadratic approximation of $V_0(z,\xi)$ is sufficient for the construction of a Lyapunov function for (Σ_0).

Corollary 5.26 *(Quadratic approximation of $V_0(z,\xi)$)*
Suppose that, in addition to the assumptions of Proposition 5.24, matrices M_l in (5.3.21) are zero, that is,

$$\psi(z,\xi) = M\xi + r(z,\xi)$$

Then a Lyapunov function for (Σ_0) is

$$V(z,\xi) = \ln(\bar{V}(z,\xi) + 1) + \int_0^{U(\xi)} \gamma(s)ds$$

where $\bar{V}(z,\xi)$ is the quadratic approximation of the Lyapunov function $V_0(z,\xi)$ given by (5.2.1). □

5.4 Stabilization of Augmented Cascades

5.4.1 Design of the stabilizing feedback laws

Lyapunov function $V_0(z,\xi)$ constructed in the preceding section will now be employed for controller design for two types of cascade systems obtained by augmenting the core system (Σ_0).

In the first type of cascade

$$(\Sigma_1)\left\{\begin{array}{l} \dot{z} = f(z) + \psi(z,\xi) + g_1(z,\xi)u \\ \dot{\xi} = a(\xi) + b(\xi)u \end{array}\right. \tag{5.4.1}$$

(Σ_0) appears as its uncontrolled part ($u = 0$). The *damping control* with V_0 is

$$u_1(z,\xi) = L_G V_0(z,\xi) = -\frac{\partial V_0}{\partial z}(z,\xi)g_1(z,\xi) - \frac{\partial V_0}{\partial \xi}(z,\xi)b(\xi) \tag{5.4.2}$$

where $G^T(z,\xi) = [g_1^T(z,\xi)\ b^T(\xi)]$. For the closed-loop system (Σ_1, u_1) the derivative $\dot{V}_0$ is

$$\dot{V}_0(z,\xi) = L_f W(z) + L_a U(\xi) - u_1^2(z,\xi) \le 0 \tag{5.4.3}$$

From Section 3.5.2 we know that, by construction, the system (Σ_1) with the output $y = u_1(z,\xi)$ is passive with the storage function V_0. Furthermore, if (Σ_1) with the output $y = u_1(z,\xi)$ is ZSD, then the feedback law $u = u_1(z,\xi)$ achieves GAS of the equilibrium $(z,\xi) = (0,0)$.

In the second type of cascade

$$(\Sigma_2)\left\{\begin{array}{l} \dot{z} = f(z) + \psi(z,\xi) + g_2(z,\xi,y)y \\ \dot{\xi} = a(\xi) + b(\xi)y \\ \dot{y} = u \end{array}\right. \tag{5.4.4}$$

(Σ_0) is the zero-dynamics subsystem with respect to the output y. For (Σ_2) the *feedback passivation* design of Section 5.4 achieves global stability and, under additional assumptions, global asymptotic stability. In the first step, the feedback transformation

$$u = -\frac{\partial V_0}{\partial z}g_2 - \frac{\partial V_0}{\partial \xi}b + v \tag{5.4.5}$$

renders the system passive from the new input v to the output y with the storage function

$$V_2(z,\xi,y) = V_0(z,\xi) + \frac{1}{2}y^2 \tag{5.4.6}$$

The additional feedback $v = -y$ results in the control law

$$u_2(z,\xi,y) = u_1(z,\xi,y) - y \tag{5.4.7}$$

where, with a slight abuse of notation, $u_1(z,\xi,y)$ denotes the expression (5.4.2) with $g_1(z,\xi)$ replaced by $g_2(z,\xi,y)$. The derivative $\dot{V}_2$ along the solutions of the closed-loop system (Σ_2, u_2) satisfies

$$\dot{V}_2 \leq L_f W(z) - L_a U(\xi) - y^2 \leq 0 \tag{5.4.8}$$

This guarantees global stabillity of the closed-loop system. Global asymptotic stability is achieved if the system (Σ_2) with input v and output y is ZSD. Assuming, without loss of generality, that $g_1(z,0) = g_2(z,0,0)$, we denote $u_1(z,0) = u_2(z,0,0) =: u_0(z)$. Then the global asymptotic stability of both (Σ_1, u_1) and (Σ_2, u_2) is achieved if $z = 0$ is the largest invariant set of $\dot{z} = f(z)$ contained in

$$E = \{z \in I\!R^{n_z} \mid L_f W(z) = 0; u_0(z) = 0\} \tag{5.4.9}$$

The next example illustrates the fact that, with the designs (5.4.2) and (5.4.7), higher-order terms of the function ψ can influence asymptotic stability only if they are linear in ξ, such as $z^2\xi$.

Example 5.27 *(Stabilization through higher-order terms)*
Consider the system

$$\begin{aligned} \dot{z} &= z\xi \\ \dot{\xi} &= -\xi + u \end{aligned} \tag{5.4.10}$$

which is in the form (Σ_1). Setting $u = 0$ and using $W(z) = z^2$, we obtain $V_0(z,\xi) = z_\infty^2 + \xi_2$. Taking the limit for $s \to \infty$ of the solution $\tilde{z}(s) = ze^{\xi(1-e^{-s})}$, we obtain the Lyapunov function

$$V_0(z,\xi) = z^2 e^{2\xi} + \xi^2$$

The control law (5.4.2) is then given by

$$u_1(z,\xi) = -\frac{\partial V_0}{\partial \xi} = -2(z^2 e^{2\xi} + \xi)$$

Because $u_1(z,0) = 0 \Rightarrow z = 0$, the set E in (5.4.9) is $z = 0$, which proves global asymptotic stability of the closed-loop system.

However, this design fails if the z-subsystem is controlled through the terms which are quadratic or higher-order in ξ. For the nonlinear system

$$\begin{aligned} \dot{z} &= \xi^3 \\ \dot{\xi} &= -\xi + u \end{aligned} \tag{5.4.11}$$

with $u = 0$, the solution $\tilde{z}(s) = z + \frac{\xi^3}{3}(1 - e^{-3s})$ results in the Lyapunov function

$$V_0(z, \xi) = (z + \frac{\xi^3}{3})^2 + \xi^2$$

The feedback control (5.4.2) is now given by

$$u_1(z, \xi) = -\frac{\partial V_0}{\partial \xi} = -2(z + \frac{\xi^3}{3})\xi^2 - 2\xi$$

and it does not achieve asymptotic stability, because $\xi = 0$ is an equilibrium manifold of the closed-loop system.

This design method fails, even though the system (5.4.11) is stabilizable. For example, the control law $u(z, \xi) = -z$ achieves global asymptotic stability as can be verified with the Lyapunov function $V(z, \xi) = \frac{z^2}{2} + \frac{\xi^4}{4}$. □

5.4.2 A structural condition for GAS and LES

We now give a GAS condition which can be verified before the design. It connects stabilizability of the Jacobian linearization of (Σ_1) with GAS of the closed-loop systems (Σ_1, u_1) and (Σ_2, u_2).

Assumption 5.28 *(Structural conditions for asymptotic stabilization)*
The subsystem Lyapunov functions $W(z)$ and $U(\xi)$ are locally quadratic, that is $\frac{\partial^2 W}{\partial z^2}(0,0) = \bar{W} > 0$, $\frac{\partial^2 U}{\partial \xi^2}(0) = \bar{U} > 0$. Furthermore, z can be partitioned into $z = (z_1, z_2)$ in such a way that Assumption 5.10 is satisfied and, for all $z = (0, z_2)$, the following holds:

$$\frac{\partial \psi}{\partial \xi}(z, 0) := M, \ g_1(z, 0) := g_0, \text{and } \frac{\partial W}{\partial z}(z) = z_2^T \bar{W}_2$$

where M and $\bar{W}_2$ are constant matrices, and g_0 is a constant vector.

Theorem 5.29 *(GAS and stabilizability of the Jacobian linearization)*
Under Assumption 5.28, (Σ_1, u_1) and (Σ_2, u_2) are globally asymptotically stable if the span of $\{\frac{\partial}{\partial z_2}\}$ lies in the stabilizable subspace of the Jacobian linearization of (Σ_1):

$$(\bar{\Sigma}_1) \begin{cases} \dot{z}_1 = F_1 z_1 + M_1 \xi + g_{01} u \\ \dot{z}_2 = F_{21} z_1 + F_2 z_2 + M_2 \xi + g_{02} u \\ \dot{\xi} = A\xi + b_0 u, \end{cases} \tag{5.4.12}$$

Proof : Using (5.4.3) and Assumption 5.10, it is sufficient to consider the invariant sets of $\dot{z}_2 = F_2 z_2$ in $E' = \{(z,\xi) = (0, z_2, 0) | \ u_0(z) = 0\}$. Using Assumption 5.28, we rewrite E' as

$$E' = \{(z,\xi) = (0,z_2,0)| \ - z_2^T \bar{W}_2 g_0 - \frac{\partial \Psi}{\partial z}(z,0) g_0 - \frac{\partial \Psi}{\partial \xi}(z,0) b_0 = 0\} \quad (5.4.13)$$

To show that $z_2 = 0$ attracts all solutions of $\dot{z}_2 = F_2 z_2$ in E', we use a state decomposition $(z,\xi) =: \zeta = (\zeta_u, \zeta_s)$ of $\bar{\Sigma}_1$ into its unstabilizable and stabilizable parts:

$$\dot{\zeta} = \bar{A}\zeta + \bar{b}u, \quad \bar{A} = \begin{pmatrix} \bar{A}_u & 0 \\ \bar{A}_{us} & \bar{A}_s \end{pmatrix}, \quad \bar{b} = \begin{pmatrix} 0 \\ \bar{b}_s \end{pmatrix}$$

Because $\frac{\partial^2 W}{\partial z^2}(0,0) > 0$, the Jacobian linearization of $\dot{z} = f(z)$ at 0 is Lyapunov stable, for otherwise $L_f W$ would not be nonpositive for all z. Hence, $(\bar{\Sigma}_1)$ is Lyapunov stable when $u \equiv 0$. Let $\bar{P} > 0$ satisfy $\bar{P}\bar{A} + \bar{A}^T \bar{P} \leq 0$. The control law $\bar{u} = -2\bar{b}^T \bar{P}\zeta$ results in

$$\begin{aligned} \dot{\zeta}_u &= \bar{A}_u \zeta_u \\ \dot{\zeta}_s &= \bar{A}_h \zeta_s + \bar{A}_1 \zeta_u, \quad A_h = A_s - 2\bar{b}_s \bar{b}_s^T \bar{P}_s \end{aligned} \quad (5.4.14)$$

with $\bar{P}_s$ being the positive definite submatrix of $\bar{P}$ corresponding to $\bar{A}_s$. Using the detectability of the pair $(\bar{b}_s^T, \bar{A}_s^T)$, we conclude that A_h is Hurwitz and, hence, any solution of (5.4.14) with $\zeta_u(0) = 0$ converges to zero. Because, by assumption, E' belongs to the stabilizable subspace of $(\bar{\Sigma}_1)$, for any initial condition in E' the solution of $(\bar{\Sigma}_1, \bar{u})$ converges to zero.

One particular choice for $\bar{P}$ results from the quadratic Lyapunov function

$$\bar{V}(z,\xi) = \frac{1}{2} z^T \bar{W} z + \bar{\Psi}(z,\xi) + \xi^T \bar{U} \xi \quad (5.4.15)$$

where, following Theorem 5.8,

$$\bar{\Psi}(z,\xi) = \int_0^\infty \bar{z}^T \bar{W} \ M \ \bar{\xi}(s) \ ds \quad (5.4.16)$$

Here $(\bar{z}(s), \bar{\xi}(s))$ is the solution of the uncontrolled system $(\bar{\Sigma}_1, u = 0)$ with the initial condition (z,ξ). The corresponding control law

$$\bar{u}(z,\xi) = -z^T \bar{W} g_0 - \frac{\partial \bar{\Psi}}{\partial z} g_0 - \frac{\partial \bar{\Psi}}{\partial \xi} b - 2\xi^T \bar{U} b_0 \quad (5.4.17)$$

achieves convergence of $z_2(t)$ to 0 along any solution of $(\bar{\Sigma}_1, \bar{u})$ starting in E'.

To complete the proof of the theorem, we will show that the control law $\bar{u}(z,\xi)$ restricted to the set $\{(z,\xi)|z_1 = 0, \xi = 0\}$ is equal to $u_0(z)$ and hence in E' both are equal to 0. Because of this, if there exists a solution of (Σ_1, u_1) which is contained in E' for all t, it is also a solution of $(\bar{\Sigma}_1, \bar{u})$, and therefore converges to zero.

In E', the control law (5.4.17) becomes

$$\bar{u}((0,z_2),0) = -z_2^T \bar{W}_2 g_0 - \frac{\partial \bar{\Psi}}{\partial z}(z,0)g_0 - \frac{\partial \bar{\Psi}}{\partial \xi}(z,0)b$$

By definition, $\Psi(z,0) = \bar{\Psi}(z,0) = 0$ for each z; it immediately follows that for all $z \in I\!R^{n_z}$

$$\frac{\partial \Psi}{\partial z}(z,0) = \frac{\partial \bar{\Psi}}{\partial z}(z,0) = 0 \tag{5.4.18}$$

Next note that, for each initial condition $(0, z_2, 0)$ and for all $s \geq 0$,

$$\tilde{z}_1(s) = \bar{z}_1(s) \equiv 0, \quad \tilde{\xi}(s) = \bar{\xi}(s) \equiv 0 \text{ and } \tilde{z}_2(s) = \bar{z}_2(s) = e^{F_2 s} z_2$$

Hence, for each initial condition $((0,z_2),0)$,

$$\frac{\partial \Psi}{\partial \xi}(z,0) = \int_0^\infty \frac{\partial^2 W}{\partial z^2}\frac{\partial \tilde{z}}{\partial \xi}\psi(\tilde{z}(s),0)\, ds + \int_0^\infty \frac{\partial W}{\partial z}\frac{\partial \psi}{\partial \xi}\frac{\partial \tilde{\xi}}{\partial \xi}(\tilde{z}(s),0)\, ds$$

The first term on the right hand side is 0 because $\psi(z,0) = 0$. Since $\frac{\partial \tilde{\xi}}{\partial \xi}(s)\Big|_{\xi=0} = e^{As}$, where $A = \frac{\partial a}{\partial \xi}(0)$, the second term becomes

$$\int_0^\infty \tilde{z}_2^T W_2 M e^{As}\, ds$$

which, using (5.4.16), is equal to $\frac{\partial \bar{\Psi}}{\partial \xi}((0,z_2),0)$. □

The set of conditions in Assumption 5.28, which allowed us to verify apriori the GAS property of the closed-loop system, restricted the form of ψ and g_1. These restrictions were introduced so that GAS can be concluded from the properties of the Jacobian linearization. That this is not always possible can be seen in the system

$$\begin{aligned}\dot{z} &= z\xi + \xi \\ \dot{\xi} &= -\xi + u\end{aligned}$$

which has controllable Jacobian linearization, but it is not stabilizible because $z = -1$ is an equilibrium of the z-subsystem for all ξ and u. The only condition of Theorem 5.29 that this system fails is that $\frac{\partial \psi}{\partial \xi}(z,0) = 1 + z$ is not constant. However, as one can see in the case of the system (5.4.10) in Example 5.27, the fact that the conditions of Theorem 5.29 are not satisfied does not mean that GAS is not achieved. Rather, it must be deduced from the Invariance Principle.

By Theorem 5.11 and Corollary 5.12, Assumption 5.28 also guarantees smoothness of the Lyapunov function and the control laws for (Σ_1) and (Σ_2). In addition, the following corollary shows that local exponential stability is achieved if the Jacobian linearization $(\bar{\Sigma}_1)$ is stabilizable. This additional property will be needed for the recursive designs in Chapter 6.

Corollary 5.30 *(LES when the Jacobian linearization is stabilizable)*
If (Σ_0) satisfies Assumptions 5.1, 5.2, and 5.28 and if the Jacobian linearization $(\bar{\Sigma}_1)$ is stabilizable, then in addition to being globally asymptotically stable, (Σ_1, u_1) and (Σ_2, u_2) are also locally exponentially stable.

Proof: We will show that the linearization of the control law (5.4.2) is a stabilizing feedback for the Jacobian linearization of (Σ_1). This implies that the Jacobian linearization of (Σ_1, u_1) is asymptotically stable and therefore exponentially stable.

First we write the approximations around $(z, \xi) = (0, 0)$ of the relevant functions by keeping the lowest order terms:

$$\begin{aligned} W(z) &= \tfrac{1}{2} z^T \bar{W} z + h.o.t., & U(\xi) &= \xi^T \bar{U} \xi + h.o.t. \\ \Psi(z,\xi) &= z^T \Psi_1 \xi + \xi^T \Psi_0 \xi + h.o.t., & \psi(z,\xi) &= M\xi + h.o.t. \\ g(z,\xi,y) &= g_0 + h.o.t, & b(\xi) &= b_0 + h.o.t \\ f(z) &= Fz + h.o.t., & a(\xi) &= A\xi + h.o.t \end{aligned}$$

where *h.o.t.* stands for "higher order terms." The linearization of the control law u_1 becomes

$$u_{1l} = z^T \bar{W} g_0 - \xi^T \Psi_1^T g_0 - z^T \Psi_1 b_0 - 2\xi^T \Psi_0 b_0 - 2\xi^T \bar{U} b_0 \tag{5.4.19}$$

For the Jacobian linearization of (Σ_1) we use the same construction as in Theorem 5.29 and design the linear control law

$$\bar{u} = z^T \bar{W} g_0 - \xi^T \Psi_{l1}^T g_0 - z^T \bar{\Psi}_1 b_0 - 2\xi^T \bar{\Psi}_0 b_0 - 2\xi^T \bar{U} b_0 \tag{5.4.20}$$

where the matrices $\bar{\Psi}_1$ and $\bar{\Psi}_0$ are obtained from

$$\bar{\Psi}(z,\xi) = \int_0^\infty \bar{z}^T(s) \bar{W} M \bar{\xi}(s)\, ds = z^T \bar{\Psi}_1 \xi + \xi^T \bar{\Psi}_2 \xi$$

As in Theorem 5.29, the control law (5.4.20) stabilizes the linearized system $(\bar{\Sigma}_1)$.

To prove that the control law (5.4.20) is identical to the linearization (5.4.19) we need to show that $\Psi_1 = \bar{\Psi}_1$ and $\Psi_0 = \bar{\Psi}_0$. This follows from the uniqueness of the solution of (5.3.24). □

5.4.3 Ball-and-beam example

The well known ball-and-beam example [35], shown in Figure 5.1, is described by

$$\begin{aligned} 0 &= \ddot{r} + G\sin\theta + \beta\dot{r} - r\dot{\theta}^2 \\ \tau &= (r^2+1)\ddot{\theta} + 2r\dot{r}\dot{\theta} + Gr\cos\theta \end{aligned} \tag{5.4.21}$$

where r is the position of the ball, θ is the angle of the beam, the control variable is the torque applied to the beam τ, G is the gravity ($G = 9.81$ for simulations), and $\beta > 0$ is the viscous friction constant ($\beta = 0.1$ for simulations).

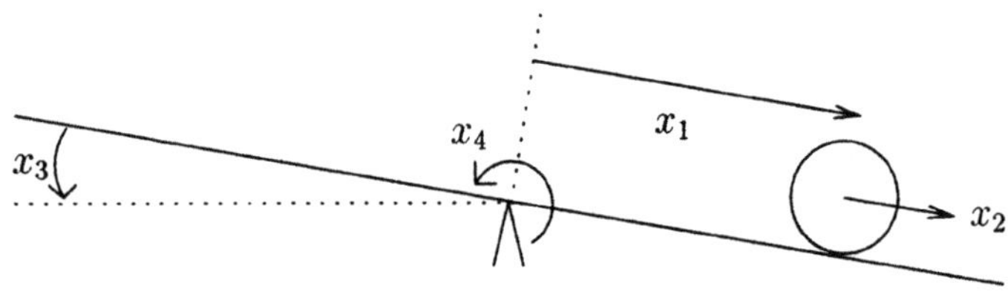

Figure 5.1: The ball-and-beam system.

If we apply the feedback transformation

$$\tau = 2r\dot{r}\dot{\theta} + Gr\cos\theta + k_1\theta + k_2\dot{\theta} + (r^2+1)u \tag{5.4.22}$$

and define $z_1 = r$, $z_2 = \dot{r}$, $\xi_1 = \theta$, $\xi_2 = \dot{\theta}$, we obtain the state equation

$$\begin{aligned} \dot{z}_1 &= z_2 \\ \dot{z}_2 &= -\beta z_2 - G\sin\xi_1 + z_1\xi_2^2 \\ \dot{\xi}_1 &= \xi_2 \\ \dot{\xi}_2 &= -k_1\xi_1 - k_2\xi_2 + u \end{aligned} \tag{5.4.23}$$

This system is in the cascade form (Σ_1). First, when $u = 0$, the ξ-subsystem is exponentially stable with the Lyapunov function $U(\xi) = \frac{1}{2}(k_1x_1^2 + x_2^2)$. Second, when $\xi = 0$, the z-subsystem is globally stable with the Lyapunov function $W(z) = \frac{1}{2}(\beta z_1 + z_2)^2 + \frac{1}{2}z_2^2$.

Because the conditions of Theorem 5.8 are satisfied, (5.4.23) is globally stable and

$$\Psi = \int_0^\infty (\beta\tilde{z}_1(s) + 2\tilde{z}_2(s))(-G\sin\tilde{\xi}_1(s) + \tilde{x}_1(s)\tilde{\xi}_2^2(s))\, ds$$

is the desired cross-term which makes $V_0 = W(x) + \Psi(x,\xi) + U(\xi)$ a Lyapunov function for the system (5.4.23).

The control law for (Σ_1) given by

$$u_1 = -\frac{\partial \Psi}{\partial \xi_2} - \frac{\partial U}{\partial \xi_2} = -\frac{\partial \Psi}{\partial \xi_2} - \xi_2 \tag{5.4.24}$$

achieves GAS as it can be shown by verifying that the consitions of Theorem 5.29 are satisfied. To evaluate u we need to compute $\frac{\partial \Psi}{\partial \xi_2}$. Among different mehods available for approximate evaluation of $\frac{\partial \Psi}{\partial \xi_2}$ we employ the on-line integration in faster than real time. For this we need $\tilde{\xi}$, $\tilde{z}$, and the variational variables $\frac{\partial \xi}{\partial \xi_2}$ and $\nu := \frac{\partial z}{\partial \xi_2}$.

We obtain

$$\tilde{\xi}(\tau) = e^{A\tau}\xi, \quad A = \begin{bmatrix} 0 & 1 \\ -k_1 & -k_2 \end{bmatrix} \tag{5.4.25}$$

and $\frac{\partial \xi}{\partial \xi_2} = [e^{A\tau}_{(12)} \; e^{A\tau}_{(22)}]^T$, where $e^{A\tau}_{(ij)}$ denotes the (i,j)-th entry of the matrix $e^{A\tau}$. The set of differential equations to be integrated on a sufficiently long interval $[0,T]$ is

$$\begin{aligned}
&\tfrac{d}{d\tau}\tilde{z}_1 = \tilde{z}_2 && \tilde{z}_1(0) = z_1 \\
&\tfrac{d}{d\tau}\tilde{z}_2 = -\beta\tilde{z}_2 - G\sin\tilde{\xi}_1 + \tilde{z}_1\tilde{\xi}_2^2 && \tilde{z}_2(0) = z_2 \\
&\tfrac{d}{d\tau}\nu_1 = \nu_2 && \nu_1(0) = 0 \\
&\tfrac{d}{d\tau}\nu_2 = -\beta\nu_2 - G\cos\tilde{\xi}_1\, e^{A\tau}_{(12)} + \nu_1\tilde{\xi}_2^2 + 2\tilde{z}_1\tilde{\xi}_2\, e^{A\tau}_{(22)} && \nu_2(0) = 0 \\
&\tfrac{d}{d\tau}\frac{\partial \Psi^*}{\partial \xi_2} = (\beta\nu_1 + 2\nu_2)(G\sin\tilde{\xi}_1 + \tilde{z}_1\tilde{\xi}_2^2) + (\beta\tilde{z}_1 + 2\tilde{z}_2) \\
&\qquad \times(G\cos\tilde{\xi}_1\, e^{A\tau}_{(12)} + \nu_1\tilde{\xi}_2^2 + 2\tilde{z}_1\tilde{\xi}_2\, e^{A\tau}_{(22)}) && \frac{\partial \Psi^*}{\partial \xi_2}(0) = 0
\end{aligned} \tag{5.4.26}$$

where, $\frac{\partial \Psi^*}{\partial \xi_2}(\tau)$, defined in Section 5.2.3, is an approximation of $\frac{\partial \Psi}{\partial \xi_2}$ obtained by truncation of the integral at the time τ. By truncating at $\tau = T$ we obtain the approximate control law

$$u_{app} = -\frac{\partial \Psi^*}{\partial \xi_2}(T; z, \xi) - \xi_2 \tag{5.4.27}$$

which is used in the simulations.

For the computer simulations, we have placed both eigenvalues of A at -2 with $k_1 = 4, k_2 = 4$. Thus $\|\tilde{\xi}(\tau)\|$ and $\|\psi(\tilde{z}(\tau), \tilde{\xi}(\tau))\|$ decay as $\tau e^{-2\tau}$. Based on this rate of decay we have set $T = 10$ seconds. A response of the closed-loop system from the initial condition $(1, 0, -1.57, 0)$ is shown in Figure 5.2. This initial condition corresponds to the upright beam with the ball at 1 unit (meter) distance below the pivot. The controller achieves an excellent control of the overshoot of the ball, but eventually the convergence of the ball position becomes slow. This is a consequence of the application of damping control,

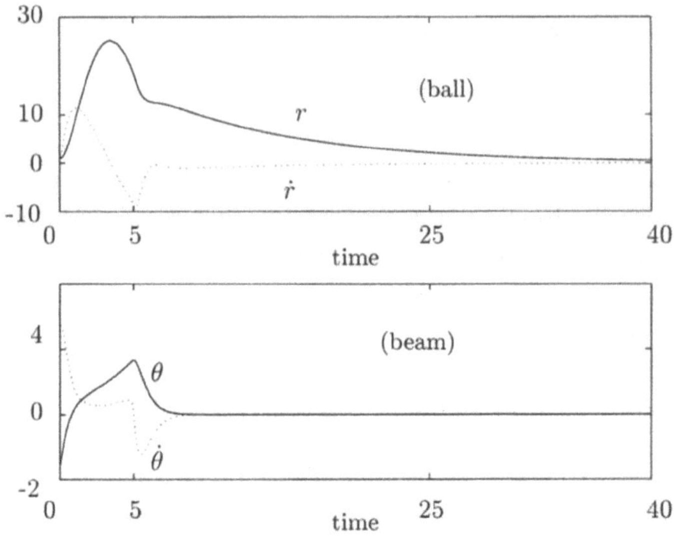

Figure 5.2: Typical transient of the designed ball-and-beam system

which, as shown in Example 3.39, prevents us from assigning a desired rate of convergence. When the states are sufficiently small so that the Jacobian linearization determines the response, the behavior of the ball position and velocity becomes dominated by the slow mode at $-\beta = -0.1$.

5.5 Lyapunov functions for adaptive control

When a nonlinear cascade system depends on an unknown parameter $\theta \in I\!R^p$, we construct the cross-term Ψ and the Lyapunov function V_0 to be parameterized by θ. Our goal is to use this construction in the adaptive controller design. As we shall see, this approach applies to systems for which other adaptive control design methods cannot be applied. A benchmark problem of this kind, proposed in [54], is the third order system

$$\begin{aligned} \dot{x}_1 &= x_2 + \theta x_3^2 \\ \dot{x}_2 &= x_3 \\ \dot{x}_3 &= u \end{aligned} \tag{5.5.1}$$

with a scalar unknown parameter θ. This system is a representative of a larger class of nonlinear systems with unknown parameters for which we will now solve the adaptive stabilization problem. The systems in this class are of the form

$$\begin{aligned} \dot{z} &= Fz + H(\xi)z + \psi^T(\xi)\theta + g^T(z,\xi,y)\theta\, y \\ \dot{\xi} &= A\xi + b^T(z,\xi,y)\theta\, y \\ \dot{y} &= \varphi^T(z,\xi,y)\theta + u \end{aligned} \tag{5.5.2}$$

where $H(0) = 0$, $\psi(0) = 0$, $z \in R^{n_z}$, $\xi \in R^{n_\xi}$, $y \in R$, $u \in R$, and $\theta \in R^p$. Output y and input u are assumed to be scalars for notational convenience. All the results apply when y and u are m-vectors. The main assumption about the cascade (5.5.2) is about the stability properties of its subsystems.

Assumption 5.31 *(Stability of subsystems)*
In (5.5.2) the matrix A is Hurwitz and $\dot{z} = Fz$ is stable, that is, there exist positive definite matrices P_F, P_A, Q_A and a positive semidefinite matrix Q_F such that

$$\begin{aligned} F^T P_F + P_F F &= -Q_F \\ A^T P_A + P_A A &= -Q_A \end{aligned}$$

If the parameters θ were known, the above assumption would make the system (5.5.2) a special case of the augmented cascade (Σ_2).

It is important to observe that for some $\theta \in R^p$ the system (5.5.2) may fail to be controllable or even stabilizable. In general, its stabilizability is restricted to $\theta \in \Omega_S \subset \mathbb{R}^p$. However, even when $\theta \notin \Omega_S$ the adaptive controller which will be designed below will achieve boundedness of all the signals. A set $\Pi \subset \Omega_S$, will be characterized in which the adaptive controller solves the state regulation problem, that is, forces all the states to converge to 0.

5.5.1 Parametric Lyapunov Functions

With the output y the zero-dynamics subsystem of (5.5.2) is

$$\begin{aligned} \dot{z} &= Fz + H(\xi)z + \psi^T(\xi)\theta \\ \dot{\xi} &= A\xi \end{aligned} \tag{5.5.3}$$

This system is in the form (Σ_0) and Assumptions 5.1 and 5.2 are satisfied. Hence the construction of the cross-term from Section 6.2 is applicable and we get

$$\Psi(z,\xi,\theta) = \int_0^\infty 2\tilde{z}^T(s;(z,\xi),0)P_F[H(\tilde{\xi}(s;\xi,0))\,\tilde{z}(s;(z,\xi),0) + \psi^T(\tilde{\xi}(s;\xi,0))\theta]\,ds \tag{5.5.4}$$

This cross-term is used in the Lyapunov function

$$V_0(z,\xi,\theta) = z^T P_F z + \Psi(z,\xi,\theta) + \xi^T P_A \xi \tag{5.5.5}$$

Its derivative along the solutions of (5.5.3) is

$$\dot{V}_0 = -z^T Q_F z - \xi^T Q_A \xi \leq 0 \tag{5.5.6}$$

Our design of the adaptive controller requires that, possibly after a reparameterization, the control law be a linear function of θ. This will be the case when the Lyapunov function $V_0(z, \xi, \theta)$ is a polynomial function of θ. The above construction satisfies this requirement because, as we now show, the cross-term Ψ is a polynomial of degree 2 in θ.

The solution of the system (5.5.3), with the initial condition (z, ξ) at time $s = 0$, is

$$\begin{aligned}
\tilde{z}(s) &= \Phi_\xi(s,0)z + \int_0^s \Phi_\xi(s,\tau)\psi^T(\tilde{\xi}(\tau))\theta \, d\tau =: \Phi_\xi(s,0)z + J^T(\xi, s)\theta \\
\tilde{\xi}(s) &= e^{As}\xi
\end{aligned} \tag{5.5.7}$$

where $\Phi_\xi(s,t)$ satisfies

$$\dot{\Phi}_\xi(s,t) = [F + H(\tilde{\xi}(s))]\Phi_\xi(s,t), \quad \Phi_\xi(t,t) = I$$

Substituting $\tilde{z}$ and $\tilde{\xi}$ into (5.5.4) we obtain Ψ as a quadratic polynomial in θ:

$$\begin{aligned}
\Psi(z,\xi) = \int_0^\infty & 2(\theta^T J(\xi,s) + z^T \Phi_\xi^T(s,0)) \times \\
& \times P_F [H(\tilde{\xi}(s))(\Phi_\xi(s,0)x + J^T(\xi,s)\theta) + \psi^T(\tilde{\xi}(s))\theta] \, ds \\
=: z^T \Psi_0(\xi) z &+ z^T \Psi_1(\xi)\theta + \theta^T \Psi_2(\xi)\theta
\end{aligned} \tag{5.5.8}$$

The coefficient matrices in the above expressions are

$$\begin{aligned}
\Psi_0(\xi) &= 2\int_0^\infty \Phi_\xi^T P_F H \Phi_\xi \, ds \\
\Psi_1(\xi) &= 2\int_0^\infty \Phi_\xi^T [(P_F H + H^T P_F) J^T + P_F \psi^T] \, ds \\
\Psi_2(\xi) &= 2\int_0^\infty J P_F [H J^T + \psi^T] \, ds
\end{aligned} \tag{5.5.9}$$

With this expression for the cross-term the Lyapunov function (5.5.5) becomes

$$V_0(z,\xi,\theta) = z^T P_F x + z^T \Psi_0(\xi) z + z^T \Psi_1(\xi)\theta + \theta^T \Psi_2(\xi)\theta + \xi^T P_A \xi \tag{5.5.10}$$

The cascade (5.5.3) satisfies Assumption 5.10 and hence Theorem 5.11 guarantees that Ψ_i's are differentiable as many times as the functions $H(\xi)$ and $\psi(\xi)$ in (5.5.2).

5.5.2 Control with known θ

Let us first assume that the parameter vector θ is known and design a controller which achieves global stability of the system (5.5.2). Because the cascade (5.5.2) is in the form (Σ_2), we employ the Lyapunov function $V_2 = V_0 + \frac{1}{2}y^2$ and the feedback passivating control law (5.4.7) which in this case is given by

$$\begin{aligned} u(z,\xi,y,\theta) &= -y - \varphi(z,\xi,y)^T\theta - \frac{\partial V_0}{\partial z} g^T(z,\xi,y)\theta - \frac{\partial V_0}{\partial \xi} b^T(z,\xi,y)\theta \\ &= -y - [\varphi^T + 2z^T(P_F + \Psi_0)g^T + 2\xi^T P_A b^T + z^T \sum_{i=1}^{n_z} z_i \frac{\partial \Psi_0^{(i)}}{\partial \xi} b^T]\theta \\ &\quad -\theta^T[\Psi_1^T g^T + \sum_{i=1}^{n_z} z_i \frac{\partial (\Psi_1^T)^{(i)}}{\partial \xi} b^T]\theta - \theta^T \sum_{i=1}^{p} \theta_i \frac{\partial \Psi_2^{(i)}}{\partial \xi} b^T \theta \end{aligned} \tag{5.5.11}$$

where, as in Section 5.3, the superscript (i) denotes the i-the column of the corresponding matrix. Because

$$\dot{V}_2 = -z^T Q_F z - \xi^T Q_A \xi - y^2 \le 0 \tag{5.5.12}$$

we conclude that without any restriction on θ, the above control law achieves global stability and the regulation of ξ and y, that is $\xi \to 0$ and $y \to 0$ as $t \to \infty$. This is true for all θ, regardless of a possible lack of stabilizability. To prove GAS of the closed-loop system (5.5.2), (5.5.11), we need one additional assumption constraining θ. The set Ω_S of θ for which (5.5.2) is stabilizable is very difficult to characterize. Instead, we apply the Invariance Principle and conclude that the solutions of the closed-loop system converge to E, the largest invariant set where $\dot{V}_2 = 0$. This motivates the following definition.

Definition 5.32 *(Addmissible set $\mathcal{P}$)*
Addmissible set $\mathcal{P}$ is the set of all $\theta \in R^p$ for which $z = 0$ is the only solution of the equations

$$\dot{z} = Fz, \quad z^T Q_F z \equiv 0, \quad \chi(z,\theta) \equiv 0 \tag{5.5.13}$$

where $\chi(z,\theta) = \varphi^T(z,0,0)\theta + u(z,0,0,\theta)$. □

One important point, later illustrated in Example 5.41, is that the equations (5.5.13) are in the closed form even when the closed-form expression for the control law is not available.

Proposition 5.33 *(GAS with θ known)*
If $\theta \in \mathcal{P}$ then the closed-loop system (5.5.2), (5.5.11) is globally asymptotically stable. □

Example 5.34 *(Benchmark problem with θ known)*
To design a controller which achieves global asymptotic stability for the benchmark system (5.5.1) we first transform (5.5.1) into the form (5.5.2) introducing the output $y = x_2 + x_3$:

$$\begin{aligned}\dot{x}_1 &= x_2 + \theta x_2^2 + \theta(y - 2x_2)y \\ \dot{x}_2 &= -x_2 + y \\ \dot{y} &= -x_2 + y + u\end{aligned} \tag{5.5.14}$$

The zero-dynamics subsystem is

$$\begin{aligned}\dot{x}_1 &= x_2 + \theta x_2^2 \\ \dot{x}_2 &= -x_2\end{aligned}$$

To construct the Lyapunov function (5.5.5) we let $P_F = 1$ and using (5.2.48) compute

$$\Psi(x_1, x_2, \theta) = \int_0^\infty \tilde{x}_1(s)(\tilde{x}_2(s) + \theta\tilde{x}_2^2(s))\, ds = -x_1^2 + (x_1 + x_2 + \frac{1}{2}\theta x_2^2)^2$$

Thus, our Lyapunov function for the zero dynamics is

$$V_0(x_1, x_2, \theta) = (x_1 + x_2 + \frac{1}{2}\theta x_2^2)^2 + x_2^2$$

When θ is known, the control law (5.5.11) is implementable and is given by

$$u = -2y - x_1 - x_2 - \theta[(x_1 + x_2)(y - x_2) + \frac{1}{2}x_2^2] - \frac{1}{2}\theta^2(y - x_2)x_2^2 \tag{5.5.15}$$

It achieves boundedness of all the states and the convergence of x_2 and y to 0. To prove assymptotic stability we need to characterize the set $\mathcal{P}$ via (5.5.13). In this case $F = 0$, $Q_F = 0$, $\varphi = 0$, so that the only nontrivial equation in (5.5.13) is

$$\chi(x_1, \theta) = u(x_1, 0, 0, \theta) = -x_1 = 0$$

Becasue $x_1 = 0$ is the only solution of (5.5.13) for all θ we have $\mathcal{P} = \mathbb{R}$. Thus the global asymptotic stability is achieved without any restriction on θ. □

Example 5.35 *(Lack of stabilizability)*
The benchmark system (5.5.1) is stabilizable for any value of the parameter θ. That this is not always the case is illustrated by

$$\begin{aligned}\dot{z} &= \xi + \theta_1 y \\ \dot{\xi} &= -\xi + \theta_2 y \\ \dot{y} &= u\end{aligned} \tag{5.5.16}$$

This linear system with two parameters is in the form (5.5.2). From $\frac{d}{dt}(z + \xi) = (\theta_1 + \theta_2)y$ it is obvious that (5.5.16) is not stabilizable for $\theta_1 + \theta_2 = 0$. Nevertheless, we are able to achieve boundedness and regulation of ξ and y for any θ_1 and θ_2.

The zero-dynamics subsystem is

$$\begin{aligned}\dot{z} &= \xi \\ \dot{\xi} &= -\xi\end{aligned}$$

With $P_F = 1$ the Lyapunov function (5.5.5) is

$$V_0(z, \xi) = z^2 + 2z\xi + 2\xi^2$$

and the control law

$$u = -2(\theta_1 + \theta_2)z - 2(\theta_1 + 2\theta_2)\xi - y$$

achieves global stability and convergence of ξ and y to 0.

In this example the only nontrivial equation in (5.5.13) is

$$\chi(z, \theta) = u(z, 0, 0, \theta) = -2z(\theta_1 + \theta_2) = 0 \tag{5.5.17}$$

If $z = 0$ is to be the only solution of (5.5.17) we must restrict θ to belong to $\mathcal{P} = \{(\theta_1, \theta_2) \in \mathbb{R}^2 : \theta_1 + \theta_2 \neq 0\}$. Thus, according to Proposition 5.33, if $\theta \in \mathcal{P}$ then the global asymptotic stability of the closed-loop system is also achieved. □

5.5.3 Adaptive Controller Design

The control law (5.5.11) is a cubic polynomial in θ. To design an adaptive version of (5.5.11) we resort to overparameterization by introducing a new parameter ϑ_l for every product of θ_i's which appears in (5.5.11), such as $\vartheta_{l_1} := \theta_i\theta_j$, $\vartheta_{l_2} := \theta_i\theta_j\theta_k$, etc. In this way we have defined the augmented vector $\Theta^T = [\theta^T \; \vartheta^T] \in \mathbb{R}^q$ where $q \leq \frac{1}{6}(p^3 + 6p^2 + 11p)$.

We rewrite the control law (5.5.11) as

$$u(z, \xi, y, \Theta) = -y - w^T(z, \xi, y)\Theta \tag{5.5.18}$$

where the function w can be derived from (5.5.11). Because the parameter vector is not known, we replace it with an estimate $\hat{\Theta}^T \triangleq [\hat{\theta}^T \; \hat{\vartheta}^T]$ and obtain the "certainty equivalence" control law

$$u(z, \xi, y, \hat{\Theta}) = -y - w^T(z, \xi, y)\hat{\Theta} \tag{5.5.19}$$

Next we modify the Lyapunov function V_2 to include the parameter estimation error $\tilde{\Theta} := \Theta - \hat{\Theta}$:

$$V_e(z, \xi, y, \theta, \tilde{\Theta}) = V_0(z, \xi, \theta) + \frac{1}{2}y^2 + \frac{1}{2}\tilde{\Theta}^T\tilde{\Theta} \tag{5.5.20}$$

Its time-derivative along the solutions of the closed-loop system is

$$\dot{V}_e(z, \xi, y, \theta, \tilde{\Theta}) = \dot{V}_0(z, \xi, \theta) + yu(z, \xi, y, \hat{\Theta}) + \tilde{\Theta}^T\dot{\tilde{\Theta}}$$

Adding and subtracting $yu(z, \xi, y, \Theta)$ from $\dot{V}_e$ and using (5.5.12) we obtain

$$\dot{V}_e = -z^T Q_F z - \xi^T Q_A \xi - y^2 + yw^T(z, \xi, y)\tilde{\Theta} + \tilde{\Theta}^T\dot{\tilde{\Theta}}$$

The parameter update law which eliminates the parameter error terms from $\dot{V}_e$ is

$$\dot{\hat{\Theta}} = w(z, \xi, y)y \tag{5.5.21}$$

The remaining expression for V_e is negative semidefinite:

$$\dot{V}_e = -z^T Q_F z - \xi^T Q_A \xi - y^2 \leq 0 \tag{5.5.22}$$

It follows by the standard argument that the adaptive controller consisting of the control law (5.5.19) and the parameter update law (5.5.21) achieves boundedness of all the states and regulation of ξ and y. Again it is not required that the system (5.5.2) be stabilizable.

Proposition 5.36 *(Stability of adaptive system)*
For any $\theta \in R^p$ the system (5.5.2) with the adaptive controller (5.5.19), (5.5.21) is globally stable and ξ and y converge to 0 as $t \to \infty$.

Additional properties can be deduced by analyzing E', the largest invariant set of the closed-loop system (5.5.2), (5.5.19), (5.5.21) where $\dot{V}_e = 0$. In general, E' is different from E and the analysis is more difficult than in Proposition 5.33.

We still want to examine whether the condition $\theta \in \mathcal{P}$ can guarantee the regulation of z, possibly with a modified adaptive controller. We will do it in two steps. First we remove the dependence of E' on $\hat{\vartheta}$ by introducing the following assumption:

Assumption 5.37 *(Restiction on uncertainties in E)*
$\varphi(z, 0, 0) = 0$ and either $\frac{\partial \psi}{\partial \xi}(0) = 0$ or $b(z, 0, 0) = 0$.

It will be clear in the sequel that this assumption is needed only for the uncertain parts of φ, ψ, and b. For example, if instead of $\psi^T(\xi)\theta$ we had $\psi_0(\xi) + \psi_1(\xi)^T\theta$, then the assumption applies only to ψ_1.

Proposition 5.38 *(Higher-order terms in θ)*
If Assumption 5.37 is satisfied then

$$\frac{\partial u}{\partial \hat{\vartheta}}(z, 0, 0, \hat{\Theta}) = 0$$

that is, the certainty equivalence control law (5.5.19) is independent of $\hat{\vartheta}$ when $\xi = 0$, $y = 0$.

Proof: Recall that ϑ stands for the terms quadratic and cubic in θ in the control law (5.5.11). It suffices to show that the functions multiplying these nonlinear terms vanish when $\xi = 0$, $y = 0$. The portion of the control law (5.5.11) which is nonlinear in θ is

$$\theta^T[\Psi_1^T g^T + \sum_{i=1}^{n_x} z_i \frac{\partial(\Psi_1^T)^{(i)}}{\partial \xi} b^T]\theta + \theta^T \sum_{i=1}^{p} \theta_i \frac{\partial \Psi_2^{(i)}}{\partial \xi} b^T \theta \tag{5.5.23}$$

Under Assumption 5.37 this expression vanishes when $\xi = 0$, $y = 0$. To see this, note that $\Psi_1(\xi)$ is at least linear in ξ because J and h are both at least linear in ξ. Also $\frac{\partial \Psi_2^{(i)}}{\partial \xi}(0) = 0$ because Ψ_2 is at least quadratic in ξ. Finally, from Assumption 5.37, either $\frac{\partial \Psi_1^{(i)}}{\partial \xi}(0) = 0$ (when $\frac{\partial \psi}{\partial \xi}(0) = 0$) or $b^T(z, 0, 0) = 0$. Therefore, when $\xi = 0$ and $y = 0$, the terms nonlinear in θ vanish and the adaptive control law (5.5.19) depends only on $\hat{\theta}$.

□

Proposition 5.38 shows that the set E' is independent of $\hat{\vartheta}$. Note that, $y = 0$ in E', in which case the estimate $\hat{\theta}$ is a constant vector denoted by $\bar{\theta}$. To achieve the regulation of z, we will make sure that $\bar{\theta} \in \mathcal{P}$. To this end we introduce a projection in the parameter update law which will keep $\hat{\theta} \in \Pi$, a closed and convex subset of $\mathcal{P}$ which need not be bounded.

Conformal with the partition of Θ into θ and ϑ, we let $w^T(z, \xi, y) = [w_1^T(z, \xi, y)\ w_2^T(z, \xi, y)]$. Then we modify the update law (5.5.21) as

$$\begin{aligned} \dot{\hat{\theta}} &= \mathrm{Proj}_\Pi\{w_1(z, \xi, y)y\} \\ \dot{\hat{\vartheta}} &= w_2(z, \xi, y)y \end{aligned} \tag{5.5.24}$$

where $\mathrm{Proj}_\Pi\{\cdot\}$ is the standard projection operator (c.f. Section 4.4 in [40]) which guarantees that the vector $\hat{\theta}(t)$ remains in the set Π.

Theorem 5.39 *(Adaptive regulation)*
If Assumption 5.37 is satisfied and if the closed and convex set $\Pi \subset \mathcal{P}$ contains θ, then the system (5.5.2) with the adaptive controller (5.5.19), (5.5.24) is globally stable and z, ξ, and y converge to 0 as $t \to \infty$.

Proof: An important property of the parameter update law (5.5.24) is that, if the set Π contains the true parameter vector θ, then

$$\tilde{\theta}^T \mathrm{Proj}_\Pi\{w_1(z,\xi,y)y\} \geq \tilde{\theta}^T w_1(z,\xi,y)y$$

Using this inequality we obtain

$$\begin{aligned} \dot{V}_e &= -z^T Q_F z - \xi^T Q_A \xi - y^2 + y w_1^T \tilde{\theta} - \tilde{\theta}^T \mathrm{Proj}_\Pi\{w_1(z,\xi,y)y\} \\ &\leq -z^T Q_F z - \xi^T Q_A \xi - y^2 \leq 0 \end{aligned}$$

By Theorem 2.20 the states of the system are uniformly bounded and y and ξ converge to 0.

Now we examine the largest invariant set E' where the following must hold:

1. $\xi = 0,\ y = 0,\ \dot{\hat{\Theta}} = 0$
2. $\hat{\theta}(t) = \bar{\theta} \in \Pi \subset \mathcal{P}$
3. $\dot{z} = Fz, \quad z^T Q_F z = 0$
4. $0 \equiv \dot{y} = u(z, 0, 0, \Theta)$.

The last item follows from Assumption 5.37 because $\dot{y} = \varphi^T\theta + u$ and φ^T vanishes when $\xi = 0$, $y = 0$. By Proposition 5.38, $u(z,0,0,\hat{\Theta})$ is independent of $\hat{\vartheta}$. Thus, $u(z,0,0,\hat{\Theta}) = \chi(z,\bar{\theta})$. Since $\bar{\theta} \in \mathcal{P}$, $z = 0$ is the only solution which satisfies items 3 and 4, which proves that the regulation of z is achieved. □

Example 5.40 *(Adaptive benchmark problem)*
Returning to the benchmark system (5.5.1) we now allow that the parameter θ be unknown. Our adaptive control law is a certainty equivalence version of the control law (5.5.15) with θ replaced by $\hat{\theta}$ and θ^2 replaced by an additional estimate $\hat{\vartheta}$:

$$u = -2y - x_1 - x_2 - \hat{\theta}[(x_1 + x_2)(y - x_2) + \frac{1}{2}x_2^2] - \frac{1}{2}\hat{\vartheta}(y - x_2)x_2^2 \qquad (5.5.25)$$

In Example 5.34 we have shown that for this problem $\mathcal{P} = \mathbb{R}$. Because the purpose of the projection in (5.5.24) was to keep $\hat{\theta}$ in $\mathcal{P}$, we conclude that in this case it is not needed. Thus, the parameter update law is

$$\begin{aligned} \dot{\hat{\theta}} &= y[(x_1 + x_2)(y - x_2) + \tfrac{1}{2}x_2^2] \\ \dot{\hat{\vartheta}} &= y(y - x_2)x_2^2 \end{aligned} \qquad (5.5.26)$$

By Proposition 5.36 the adaptive controller (5.5.25), (5.5.26) achieves boundedness of x_1, x_2, y, $\hat{\theta}$, $\hat{\vartheta}$ and the regulation of x_2 and y. From (5.5.2) and (5.5.14) we conclude that Assumption 5.37 holds because $\varphi = -x_2 + y$ and $b = 1$ do not include parametric uncertainties. Since $\mathcal{P} = \mathbb{R}$, Theorem 5.39 establishes that the regulation of x_1 is also achieved. □

Example 5.41 *(Adaptive design features)*
Other prominent features of this adaptive design will become apparent on the following nonlinear system

$$\begin{aligned} \dot{z} &= z\xi + \theta_1 \sin^2 \xi + z^2 y^2 \\ \dot{\xi} &= -\xi + \theta_2 y \\ \dot{y} &= \theta_3 a(y) z^2 + u \end{aligned} \tag{5.5.27}$$

Selecting $W(z) = z^2$, the cross-term Ψ in the Lyapunov function for the zero-dynamics subsystem

$$\begin{aligned} \dot{z} &= z\xi + \theta_1 \sin^2 \xi \\ \dot{\xi} &= -\xi \end{aligned} \tag{5.5.28}$$

is

$$\Psi(z,\xi) = \int_0^\infty 2\theta_1 \tilde{z}\ \sin^2 \tilde{\xi}\ ds = -z^2 + \left(ze^\xi + \theta_1 \rho(\xi)\right)^2 \tag{5.5.29}$$

where

$$\rho(\xi) = \int_0^\xi \frac{e^\mu}{\mu}\ \sin^2 \mu\ d\mu$$

Even though this integral cannot be evaluated in closed form, it globally defines an analytic function which can be either precomputed or generated on-line by integration.

With (5.5.29) the Lyapunov function for the zero-dynamics subsystem (5.5.28) is

$$V(z,\xi) = \left(ze^\xi + \theta_1 \rho(\xi)\right)^2 + \xi^2 \tag{5.5.30}$$

which yields the control law

$$\begin{aligned} u(z,\xi,\theta) &= -\theta_3 a(y) z^2 - y - \tfrac{\partial V}{\partial z} z^2 y - \tfrac{\partial V}{\partial \xi}\theta_2 \\ &= -\theta_3 a(y) z^2 - y - 2e^\xi \left(ze^\xi + \theta_1 \rho(\xi)\right)\left(z^2 y + \theta_2 z + \theta_1\theta_2 \tfrac{\sin^2 \xi}{\xi}\right) - 2\theta_2 \xi \end{aligned} \tag{5.5.31}$$

The equations (5.5.13) reduce to

$$\dot{z} = 0, \quad \chi(z,0,0,\theta) = -2\theta_2 z^2$$

and the admissible set is

$$\mathcal{P} = \{\theta \in \mathbb{R}^3 : \theta_2 \neq 0\} \tag{5.5.32}$$

Note that for $\theta_2 = 0$ the system (5.5.27) is not stabilizable.

Case 1: If $a(0) = 0$, Assumption 5.37 is satisfied, so the adaptive controller (5.5.19), (5.5.24) achieves boundedness of the signals and regulation of (z, ξ, y) provided that we can find an appropriate closed and convex set Π for the projection. From (5.5.32) we conclude that θ_2 is the only parameter which requires projection. For this we need to know the sign of θ_2 and a lower bound on $|\theta_2|$. The resemblance to linear adaptive systems, where the sign of the high-frequency gain is a standard assumption, is not accidental. The projection in our adaptive design serves the same purpose: to avoid the set of parameter values for which the system cannot be stabilized.

Case 2: If $a(0) \neq 0$ Assumption 5.37 is not satisfied. Nevertheless, through Proposition 5.36, the adaptive controller (5.5.19), (5.5.21) guarantees boundedness of the signals and regulation of ξ and y. To guarantee the regulation of z we need that

$$(\theta_3 - \hat{\theta}_3)a(0) - \theta_2 \neq 0$$

The above expression involves the estimates of θ (and, in general, may also involve $\hat{\vartheta}$) and is less helpful in the determination of the projection set Π. With Assumption 5.37 we avoided this difficulty and determined the set Π using (5.5.13). □

5.6 Summary

For the control design methods presented in this book it is crucial that a Lyapunov function be known. In this chapter we have developed methods for its construction. We have restricted our attention to a cascade which consists of a stable z-subsystem, GAS and LES ξ-subsystem, and an interconnection term ψ.

We have presented two basic constructions: composite Lyapunov functions and Lyapunov functions with a cross-term. The first construction method is based on a specific choice of nonlinear weights, so that the indefinite term in $\dot{V}$ is dominated by the negative terms. For this, the interconnection $\psi(z, \xi)$ must be of second or higher order in ξ. A change of coordinates needed to remove the terms linear in ξ exists when $\dot{z}$ is linear in z and a nonresonance condition is satisfied. In this case the change of coordinates is obtained by solving a set of linear algebraic equations.

To encompass a wider class of systems, we have constructed a Lyapunov function with a cross-term Ψ. We have proven that the cross-term is differentiable and that the resulting Lyapunov function is positive definite and radially unbounded. In special cases, the cross-term Ψ can be computed explicitely. In general, numerical computation is required. An alternative approach is to evaluate $\frac{\partial\Psi}{\partial z}(z,\xi)$ and $\frac{\partial\Psi}{\partial\xi}(z,\xi)$, and hence the control law, in real time at any point (z,ξ), without the need to precompute and store the data. This approach is straightforward to implement, as illustated by the ball-and-beam example.

To avoid the computational burden associated with the evaluation of the cross-term and its partial derivatives, a relaxed cross-term construction is developed, which, in contrast to the composite Lyapunov construction, is not restricted by the nonresonance conditions.

The Lyapunov constructions developed for the basic (Σ_0) are employed to design stabilizing control laws for more general systems obtained by augmenting the cascade. The ZSD property required by the control laws depends on the the cross-term. We have given structural conditions under which the ZSD property is a priori guaranteed.

We have also presented a construction of the parameterized cross-term for adaptive control of cascades with unknown parameters. As illustrated by the benchmark problem, our adaptive design applies to nonlinear systems which are not feedback linearizable.

5.7 Notes and references

Lyapunov constructions for cascades consisting of a GS subsystem and a GAS/LES subsystem have appeared recently in the work by Mazenc and Praly [75] and by the authors [46].

Mazenc and Praly pusrsued the composite approach and introduced the exact and relaxed decoupling change of coordinates presented in Section 5.3.

The cross-term construction presented in Section 5.2 was introduced by the authors in [46]. It removes nonresonance and linearity assumptions of the composite approach.

The extension of the cross-term construction to the adaptive case and the solution of the adaptive benchmark problem in Section 5.5 are due to the authors [47].

Chapter 6

Recursive designs

Feedback passivation designs, which have been successful for the cascade structures in Chapters 4 and 5, will now be extended to larger classes of nonlinear systems. The common idea of the two main recursive procedures in this chapter, *backstepping* and *forwarding*, is to apply a passivation design to a small part of the system, and then to reapply it step-by-step by augmenting the subsystem at each step. The design is completed when the augmentations recover the whole system.

Backstepping and forwarding, presented in Section 6.1 and 6.2, complement each other: backstepping is applicable to the lower-triangular, and forwarding to the upper-triangular systems. Backstepping employs an analytic expression for the time-derivative of the control law designed at the preceding step. In forwarding, this operation is integration. The two procedures proceed in reverse order. Backstepping starts with the system equation (integrator) which is the farthest from the control input and reaches the control input at the last step. Forwarding, on the other hand, starts from the input and moves forward.

Both procedures construct a passivating output and a storage function to be used as a Lyapunov function. They accomplish the same task by removing two different obstacles to passivation: backstepping removes the relative degree one obstacle, while forwarding removes the minimum phase obstacle.

In addition to backstepping and forwarding, we also present a family of simplified designs. Instead of analytically implementing the derivatives used in backstepping, a high-gain design dominates them by increasing the feedback gains which, in turn, enforce a hierarchy of time scales and invariant manifolds. The flexibility provided by backstepping can be used to avoid cancellations and increase robustness, as in our control laws based on domination of destabilizing uncertainties. Close to the origin, such control laws are "softer" than their linear high-gain simplifications. They may exhibit a high-gain behavior only for large signals.

Simplifications of forwarding, such as nested saturation designs introduced by Teel [109], also involve a hierarchy of time scales and invariant manifolds. The time scales of nested saturation designs are slower at each step, while the time scales of high-gain designs are faster at each step.

In Section 6.3, we consider *interlaced systems*, which can be designed by combining the steps of forwarding and backstepping. This is the largest class of systems which can be globally stabilized with restrictions only on the structure of their feedback and feedforward interconnections.

6.1 Backstepping

6.1.1 Introductory example

Backstepping and simplified high-gain designs will be introduced for the following strict-feedback system

$$\begin{aligned} \dot{x}_1 &= x_2 + \theta x_1^2 \\ \dot{x}_2 &= x_3 \\ \dot{x}_3 &= u \end{aligned} \tag{6.1.1}$$

where θ is an uncertain parameter known to belong to the interval $\theta \in [-1, 1]$. This system is represented by the block-diagram in Figure 6.1 which shows a

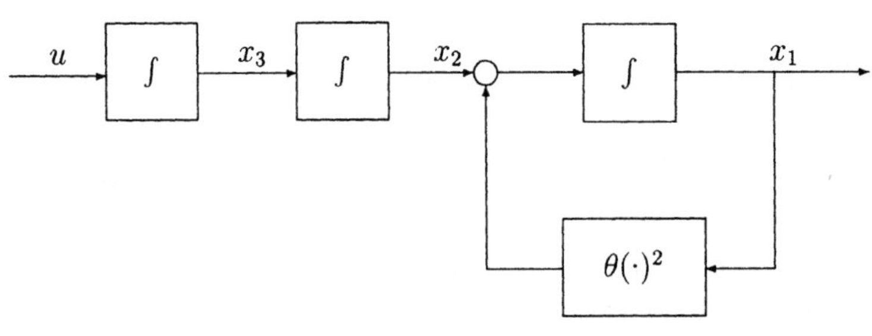

Figure 6.1: The block-diagram of a strict-feedback system.

feedback loop and the absence of feedforward paths other than the integrator chain. For $u = 0$ the system exhibits two types of instability: a linear instability due to the double integrator (x_2, x_3), and a more dramatic nonlinear instability occurring in the subsystem $\dot{x}_1 = \theta x_1^2$. Our goal is to achieve global asymptotic stability of this system by a systematic passivation design. To apply a passivation design from Chapter 4, we need to find a passivating output and a storage function to be used as Lyapunov function. We will accomplish this by a recursive construction.

Recursive passivating outputs. The two requirements of a passivating output are: first, relative degree one, and second, weak minimum phase. For an output of (6.1.1) to be relative degree one, it must be a function of x_3; thus we let $y_3 = x_3 - \alpha_2(x_1, x_2)$. Next we need to select $\alpha_2(x_1, x_2)$ to satisfy the minimum phase requirement, that is the GAS property of the zero dynamics. Setting $y_3 \equiv 0$ shows that the zero-dynamics subsystem is

$$\begin{aligned} \dot{x}_1 &= x_2 + \theta x_1^2 \\ \dot{x}_2 &= \alpha_2(x_1, x_2) \end{aligned} \tag{6.1.2}$$

For this subsystem we must find a stabilizing "control law" $\alpha_2(x_1, x_2)$, that is, we are again facing a feedback stabilization problem, However, and this is extremely significant, this new stabilization problem is for a lower order subsystem of the original third order system (6.1.1). The original problem is thus reduced to the stabilization of the second order subsystem

$$\begin{aligned} \dot{x}_1 &= x_2 + \theta x_1^2 \\ \dot{x}_2 &= x_3 \end{aligned} \tag{6.1.3}$$

in which x_3 is the "control". To solve this lower order problem we need to construct a new relative degree one passivating output $y_2 = x_2 - \alpha_1(x_1)$ and design $\alpha_1(x_1)$ to achieve GAS of the zero-dynamics subsystem

$$\dot{x}_1 = \alpha_1(x_1) + \theta x_1^2 \tag{6.1.4}$$

Once more the problem has been reduced, now to the stabilization of the first order subsystem

$$\dot{x}_1 = x_2 + \theta x_1^2 \tag{6.1.5}$$

in which x_2 is the "control", and $y_1 = x_1$ is the output.

Recursive passivating controls: backstepping. Our definitions of passivating outputs y_1, y_2, and y_3 proceeded in the *bottom-up* direction: from y_3, to y_2, to y_1. These outputs are to be obtained by constructing the functions $\alpha_1(x_1)$ and $\alpha_2(x_1, x_2)$, each playing the part of a "control law": $\alpha_1(x_1)$ for x_2 as a "virtual control" of (6.1.5), and $\alpha_2(x_1, x_2)$ for x_3 as a "virtual control" of (6.1.3). This shows that the recursive design procedure must proceed in the *top-down* direction, by first designing $\alpha_1(x_1)$, then $\alpha_2(x_1, x_2)$, and finally $\alpha_3(x_1, x_2, x_3)$ for the actual control u. In this top-down direction, we start from the scalar subsystem (6.1.5), then augment it by one equation to (6.1.3), and again by one more equation to the original system (6.1.1). On a block-diagram we move "backward" starting with the integrator farthest from the control input. Hence the term *backstepping*.

Let us now reinterpret the construction of the passivating outputs as a backstepping construction of the "control laws" $\alpha_1(x_1)$, $\alpha_2(x_1,x_2)$, and $\alpha_3(x_1,x_2,x_3)$. In the first step, the subsystem (6.1.5), with the output y_1, and the input x_2, is rendered passive by the "control law" $\alpha_1(x_1)$. At the second step, the subsystem (6.1.3) with the output $y_2 = x_2 - \alpha_1(x_1)$ and the input x_3 is rendered passive by the "control law" $\alpha_2(x_1,x_2)$. At the third and final step, the original system (6.1.1) with the output $y_3 = x_3 - \alpha(x_1,x_2)$ and the input u is rendered passive and GAS by the control law $u = \alpha_3(x_1,x_2,x_3)$. At each step a Lyapunov function is constructed which also serves as a storage function.

An interpretation is that backstepping circumvents the relative degree obstacle to passivation. For the output $y = x_1$, the original system has relative degree three. However, at each design step, the considered subsystem has relative degree one with the zero dynamics rendered GAS at the preceding step.

We now present the design steps in more detail.

First step. At this step we design $\alpha_1(x_1)$ to stabilize (6.1.5). If θ were known, this problem would be very simple, but even then we would not use a cancellation control law $\alpha_1(x_1) = -x_1 - \theta x_1^2$, because it would lead to nonrobustness with respect to small variations of θ. Instead, we apply *domination*. Knowing that $\theta \in [-1,1]$, we proceed with a design in which $\alpha_1(x_1)$ *dominates* the term θx_1^2. One such design is $\alpha_1(x_1) = -x_1 - x_1^3$. It achieves GAS of (6.1.5) for $|\theta| < 2$. With this $\alpha_1(x_1)$, and $V_1 = \frac{1}{2}x_1^2$ as a Lyapunov function, the derivative $\dot{V}_1$ of V_1 for (6.1.5) is

$$\dot{V}_1|_{x_2=\alpha_1} = -x_1^2(1 - \theta x_1 + x_1^2) \leq -\frac{1}{2}x_1^4 - \frac{1}{2}x_1^2 \qquad (6.1.6)$$

With $\alpha_1(x_1)$ and $V_1 = \frac{1}{2}x_1^2$ constructed, the first step is completed.

Second step. Using $y_2 = x_2 - \alpha_1(x_1)$ as the output for (6.1.3) with x_3 as the input, we rewrite (6.1.3) as

$$\begin{aligned} \dot{x}_1 &= \alpha_1 + \theta x_1^2 + y_2 \\ \dot{y}_2 &= x_3 + (1+3x_1^2)(\alpha_1 + \theta x_1^2 + y_2) \end{aligned} \qquad (6.1.7)$$

where we have substituted $x_2 = y_2 + \alpha_1(x_1)$ and used the analytical expression for

$$\dot{\alpha}_1 = \frac{\partial \alpha_1}{\partial x_1}\dot{x}_1 = -(1+3x_1^2)(\alpha_1 + \theta x_1^2 + y_2) \qquad (6.1.8)$$

We now proceed to find a Lyapunov function V_2 for (6.1.7). Because y_2 is a passivating output, a possible choice is the storage function $V_2 = V_1 + \frac{1}{2}y_2^2$ whose derivative is

$$\dot{V}_2 = \dot{V}_1 + y_2\dot{y}_2 = \dot{V}_1\Big|_{y_2=0} + \frac{\partial V_1}{\partial x_1}y_2 + y_2(x_3 + (1+3x_1^2)(\alpha_1 + \theta x_1^2 + y_2)) \quad (6.1.9)$$

The key property of this expression is that all the potentially indefinite terms appear multiplied by y_2. Hence, our virtual control $x_3 = \alpha_2(x_1, x_2)$ can be chosen to make $\dot{V}_2$ negative definite. A possible design is

$$\alpha_2 = -y_2 - (1+3x_1^2)y_2 - x_1 - (1+3x_1^2)(\alpha_1 + \theta x_1^2) \quad (6.1.10)$$

but its last term is not implementable because it cancels a nonlinearity which contains the uncertain parameter θ. Instead of (6.1.10) we proceed with *domination.*

To dominate the θ-term in $\dot{V}_2$, we can use the control law

$$\alpha_2 = -(2+3x_1^2)y_2 - 2(2+3x_1^2)^2(1+x_1^2)y_2 \quad (6.1.11)$$

which, substituted into (6.1.9), yields

$$\begin{aligned} \dot{V}_2 &= -x_1^2(1-\theta x_1 + x_1^2) + y_2[(1+3x_1^2)(1-\theta x_1 + x_1^2) + x_1] - \\ &\quad -2y_2^2(2+3x_1^2)^2(1+x_1^2) \\ &\leq -\frac{1}{2}\dot{V}_1\Big|_{y_2=0} - (2+3x_1^2)(1+x_1^2)y_2^2 \end{aligned} \quad (6.1.12)$$

We have dominated the θ-term in $\dot{V}_2$ by "completing the squares", which results in a rapidly growing nonlinear gain in the virtual control law (6.1.11). As shown in [23], such a "hardening" of the nonlinearities in the virtual control laws at each consecutive step is due to the quadratic form of the Lyapunov functions. Here, we can avoid the hardening of the control law by noting that the θ-term in $\dot{V}_2$ is

$$-y_2(1+3x_1^2)(x_1 + x_1^3 - \theta x_1^2) \quad (6.1.13)$$

This term has the form $-y_2\frac{\partial \tilde{V}_1}{\partial x_1}$, where $\tilde{V}_1$ is the positive definite function

$$\tilde{V}_1(x_1) = \int_0^{x_1}(1+3s^2)(s+s^3-\theta s^2)ds$$

Thus, when we use $\tilde{V}_1$ instead of $V_1 = \frac{1}{2}x_1^2$, the term (6.1.13) does not appear in $\dot{\tilde{V}}_1$. As a consequence, for the modified Lyapunov function $\tilde{V}_2 = \tilde{V}_1 + \frac{1}{2}y_2^2$, we obtain

$$\dot{\tilde{V}}_2 = \frac{\partial \tilde{V}_1}{\partial x_1}(-x_1 - x_1^3 + \theta x_1^2) + y_2(\alpha_2 + (1+3x_1^2)y_2) \quad (6.1.14)$$

A control law rendering $\dot{\tilde{V}}_2$ negative definite is

$$\alpha_2 = -y_2 - (1 + 3x_1^2)y_2 \tag{6.1.15}$$

We have thus designed a control law with a gain margin $[1, \infty)$, which also uses less effort than the control law (6.1.11). The construction of a Lyapunov function like $\tilde{V}_1(x_1)$ is always applicable to second-order systems and is one of the flexibilities of backstepping.

Third step. With $\alpha_2(x_1, y_2)$ in (6.1.15), we have constructed a passivating output $y_3 = x_3 - \alpha_2$ for the full system (6.1.1), which in the new coordinates (x_1, y_2, y_3), is given by

$$\begin{array}{rcl} \dot{x}_1 & = & \alpha_1 + \theta x_1^2 + y_2 \\ \dot{y}_2 & = & \alpha_2(x_1, x_2) + y_3 - \dot{\alpha}_1(x_1, y_2) \\ \dot{y}_3 & = & u - \dot{\alpha}_2(x_1, y_2, y_3) \end{array} \tag{6.1.16}$$

In this system the explicit expressions for $\dot{\alpha}_1$ and $\dot{\alpha}_2$ are known. A Lyapunov function is $V_3 = \tilde{V}_2 + \frac{1}{2}y_3^2$ and its derivative for (6.1.16) is

$$\dot{V}_3 = \dot{\tilde{V}}_2|_{y_3=0} + y_3(u - \dot{\alpha}_2 + y_2) \tag{6.1.17}$$

It is clear that a control law $u = \alpha_3(x, y_1, y_2)$ can be designed to make $\dot{V}_3$ negative definite. This control law will necessarily contain a term to dominate the θ-dependent part of $\dot{\alpha}_2$.

Exact and robust backstepping. With the just completed backstepping design we have achieved GAS of the nonlinear system (6.1.1) with an uncertain parameter θ. The presence of the uncertainty prevented us from using a simpler cancellation control law. For $\theta = 1$ such a cancellation control law would be $\alpha_1(x_1) = -x_1 - x_1^2$ and the first equation would have become $\dot{x}_1 = -x_1 + y_2$. We will refer to this idealized form of backstepping as *exact backstepping.* In the presence of uncertainties, such as $\theta \in [-1, 1]$, we are forced to use *robust backstepping.* Then the "control laws" α_1 and α_2, as well as the true control law $u = \alpha_3$ contain terms constructed to dominate the uncertainties.

The above example shows that *robust backstepping* is more complicated and more "nonlinear" than exact backstepping. The complexity of backstepping control laws is considerable even in the case of exact backstepping. It grows with the number of steps primarily because the analytical expressions of the time-derivatives such as $\dot{\alpha}_1$ and $\dot{\alpha}_2$ are increasingly complex. This motivates various simplifications, some of which we will explore later. We first present the general backstepping procedure for strict-feedback systems.

6.1.2 Backstepping procedure

A family of backstepping designs can be constructed by recursive applications of different versions of the same basic step: the augmentation by one equation of the subsystem already made passive by a "virtual control". The basic step, already presented in Proposition 4.25, is now given in a different form.

Proposition 6.1 *(Backstepping as recursive feedback passivation)*
Assume that for the system

$$\dot{z} = f(z) + g(z)u, \tag{6.1.18}$$

a C^1 feedback transformation $u = \alpha_0(z) + v_0$ and a C^2 positive definite, radially unbounded storage function $W(x)$ are known such that this system is passive from the input v_0 to the output $y_0 = (L_gW)^T(z)$, that is $\dot{W} \leq y_0^T v_0$.

Then the augmented system

$$\begin{array}{rcl} \dot{z} &=& f(z) + g(z)\xi \\ \dot{\xi} &=& a(z,\xi) + b(z,\xi)u, \end{array} \tag{6.1.19}$$

where $b^{-1}(z,\xi)$ exists for all (z,ξ), is (globally) feedback passive with respect to the output $y = \xi - \alpha_0(z)$ and the storage function $V(z,y) = W(z) + \frac{1}{2}y^Ty$. A particular control law ("exact backstepping") which achieves passivity of (6.1.19) is

$$u = b^{-1}(z,\xi)(-a(z,\xi) - y_0 + \frac{\partial \alpha_0}{\partial z}(f(z) + g(z)\xi) + v) \tag{6.1.20}$$

The system (6.1.19) with (6.1.20) is ZSD for the input v if and only if the system (6.1.18) is ZSD for the input v_0. Moreover, if $W(z(t))$ is strictly decreasing for (6.1.18) with $u = \alpha_0(z)$, then $W(z(t)) + \frac{1}{2}y^T(t)y(t)$ is strictly decreasing for (6.1.19) with $v = -y$.

Proof: Substituting $\xi = y + \alpha_0(z)$, we rewrite (6.1.19) as

$$\begin{array}{rcl} \dot{z} &=& f(z) + g(z)(\alpha_0(z) + y) \\ \dot{y} &=& a(z, y + \alpha_0(z)) + b(z, y + \alpha_0(z))u - \dot{\alpha}_0(z,y), \end{array} \tag{6.1.21}$$

After the feedback transformation (6.1.20), this system becomes

$$\begin{array}{rcl} \dot{z} &=& f(z) + g(z)(\alpha_0(z) + y) \\ \dot{y} &=& -y_0 + v, \end{array} \tag{6.1.22}$$

The passivity property from y to v is established with the storage function $V = W(z) + \frac{1}{2}y^Ty$. Its time-derivative satisfies

$$\dot{V} = \dot{W} + y^T(-y_0 + v) \leq y^T v$$

where we have used the passivity assumption $\dot{W} \leq y_0^T v_0$ and the fact that $v_0 = y$.

To verify the ZSD property of (6.1.22), we set $y \equiv v \equiv 0$ which implies $y_0 \equiv 0$. Hence, the system (6.1.22) is ZSD if and only if $z = 0$ is attractive conditionally to the largest invariant set of $\dot{z} = f(z) + g(z)\alpha_0(z)$ in the set where $y_0 = (L_g W)^T = 0$. This is equivalent to the ZSD property of the original system (6.1.18) for the input v_0 and the output y_0.

Finally, with the control law $v = -y$, we obtain $\dot{V} = \dot{W}|_{y=0} - y^T y$ which is negative definite if and only if $\dot{W}|_{y=0} < 0$ for all $z \neq 0$.

□

In Proposition 6.1 a new passivating output y is constructed from the previous passivating control law $\alpha_0(z)$, and the new storage function is obtained by adding $y^T y$ to the old storage function. Moreover, the ZSD property is preserved in the augmented system.

Strict-feedback systems. Because Proposition 6.1 ensures that the augmented system inherits the properties of the original system, we can use it at each step of a recursive design procedure for a system which is an augmentation of the z-subsystem in (6.1.19) by a lower-triangular ξ-subsystem:

$$
\begin{aligned}
\dot{z} &= f(z) + g(z)\xi_1 \\
\dot{\xi}_1 &= a_1(z, \xi_1) + b_1(z, \xi_1)\xi_2 \\
\dot{\xi}_2 &= a_2(z, \xi_1, \xi_2) + b_1(z, \xi_1, \xi_2)\xi_3 \\
&\vdots \\
\dot{\xi}_{n-1} &= a_{n-1}(z, \xi_1, \ldots, \xi_{n-1}) + b_{n-1}(z, \xi_1, \ldots, \xi_{n-1})\xi_n \\
\dot{\xi}_n &= a_n(z, \xi_1, \ldots, \xi_n) + b_n(z, \xi_1, \ldots, \xi_n)u, \quad \xi_i \in I\!R^q, \ i = 1, \ldots, n
\end{aligned}
\tag{6.1.23}
$$

The systems in the lower-triangular configuration (6.1.23) are called *strict-feedback systems*, because every interconnection in the system is a *feedback* connection from the states located farther from the input. Assuming that the z-subsystem satisfies Proposition 6.1 and that every $b_i(z, \xi_1, \ldots, \xi_i)$ is invertible for all $(z, \xi_1, \ldots, \xi_i)$, the system (6.1.23) with the output $y_1 = \xi_1 - \alpha_0(z)$ has relative degree n. We will recursively reduce the relative degree to one by proceeding as in the introductory example. For $y_n = \xi_n - \alpha_{n-1}(z, \xi_1, \ldots, \xi_n)$ to be a passivating output for the whole system, the virtual control law $\xi_n = \alpha_{n-1}(z, \xi_1, \ldots, \xi_{n-1})$ must be a passivating feedback for the zero-dynamics subsystem consisting of (6.1.23) minus the last equation. Likewise, $y_{n-1} = \xi_{n-1} - \alpha_{n-2}(z, \xi_1, \ldots, \xi_{n-2})$ will be a passivating output for this subsystem if α_{n-2} is a passivating feedback for its zero-dynamics subsystem. Continuing this process upward, we end up with the recursive expressions for passivating outputs:

$$y_i = \xi_i - \alpha_{i-1}(z, \xi_1, \ldots, \xi_{i-1})$$

In the presence of uncertainties, α_i's must be constructed to dominate them ("robust backstepping"). If, in the absence of uncertainties, $\alpha_i' s$ are constructed employing some cancellations ("exact backstepping"), then the backstepping recursion is

$$\begin{aligned} y_i &= \xi_i - \alpha_{i-1}(z, \xi_1, \ldots, \xi_{i-1}) \\ \alpha_i(z, \xi_1, \ldots, \xi_i) &= b_i^{-1}(-a_i - y_{i-1} + \dot{\alpha}_{i-1} - y_i), \quad i = 2, \ldots, n \end{aligned} \tag{6.1.24}$$

In these expressions, the time-derivatives $\dot{\alpha}_i$ are evaluated as explicit functions of the state variables, that is, $\dot{\alpha}_0 = \frac{\partial \alpha_1}{\partial z}(f + g\xi_1)$, $\dot{\alpha}_1 = \frac{\partial \alpha_1}{\partial z}(f + g\xi_1) + \frac{\partial \alpha_1}{\partial \xi_1}(a_1 + b_1\xi_2)$, etc.

Construction of a CLF. Proposition 6.1 ensures another important feature of backstepping. It guarantees that, if at the first step the strict negativity of $\dot{W}$ is achieved with $u = \alpha_0(z)$, then this property is propagated through each step of backstepping. Because of it, the final storage function $V(z, \xi) = W + \sum_{i=1}^{n} y_i^T y_i$ is a CLF for (6.1.23) and can be used to design a control law with desirable stability margins. In this way, the n steps of backstepping can be seen as construction of the CLF $V(z, \xi) = W + \sum_{i=1}^{n} y_i^T y_i$ and of the new coordinates $y_1, \ldots, y_n$. Even in exact backstepping, this construction involves no cancellation in the control law until the last step. At the last step, instead of the passivating control law

$$u = \alpha_n(z, \xi_1, \ldots, \xi_n) + v \tag{6.1.25}$$

which requires cancellations and does not possess a guaranteed stability margin, we can design the control law

$$u_S(x, \xi) = \begin{cases} -\left(c_0 + \frac{L_F V + \sqrt{(L_F V)^2 + ((L_G V)^T L_G V)^2}}{(L_G V)^T L_G V}\right) L_G V &, \; L_G V \neq 0 \\ 0 &, \; L_G V = 0 \end{cases} \tag{6.1.26}$$

where

$$\begin{aligned} G(z, \xi) &= (0, \ldots, 0, b_n(z, \xi))^T, \\ F(z, \xi) &= (f(z) + g(z)\xi_1, a_1(z, \xi_1) + b_1(z, \xi_1)\xi_2, \ldots, a_n(z, \xi_1, \ldots, \xi_n))^T \end{aligned}$$

According to Proposition 3.44, the control law (6.1.26) minimizes a cost functional of the form $J = \int_0^\infty (l(x) + u^T R(x) u) dt$. This means that it achieves a sector margin $(\frac{1}{2},\infty)$. It may also serve as the starting point of a domination redesign to achieve a disk margin $D(\frac{1}{2})$.

Removing the relative degree obstacle. The absence of any feedforward connection in the system (6.1.23) is crucial for recursive backstepping: it guarantees that the relative degree of ξ_i is $r_i = n - i + 1$ for each i. Because of this property, the relative degree one requirement of feedback passivation is met at step i, not with respect to the true input u but rather with respect to the virtual input ξ_{i+1}. Only the output y_n is relative degree one with respect to the true input u.

In all the passivation designs of cascade systems in Chapter 4, we have required that a *relative degree one* output can be factored out of the interconnection term. Using the above backstepping procedure, this restriction is now removed and replaced by a right-invertibility assumption.

Proposition 6.2 *(Feedback passivation with backstepping)*
Suppose that for the cascade

$$\begin{aligned} \dot{z} &= f(z) + \psi(z,\xi) \\ \dot{\xi} &= A\xi + Bu \end{aligned} \qquad (6.1.27)$$

there exists an output $y = C\xi$ such that
(i) the interconnection $\psi(z,\xi)$ can be factored as $\psi(z,\xi) = \tilde{\psi}(z)y$;
(ii) the system (A, B, C) is right-invertible and minimum phase.

If $\dot{z} = f(z)+\tilde{\psi}(z)k(z)$ is GAS and if $W(z)$ is a positive definite and radially unbounded function such that $L_{f+\tilde{\psi}k}W \leq 0$, then global stabilization of the cascade (6.1.27) is achieved by recursive backstepping starting with the virtual control law $y = k(z)$. □

Proof: We assume without loss of generality that the system (A, B, C) is in the normal form

$$\begin{aligned} \dot{\xi}_0 &= A_0\xi_0 + B_0 y \\ y^{(r)} &= u \end{aligned} \qquad (6.1.28)$$

where $r \geq 1$ is the relative degree of the system. (This form may involve adding integrators for some control components.) If $r = 1$, Proposition 4.25 yields a globally stabilizing feedback $u = \alpha_0(\xi_0, y, z)$. If $r > 1$, α_0 is a virtual control law which can be backstepped through $r - 1$ integrators by a repeated application of Proposition 6.1.

□

6.1.3 Nested high-gain designs

As can be seen from our introductory example, the recursive formula (6.1.24) for α_i generates analytical expressions of increasing complexity, primarily due to the dependence of α_i on the time-derivative $\dot{\alpha}_{i-1}$. After a couple of recursive steps, the expression for α_i may become discouragingly complicated and motivate the designer to seek some simplifications. One such simplification, proposed in [36], is to use an approximately differentiating filter, that is to replace $\dot{\alpha}_{i-1}$ by $\frac{s}{\tau_i s+1}(\alpha_{i-1})$, where τ_i is a small time constant. Another possibility, discussed here, is to dominate $\dot{\alpha}_{i-1}$ by a linear high-gain feedback. This simplified design is of interest because it reveals the underlying geometry of backstepping. To illustrate its main features, we return to our introductory example (6.1.1) and consider its (x_1, x_2)-subsystem

$$\begin{aligned} \dot{x}_1 &= x_2 + \theta x_1^2 \\ \dot{x}_2 &= x_3 \end{aligned} \tag{6.1.29}$$

in which we treat x_3 as the control variable. As before, in the first step we design $\alpha_1(x_1) = -x_1 - x_1^3$ as the control law for the virtual control x_2. We proceed to introduce $y_2 = x_2 - \alpha_1(x_1) = x_2 + x_1 + x_1^3$ and rewrite (6.1.29) as

$$\begin{aligned} \dot{x}_1 &= -x_1 - x_1^3 + \theta x_1^2 + y_2 \\ \dot{y}_2 &= x_3 - \dot{\alpha}_1 = x_3 + (1 + 3x_1^2)(-x_1 - x_1^3 + \theta x_1^2 + y_2) \end{aligned} \tag{6.1.30}$$

Although in this case the expression for $\dot{\alpha}_1$ is not very complicated, let us avoid it by using the high-gain feedback

$$x_3 = \alpha_2 = -k y_2 \tag{6.1.31}$$

to obtain

$$\begin{aligned} \dot{x}_1 &= -x_1 - x_1^3 + \theta x_1^2 + y_2 \\ \dot{y}_2 &= -k y_2 - \dot{\alpha}_1 \end{aligned} \tag{6.1.32}$$

While *global* stabilization cannot be guaranteed, we expect that with larger values of k we can increase the region in which $k y_2$ dominates $\dot{\alpha}_1$ and therefore results in larger regions of attraction. Differentiating the Lyapunov function $\tilde{V}_2$ of Section 6.1.1, we obtain

$$\dot{\tilde{V}}_2 \le -y_2^2(k - (1 + 3x_1^2))$$

An estimate of the region of attraction is thus given by the largest level set of $\tilde{V}_2$ in which $x_1^2 < \frac{k-1}{3}$. It is clear that this region expands as $k \to \infty$.

Geometry: invariant manifolds. The high-gain feedback $\alpha_2 = -k y_2$ enforces a fast decay of y_2, that is, a rapid convergence of x_2 to its desired value

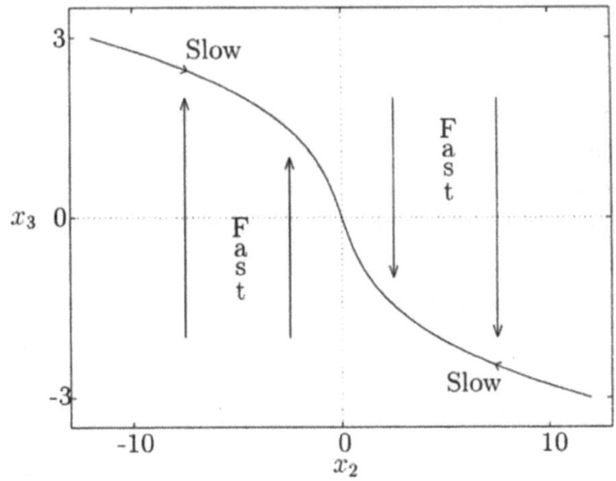

Figure 6.2: The fast convergence towards the desired manifold $x_2 = \alpha(x_1)$ enforced by the high-gain k

$x_2 = \alpha_1(x_1)$. As a result, the virtual control law $x_2 = \alpha_1(x_1)$ is approximately implemented after a fast transient. Geometrically, this means that the solutions of the feedback system converge rapidly to the desired manifold $x_2 = \alpha(x_1)$. The design creates a time-scale separation between the convergence *to* the manifold, which is *fast*, and the convergence to the origin *in* the manifold, which is *slow*. This qualitative description of the solutions, illustrated in Figure 6.2, holds in a bounded region whose size increases with the gain k.

Defining the small parameter $\epsilon = \frac{1}{k}$, the feedback system (6.1.30) is in the standard singularly perturbed form

$$\begin{aligned} \dot{x}_1 &= -x_1 - x_1^3 + \theta x_1^2 + y_2 \\ \epsilon \dot{y}_2 &= -y_2 - \epsilon \dot{\alpha}_1 \end{aligned} \tag{6.1.33}$$

For ϵ sufficiently small (k large), singular perturbation theory guarantees that (6.1.33) possesses an invariant manifold $y_2 = h(x, \epsilon)$. Differentiating this expression with respect to time and using $\dot{x}_1$ and $\dot{y}_2$ from (6.1.33) we obtain the manifold PDE

$$-h(x_1, \epsilon) - \epsilon \dot{\alpha}_1(x_1, h(x_1, \epsilon)) = \epsilon \frac{\partial h}{\partial x_1}(-x_1 - x_1^3 + \theta x_1^2 + h(x_1, \epsilon)) \tag{6.1.34}$$

For an $\mathcal{O}(\epsilon^n)$-approximation of its solution $h(x_1, \epsilon)$, we substitute $h(x_1, \epsilon) = h_0(x_1) + \epsilon h_1(x_1) + \ldots + \epsilon^{n-1} h_{n-1}(x_1) + \mathcal{O}(\epsilon^n)$ in (6.1.34) and equate the terms of the like powers in ϵ. In particular, the zeroth-order approximation is $y_2 = 0$. Since $y_2 = x_2 - \alpha_1(x_1)$ this means that the invariant manifold $y_2 = h(x_1, \epsilon)$ is ϵ-close to the *desired* manifold $x_2 = \alpha_1(x_1)$. We say in this case that $x_2 = \alpha_1(x_1)$ is near-invariant. In the limit as $\epsilon = \frac{1}{k} \to 0$, the system (6.1.33) reduced to the manifold $x_2 = \alpha(x_1)$ is the GAS zero-dynamics subsystem $\dot{x}_1 = -x_1 - x_1^3 + \theta x_1^2$.

It is of interest to compare the high-gain feedback $\alpha_2 = -ky_2$ with the previously designed backstepping control law $\alpha_2 = -(2 + 3x_1^2)y_2$. A simplification is obvious: the nonlinear "gain" $(2 + 3x_1^2)$ is replaced by the constant gain k which, if high, would make $x_2 = \alpha_1(x_1)$ near-invariant. In contrast, the backstepping "gain" is high only for large x_1^2 where it is needed to dominate uncertainties. Near $x_1 = 0$ this nonlinear gain is low. Thus, in the backstepping design, the manifold $x_2 = \alpha_1(x_1)$ is not invariant and it serves only to construct a passivating output and a Lyapunov function. As a consequence, the backstepping control law does not need large gains which would render $x_2 = \alpha_1(x_1)$ near-invariant even near $x_1 = 0$.

The high-gain feedback design extends to more general systems involving chains of intergrators.

Proposition 6.3 *(High-gain design)*
Assume that for the system

$$\dot{z} = f(z) + g(z)u,$$

a C^1 control law $u = \alpha_0(z)$ achieves GAS and LES of $z = 0$, and consider the augmented system

$$\begin{aligned} \dot{z} &= f(z) + g(z)\xi_1 \\ \dot{\xi}_1 &= \xi_2, \\ &\vdots \\ \dot{\xi}_n &= u \end{aligned} \tag{6.1.35}$$

Let $p(s) = \lambda^n + a_{n-1}\lambda^{n-1} + \ldots + a_1\lambda + a_0$ be an arbitrary Hurwitz polynomial. Then the feedback

$$u = -k(a_{n-1}\xi_n + k(a_{n-2}\xi_{n-1} + k(\ldots + k(a_1\xi_2 + ka_0(\xi_1 - \alpha_0(z))\ldots) \tag{6.1.36}$$

achieves *semiglobal* stabilization of $(z, \xi) = (0, 0)$, that is, for any compact neighborhood Ω of $(z, \xi) = (0, 0)$, there exists k^* such that for all $k \geq k^*$, the region of attraction contains Ω.

Proof: Let $\zeta_1 = \xi_1 - \alpha_0(z)$ and introduce the scaling

$$\zeta_i = \frac{\xi_i}{k^{i-1}}, \; i = 2, \ldots, n$$

In these new coordinates, the closed-loop system is

$$\begin{aligned} \dot{z} &= F(z) + g(z)\zeta_1 \\ \dot{\zeta} &= kA\zeta + e_1\alpha_0(z,\zeta_1), \quad e_1^T = (1,0,\ldots,0) \end{aligned} \tag{6.1.37}$$

where $\dot{z} = F(z) = f(z) + \alpha_0(z)g(z)$ is LES and GAS, and the matrix A is Hurwitz with characteristic polynomial $p(s)$.

Let $W(z)$ be a radially unbounded positive definite function, locally quadratic, such that $L_F W(z) < 0$ for all $z \neq 0$, and let $P > 0$ be the solution of the Lyapunov equation $A^T P + PA = -I$. For the system (6.1.37) we employ the composite Lyapunov function

$$V(z,\zeta) = W(z) + \zeta^T P \zeta \tag{6.1.38}$$

Let Ω be a desired compact region of attraction of $(z,\xi) = (0,0)$ and let $R > 0$ be such that

$$\forall (z,\xi) \in \Omega; \ \|(z,\xi)\| \leq R$$

Assuming without loss of generality that $k \geq 1$, we have $\|(z,\xi)\| \leq R \Rightarrow \|(z,\zeta)\| \leq R$. Because V is radially unbounded, we can pick a level set V_R such that $\|(z,\zeta)\| \leq R \Rightarrow V(z,\zeta) \leq V_R$. By construction, Ω is included in the compact region

$$\Omega_R = \{(z,\zeta) | V(z,\zeta) \leq V_R\}$$

We will now show that k can be chosen large enough such that $\dot{V}$ is negative definite in Ω_R, which is therefore included in the region of attraction of $(z,\xi) = (0,0)$.

The time-derivative of V is

$$\dot{V} = L_F W(z) + \zeta_1(L_g W + \alpha_0(z,\zeta_1)) - k\zeta^T\zeta \tag{6.1.39}$$

Because $\dot{z} = F(z)$ is LES and Ω_R is compact, there exist constants $c_1 > 0$ and $c_2 > 0$ such that for all $(z,\zeta) \in \Omega_R$,

$$L_F W(z) \leq -c_1\|z\|^2 \ \text{and} \ |L_g W + \alpha_0(z,\zeta_1)| \leq c_2\|z\|$$

Completing the squares in (6.1.39) and selecting $k > \frac{c_1^2}{c_2}$ proves that $\dot{V} < 0$ in Ω_R. □

Strict-feedback systems. For the strict-feedback system

$$\begin{aligned} \dot{z} &= f(z) + g(z)\xi_1 \\ \dot{\xi}_1 &= \xi_2 + a_1(z,\xi_1) \\ \dot{\xi}_2 &= \xi_3 + a_2(z,\xi_1,\xi_2) \\ &\vdots \\ \dot{\xi}_n &= u + a_n(z,\xi_1,\ldots,\xi_n), \quad \xi_i \in I\!R^q, \ i = 1,\ldots,n \end{aligned} \tag{6.1.40}$$

the proof of Proposition 6.3 is easily adapted to the case when the nonlinearities satisfy a linear growth assumption in $(\xi_2, \ldots, \xi_n)$, that is, when there exist continuous functions γ_{ij} such that

$$\|a_i(z, \xi_1, \ldots, \xi_i)\| \leq \gamma_{i1}(\|(z, \xi_1)\|) + \gamma_{i2}(\|(z, \xi_1)\|)\|(\xi_2, \ldots, \xi_i)\|, \; i = 2, \ldots, n \tag{6.1.41}$$

With this growth restriction, an increase of the controller gain k is sufficient to dominate the nonlinearities in any prescribed region. However, if the growth of a_i's is not restricted, a prescribed region of attraction can no longer be guaranteed with the control law (6.1.36). Worse yet, the actual size of the region of attraction may shrink in certain directions as k increases. An example, adapted from [58], illustrates this phenomenon.

Example 6.4 *(Vanishing region of attraction)*
In the strict-feedback system

$$\begin{aligned} \dot{z} &= -z + \xi_1 z^2 \\ \dot{\xi}_1 &= \xi_2 \\ \dot{\xi}_2 &= u + \xi_2^3 \end{aligned} \tag{6.1.42}$$

the z-subsystem is globally stabilized by the virtual control law $\xi_1 = \alpha_0(x) = 0$. With this $\alpha_0(x)$, the control law (6.1.36) reduces to the high-gain linear feedback

$$u = -k^2\xi_1 - k\xi_2$$

Using the scaling transformation

$$\tau = k\, t, \;\; \zeta_1 = k^{\frac{1}{2}}\xi_1, \;\; \zeta_2 = k^{-\frac{1}{2}}\xi_2$$

we rewrite the closed-loop ξ-subsystem as

$$\begin{aligned} \tfrac{d}{d\tau}\zeta_1 &= \zeta_2 \\ \tfrac{d}{d\tau}\zeta_2 &= -\zeta_1 - \zeta_2 + \zeta_2^3 \end{aligned} \tag{6.1.43}$$

Recognizing this system as a reversed-time version of the Van der Pol's equation, we conclude that its equilibrium $(\zeta_1, \zeta_2) = (0, 0)$ is surrounded by an unstable limit cycle which is the boundary of the region of attraction. All the initial conditions outside this region, such as those satisfying

$$\zeta_1^2 + \zeta_2^2 > 3^2 \tag{6.1.44}$$

lead to unbounded solutions. In the original coordinates, the instability condition (6.1.44) is

$$k\xi_1^2 + \frac{1}{k}\xi_2^2 > 3$$

In particular, every initial condition $(\xi_1(0), 0)$ leads to an unbounded solution if $\xi_1(0) > \frac{3}{\sqrt{k}}$. We conclude that, as $k \to \infty$, the region of attraction of the system (6.1.43) shrinks to zero along the axis $\xi_2 = 0$. This shows that the control law (6.1.36) does not achieve semiglobal stabilization for a general strict-feedback system. □

The shrinking of the region of attraction as the controller gain increases is a consequence of peaking. We have shown in Chapter 4 that a fast convergence of ξ_1 implies that its derivatives $\xi_2, \ldots, \xi_n, u$, peak with increasing exponents. If a destabilizing nonlinearity is multiplied by a peaking state, a higher gain is needed to counteract its effect. On the contrary, the higher the gain, the more destabilizing is the effect of peaking. With a sufficient growth of the nonlinearities, this will cause the region of attraction to shrink.

To achieve larger regions of attraction for the strict-feedback system (6.1.40), we replace the control law (6.1.36) with the more general expression

$$u = -k_n(\xi_n + k_{n-1}(\xi_{n-1} + k_{n-2}(\ldots + k_2(\xi_2 + k_1(\xi_1 - \alpha_0(z))\ldots) \qquad (6.1.45)$$

In this control law we can increase not only the gains but also their separation. The existence of a suitable set of parameters $\{k_1, \ldots, k_n\}$ to guarantee a prescribed region of attraction is asserted by a recursive application of the following result by Teel and Praly [112], quoted without proof.

Proposition 6.5 (*Semiglobal backstepping*)
Assume that for the system

$$\dot{z} = f(z) + g(z)u,$$

a C^1 control law $u = \alpha_{k_1}(z)$ achieves semiglobal asymptotic stability of $z = 0$, that is, the region of attraction can be arbitrarily increased by increasing the parameter k_1.

If, in addition, $u = \alpha_{k_1}(z)$ achieves LES of $z = 0$, then for the augmented system

$$\begin{aligned} \dot{z} &= f(z) + g(z)\xi \\ \dot{\xi} &= a(z, \xi) + u \end{aligned} \qquad (6.1.46)$$

semiglobal stabilization of $(z, \xi) = (0, 0)$ is achieved with the control law

$$u = -k_2(\xi - \alpha_{k_1}(z)) \qquad (6.1.47)$$

□

Instead of the simple sum $W(z) + y^T y$ used in Proposition 6.3, the proof of this result employs a composite Lyapunov function of the form

$$V(z, y) = c\frac{W(z)}{c+1-W(z)} + \mu\frac{y^T y}{\mu+1-y^T y} \tag{6.1.48}$$

where the constants c and μ can be adjusted for the prescribed region of attraction, as illustrated on the system considered in Example 6.4.

Example 6.6 (*Semiglobal stabilization with sufficient separation of the gains*) We return to the strict-feedback system

$$\begin{aligned} \dot{z} &= -z + \xi_1 x^2 \\ \dot{\xi}_1 &= \xi_2 \\ \dot{\xi}_2 &= u + \xi_2^3 \end{aligned} \tag{6.1.49}$$

to achieve its stabilization in a prescribed region of attraction $\|(z, \xi_1, \xi_2)\|^2 \leq R$. We first consider the subsystem

$$\begin{aligned} \dot{z} &= -z + \xi_1 x^2 \\ \dot{\xi}_1 &= \xi_2 \end{aligned}$$

for which semiglobal stabilization is achieved with the virtual control law $\xi_2 = \alpha_1(z, \xi_1) = -k_1\xi_1$. With this control law, the time-derivative of $W = z^2 + \xi_1^2$ is

$$\dot{W} = -2z^2 + 2\xi_1 z^3 - 2k_1\xi_1^2$$

which is negative in the set where $z^4 < 4k_1$. Hence we choose $k_1 = \frac{R^2+1}{4}$ to include the level set $W(z, \xi_1) = R$ in the region of attraction.

To stabilize the complete system (6.1.49), we apply one step of semiglobal backstepping, which yields the linear control law

$$u = -k_2 y_2 = -k_2(\xi_2 + k_1\xi_1) \tag{6.1.50}$$

The gain k_2 of (6.1.50) is determined with the help of the Lyapunov function

$$V((z, \xi_1), y) = \mu_1\frac{W(z, \xi_1)}{\mu_1 + 1 - W(z, \xi_1)} + \mu_2\frac{y_2^2}{\mu_2 + 1 - y_2^2}$$

where $\mu_1 = R$ and $\mu_2 = (1 + k_1)^2 R$ are chosen to satisfy

$$\|(z, \xi_1, \xi_2)\|^2 \leq R \Rightarrow V((z, \xi_1), y_2) \leq \mu_1^2 + \mu_2^2$$

We now show that with k_2 large enough we can render $\dot{V}$ negative definite in the region where $V \leq \mu_1^2 + \mu_2^2 + 1$. Differentiating V yields

$$\dot{V} = \frac{\mu_1(\mu_1+1)}{(\mu_1+1-W)^2}\dot{W} + \frac{\mu_2(\mu_2+1)}{(\mu_2+1-y_2^2)^2}(-2(k_2-k_1)y_2^2 + 2y_2(k_1^2\xi_1 + (y_2+k_1\xi_1)^3) \tag{6.1.51}$$

When $V \le \mu_1^2 + \mu_2^2 + 1$, we can use the bounds

$$c_{1m} := \frac{\mu_1}{\mu_1+1} \le \frac{\mu_1(\mu_1+1)}{(\mu_1+1-W)^2} \le \frac{(\mu_1^2+\mu_2^2+1+\mu_1)^2}{\mu_1(\mu_1+1)} =: c_{1M}$$
$$c_{2m} := \frac{\mu_2}{\mu_2+1} \le \frac{\mu_2(\mu_2+1)}{(\mu_2+1-y_2^2)^2} \le \frac{(\mu_1^2+\mu_2^2+1+\mu_2)^2}{\mu_2(\mu_2+1)} =: c_{2M}$$

and obtain

$$\dot{V} \le -c_{1m}(x^2+k_1\xi_1^2)+2|y_2|(|\xi_1|(c_{1M}+c_{2M}k_1^2)+c_{2M}|y_2+k_1\xi_1|^3)-2c_{2m}(k_2-k_1)y_2^2 \tag{6.1.52}$$

It is clear that with k_2 large enough the negative terms dominate the cross-term and render $\dot{V}$ negative definite in the region where $V \le \mu_1^2+\mu_2^2+1$. Hence the region of attraction contains the prescribed compact set $\|(x,\xi_1,\xi_2)\|^2 \le R$. □

The above example shows how the gains needed to achieve a prescribed region of attraction can be estimated from a Lyapunov function (6.1.48). However, it also points to two practical difficulties of control laws with several nested high-gains such as (6.1.45): first, excessive gains may be needed for prescribed regions of attractions, and second, the simplification of the backstepping design is lost in the analysis required to determine these gains.

The situation is more favorable when several time scales are already present in the system and the desired manifolds can be created without excessive gains. This is the case with the following VTOL aircraft example.

Example 6.7 (VTOL aircraft)
The model

$$\begin{aligned} \ddot{x} &= -u_1 \sin\theta + \epsilon u_2 \cos\theta \\ \ddot{y} &= u_1 \cos\theta + \epsilon u_2 \sin\theta - 1 \\ \ddot{\theta} &= u_2 \end{aligned} \tag{6.1.53}$$

has been employed by Hauser, Sastry, and Meyer [34] for the motion of a VTOL (vertical take off and landing) aircraft in the vertical (x,y)-plane. The parameter $\epsilon > 0$ is due to the "sloped" wings and is very small, $\epsilon \ll 1$. We will thus base our design on the model (6.1.53) with $\epsilon = 0$ and then select the controller parameters to take into account the effects of $\epsilon \neq 0$.

In this case study we first consider backstepping and then proceed with a linear high-gain approximation of backstepping. Finally, recognizing that the model (6.1.53) with $\epsilon > 0$ is a *peaking cascade*, we select the controller parameters to reduce peaking to an acceptable level. Compared with the *dynamic extension* design of [34, 74], ours is a *dynamic reduction* design, based on singular perturbations [27].

Backstepping. For model (6.1.53) with $\epsilon = 0$, the backstepping idea is to use u_1 and θ as the first pair of controls and assign the independent linear dynamics to x and y:

$$\begin{array}{rclclcl} \ddot{x} & = & -u_1 \sin\theta & = & -k_{11}x - k_{12}\dot{x} & := & v_1(x, \dot{x}) \\ \ddot{y} & = & u_1 \cos\theta - 1 & = & -k_{21}y - k_{22}\dot{y} & := & v_2(y, \dot{y}) \end{array} \tag{6.1.54}$$

This will be achieved if u_1 and θ satisfy

$$u_1 = \left[v_1^2(x, \dot{x}) + (v_2(y, \dot{y}) + 1)^2\right]^{\frac{1}{2}} \tag{6.1.55}$$

$$\theta = \alpha(x, \dot{x}, y, \dot{y}) = \arctan\left(\frac{v_1(x, \dot{x})}{v_2(y, \dot{y}) + 1}\right) \tag{6.1.56}$$

Since u_1 is an actual control variable, we can use (6.1.55) as its control law. This is not the case with θ, which can not satisfy (6.1.56), because it is a state variable. To proceed with backstepping, we introduce the error variable $\xi = \theta - \alpha$, select a Lyapunov function and design a control law for u_2 in the subsystem

$$\ddot{\xi} = u_2 - \ddot{\alpha} \tag{6.1.57}$$

The lengthy expression for $\ddot{\alpha}$ as a function of $x, \dot{x}, y, \dot{y}, \xi, \dot{\xi}$, is obtained by twice differentiating (6.1.56). Either backstepping or cascade designs make use of this complicated expression.

High-gain design ("dynamic reduction"). A simpler approach is to approximately implement (6.1.56). With a high-gain control law for $\ddot{\theta} = u_2$ we will create an attractive invariant manifold near $\theta = \alpha$ to which the states will converge after a fast transient.

Using standard trigonometric identities we rewrite the model (6.1.53) with $\epsilon = 0$, in the cascade form

$$\begin{array}{rcl} \ddot{x} & = & v_1 - 2v_1 \sin(\frac{\theta+\alpha}{2}) \sin(\frac{\theta-\alpha}{2}) \\ \ddot{y} & = & v_2 - 2(v_2 + 1) \cos(\frac{\theta+\alpha}{2}) \sin(\frac{\theta-\alpha}{2}) \\ \ddot{\theta} & = & u_2 \end{array} \tag{6.1.58}$$

The interconnection term is zero at $\theta - \alpha = 0$, which makes it obvious that the desired dynamics of x and y are achieved if $\theta = \alpha$. To enforce $\theta = \alpha$ we use the high-gain controller

$$u_2 = -k_1 k^2 (\theta - \alpha) - k_2 k \dot{\theta} \tag{6.1.59}$$

where $0 < k_1 \leq 1$, $0 < k_2 \leq 1$, and k is the high gain proportional to the magnitude of the eigenvalues of the θ-subsystem:

$$\frac{\ddot{\theta}}{k^2} + k_2 \frac{\dot{\theta}}{k} + k_1 \theta = k_1 \alpha \tag{6.1.60}$$

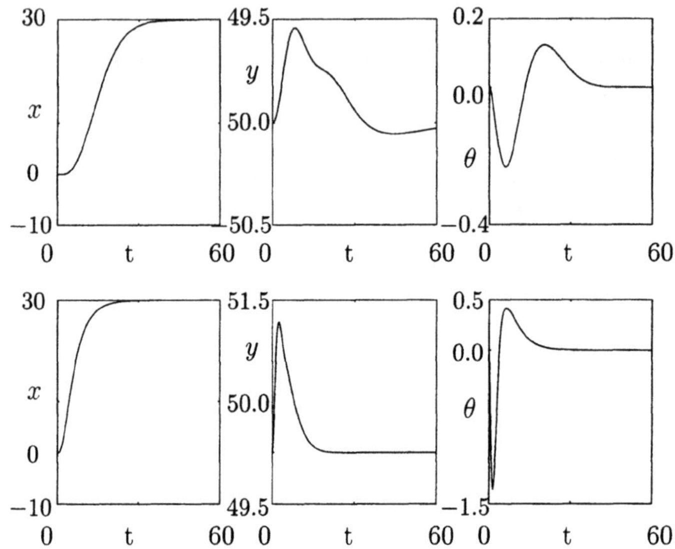

Figure 6.3: The response of the VTOL aircraft to the 30-unit step command for x: S-design is above and F-design is below.

Clearly, if $k \to \infty$ then $\theta \to \alpha$ and the off-manifold transients decay exponentially as e^{-kt}. The conditions of Proposition ?? are satisfied and the control law (6.3) achieves semiglobal stabilization of (6.1.58).

Controller calibration to limit peaking. We now consider the fact that the feedback control laws (6.1.55) and (6.1.59) will be applied to the model (6.1.53) with $\epsilon > 0$, say $\epsilon = 0.1$. It is clear from (6.1.59) that the high-gain control u_2 is initially peaking with k^2, because, in general, $\theta - \alpha \neq 0$ at $t = 0$. This means that the neglected ϵ-terms in (6.1.53) will be large, unless $\epsilon k^2 \ll 1$, which severely restricts the value of k. The time-scale separation between the slow x- and y-dynamics, and the fast θ-dynamics, can still be enforced by slowing down x and y, rather than speeding up θ. This can be accomplished by lowering the gains k_{11}, k_{12}, k_{21}, k_{22}, in (6.1.54), while retaining the slow manifold geometry which is due to the singular perturbation form of the designed system. By selecting two sets of values for k_{11}, k_{12}, k_{21}, k_{22}, k_1, k_2, and k, we assign two sets (S and F) of locations of the eigenvalues $\lambda_{x,y}$ and λ_θ:

$$\begin{array}{llll} S: & \lambda_{x,y} = -0.08 \pm j0.06, & \lambda_\theta = -0.4 \pm j0.3 \\ F: & \lambda_{x,y} = -0.3, -0.3 & \lambda_\theta = -2.4 \pm j1.8 \end{array}$$

For both sets, the λ_θ's are about 5–8 times larger than the λ_{xy}'s.

The responses in Figure 6.3 are for a transfer of x from 0 to 30 which, ideally, should not disturb y. In the faster transfer (F), the peak in y is about three times larger than in the slower transfer (S). In both cases, the effect of peaking is small and may be practically acceptable. As ϵ increases, so does the effect of peaking, and an alternative design may be required. □

6.2 Forwarding

6.2.1 Introductory example

The main ideas of forwarding will be introduced for the following *strict-feedforward* system

$$\begin{aligned} \dot{x}_1 &= x_2 + x_3^2 + x_2 u \\ \dot{x}_2 &= x_3 - x_3^2 u \\ \dot{x}_3 &= u \end{aligned} \tag{6.2.1}$$

represented by the block-diagram in Figure 6.4.

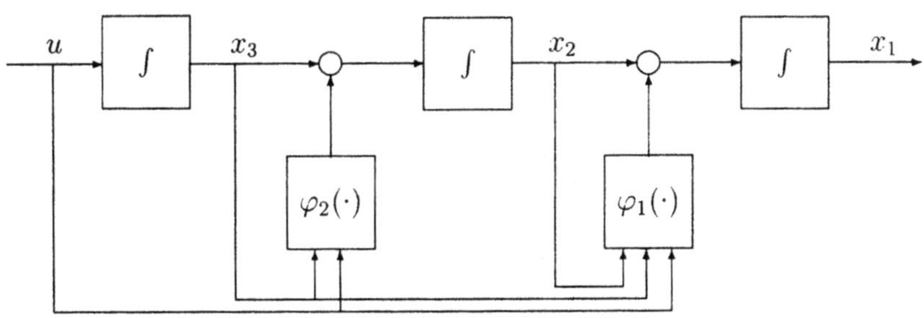

Figure 6.4: Block-diagram of a strict-feedforward system.

The block-diagram is characterized by the absence of feedback paths. This property excludes severe instabilities such as $\dot{x}_1 = x_1^2$, which may appear in strict-feedback systems, see Section 6.1.1. The solutions exist for all $t \geq 0$ as can be seen by rewriting (6.2.1) in the integral form:

$$\begin{aligned} x_3(t) &= x_3(0) + \textstyle\int_0^t u(s)ds \\ x_2(t) &= x_2(0) + \textstyle\int_0^t (x_3(s) + x_3^2(s)u(s))ds \\ x_1(t) &= x_1(0) + \textstyle\int_0^t (x_2(s) + x_3^2(s) + x_2(s)u(s))ds \end{aligned} \tag{6.2.2}$$

However, it is important to observe that, when $u = 0$ this system is unstable due to the triple integrator. Hence, some of its solutions grow unbounded as

$t \to \infty$. Our goal is to achieve global asymptotic stability by a systematic design procedure.

In backstepping, we have exploited the *lower-triangular* configuration of strict-feedback systems to develop a "top-down" recursive procedure. In a dual way, we will now exploit the *upper-triangular* configuration of the system (6.2.1) to develop a "bottom-up" recursive procedure.

Forwarding design, first step. In the first step we stabilize the *last* equation of (6.2.1), that is, $\dot{x}_3 = u$. For this passive system, a storage function is $V_3 = \frac{1}{2}x_3^2$ and a stabilizing feedback is $u = \alpha_3(x_3) = -x_3$.

With $u = -x_3$ we augment $\dot{x}_3 = -x_3$ by the x_2-equation and write the augmented system in the cascade form:

$$\begin{aligned} \dot{x}_2 &= 0 + \psi_2(x_3) \\ \dot{x}_3 &= -x_3 \end{aligned} \tag{6.2.3}$$

where $\psi_2(x_3) = x_3 - x_3^2\alpha_3(x_3) = x_3 + x_3^3$ is the interconnection term. In this cascade $\dot{x}_2 = 0$ is stable and $\dot{x}_3 = -x_3$ is GAS and LES. Therefore, the cascade is globally stable and the cross-term constructions of Chapter 5 are applicable. We apply it to construct a Lyapunov function V_2 for the augmented system (6.2.3) given the Lyapunov function V_3 for the system $\dot{x}_3 = -x_3$. From Section 5.3.1 we get

$$\begin{aligned} V_2 &= V_3 + \frac{1}{2}x_2^2 + \int_0^\infty \tilde{x}_2(s)\psi(\tilde{x}_3(s))ds \\ &= V_3 + \frac{1}{2}\lim_{s\to\infty} \tilde{x}_2^2(s) \\ &= V_3 + \frac{1}{2}(x_2 + x_3 + \frac{x_3^3}{3})^2 \end{aligned} \tag{6.2.4}$$

By construction, the time-derivative of V_2 satisfies

$$\dot{V}_2\Big|_{u=\alpha_3(x_3)} = \dot{V}_3\Big|_{u=\alpha_3(x_3)} = -x_3^2 \tag{6.2.5}$$

Although the control law $u = \alpha_3(x_3)$ has not achieved *asymptotic* stability of the augmented system (6.2.6), it allowed us to construct a Lyapunov function V_2 whose derivative for the subsystem

$$\begin{aligned} \dot{x}_2 &= x_3 - x_3^2 u \\ \dot{x}_3 &= u \end{aligned} \tag{6.2.6}$$

can be rendered negative by feedback $u = \alpha_2(x_2, x_3) = \alpha_3(x_3) + v_2$:

$$\dot{V}_2\Big|_{u=\alpha_3(x_3)+v_2} = \dot{V}_2\Big|_{u=\alpha_3(x_3)} + x_3 v_2 + (x_2 + x_3 + \frac{x_3^3}{3})(1 + x_3^2)v_2 \tag{6.2.7}$$

To make $\dot{V}_2$ negative we let

$$v_2 = -(x_2 + x_3 + \frac{x_3^3}{3})(1 + x_3^2) \tag{6.2.8}$$

and obtain

$$\dot{V}_2 = -x_3^2 + x_3 v_2 - v_2^2 < 0$$

so that the control law $\alpha_2(x_2, x_3) = -x_3 + v_2$ achieves GAS/LES of the equilibrium $(x_2, x_3) = (0, 0)$ of (6.2.6).

Optimality of forwarding. The optimality of the forwarding design is demonstrated by rewriting (6.2.7) as

$$\dot{V}_2 = \dot{V}_2\Big|_{u=u_3(x_3)} + (L_g V_2)v_2 = -(L_g V_3)^2 + (L_g V_2)v_2 \tag{6.2.9}$$

where $g^T(x) = (x_2, x_3^2, 1)$ is the control vector field of the system (6.2.1). From (6.2.8) we see that $v_2 = -L_g V_2 + L_g V_3$, which gives

$$\dot{V}_2 = -(L_g V_3)^2 + (L_g V_2)(L_g V_3) - (L_g V_2)^2 \leq \frac{1}{2}(L_g V_3)^2 - \frac{1}{2}(L_g V_2)^2 \tag{6.2.10}$$

and

$$\alpha_2 = \alpha_3 + v_2 = -L_g V_3 - (L_g V_2 - L_g V_3) = -L_g V_2 \tag{6.2.11}$$

This proves that, with respect to the output $y_2 = L_g V_2$, the system (6.2.6) is OFP$(-\frac{1}{2})$ and that V_2 is a storage function. Hence, using the results of Section 3.4, we conclude that for the subsystem (6.2.6), the control law

$$u = \alpha_2(x_2, x_3) = -x_3 - (x_2 + x_3 + \frac{x_3^3}{3})(1 + x_3^2) \tag{6.2.12}$$

minimizes a cost functional of the form

$$J = \int_0^\infty (l(x_2, x_3) + u^2)\, dt, \quad l \geq 0$$

and has a disk margin $D(\frac{1}{2})$. This property will be propagated through each step of forwarding.

Forwarding design, second step. Having completed the design of a stabilizing control law for the second-order subsystem (6.2.6), we proceed to the stabilization of the full third-order system

$$\begin{aligned} \dot{x}_1 &= x_2 + x_3^2 + x_2 u \\ \dot{x}_2 &= x_3 - x_3^2 u \\ \dot{x}_3 &= u \end{aligned} \tag{6.2.13}$$

With $u = \alpha_2(x_2, x_3)$ in (6.2.12), this system has the cascade form

$$\begin{aligned} \dot{x}_1 &= 0 + \psi_1(x_2, x_3) \\ \dot{x}_2 &= x_3 - x_3^2\alpha_2(x_2, x_3) \\ \dot{x}_3 &= \alpha_2(x_2, x_3) \end{aligned} \tag{6.2.14}$$

where $\psi_1(x_2, x_3) = x_2 + x_3^2 + x_2\alpha_2(x_2, x_3)$ is the interconnection term. For this globally stable cascade, the cross-term construction of Chapter 5 yields the Lyapunov function

$$\begin{aligned} V_1 &= V_2 + \frac{1}{2}x_1^2 + \int_0^\infty \tilde{x}_1(s)\psi_1(\tilde{x}_2(s), \tilde{x}_3(s))ds \\ &= V_2 + \frac{1}{2}\lim_{s\to\infty} \tilde{x}_1^2(s) \\ &= V_2 + \frac{1}{2}(x_1 + \phi(x_2, x_3))^2 \end{aligned} \tag{6.2.15}$$

In contrast to the explicit construction of V_2 in (6.2.4), we no longer have a closed-form expression for

$$\phi_1(x_2, x_3) = \int_0^\infty \left(\tilde{x}_2(s) + \tilde{x}_3^2(s) + \tilde{x}_2 u_2(\tilde{x}_2(s), \tilde{x}_3(s))\right) ds \tag{6.2.16}$$

This function has to be evaluated numerically or approximated analytically. By construction, the time-derivative of V_1 satisfies

$$\dot{V}_1\Big|_{u=\alpha_2(x_2,x_3)} = \dot{V}_2\Big|_{u=\alpha_2(x_2,x_3)} \leq -\frac{1}{2}(L_gV_2)^2 \tag{6.2.17}$$

and hence,

$$\dot{V}_1\Big|_{u=\alpha_2(x_2,x_3)+v_1} = \dot{V}_1\Big|_{u=\alpha_2(x_2,x_3)} + L_gV_1v_1 \leq -\frac{1}{2}(L_gV_2)^2 + L_gV_1v_1 \tag{6.2.18}$$

By choosing $v_1 = -L_gV_1 + L_gV_2$ we obtain $u = \alpha_1(x_1, x_2, x_3) = -L_gV_1$ and $\dot{V}_1 \leq -\frac{1}{2}(L_gV_1)^2$. The control law u_1 achieves GAS for the system (6.2.1) because it can be verified that $\dot{V}_1$ is negative definite.

The disk margin $D(\frac{1}{2})$ of the control law is thus preserved in the forwarding recursion. The control law $u_1(x_1, x_2, x_3)$ requires the partial derivatives of the function $\phi_1(x_2, x_3)$, which can be precomputed or evaluated on-line.

Instead of $v_1 = -L_gV_1 + L_gV_2$, we could have used $v_1 = -L_gV_1$ to make $\dot{V}_1$ in (6.2.18) negative definite. The choice $v_i = -L_gV_i$ at each step of forwarding results in the optimal value function $V = V_1 + V_2 + V_3$.

Reducing the complexity. Because of the integrals like (6.2.16), the complexity of forwarding control laws is considerable. A possible simplification is to use the relaxed constructions of Chapter 5 for the successive Lyapunov functions constructed at each step. Because of the nonlinear weighting of such Lyapunov functions, the resulting design is akin to the designs using nested saturations introduced by Teel [109]. We will discuss such simplified designs in Section 6.2.4.

6.2.2 Forwarding procedure

To present the forwarding procedure we start from a system

$$\dot{\xi} = a(\xi) + b(\xi)u \tag{6.2.19}$$

which, by assumption, is OFP($-\frac{1}{2}$) with an already constructed storage function $U(\xi)$ and the output $y = L_gU(\xi)$. To make the procedure recursive, we want to achieve the same OFP($-\frac{1}{2}$) property for the augmented system

$$\begin{aligned} \dot{z} &= f(z) + \psi(z,\xi) + g(z,\xi)u \\ \dot{\xi} &= a(\xi) + b(\xi)u \end{aligned} \tag{6.2.20}$$

The construction of the new storage function makes use of the cascade results of Chapter 5 valid under assumptions which we now collect for convenience in a single *forwarding assumption.*

Assumption 6.8 *(Forwarding assumption)*

(i) $\dot{z} = f(z)$ is GS, with a Lyapunov function $W(z)$ which satisfies Assumption 5.2.

(ii) The functions $\psi(z,\xi)$ and $g(z,\xi)$ satisfy a linear growth assumption in z, *Assumption 5.1.*

(iii) The function $f(z)$ has the form

$$f(z) = \begin{pmatrix} f_1(z_1) \\ F_2z_2 + f_2(z_1, z_2) \end{pmatrix}, \quad z = \begin{pmatrix} z_1 \\ z_2 \end{pmatrix} \tag{6.2.21}$$

where $f_2(0, z_2) = 0$, $\dot{z}_1 = f_1(z_1)$ is GAS, and $\dot{z}_2 = F_2z_2$ is GS, *Assumption 5.10.*

(iv) $W(z_1, z_2)$ is locally quadratic, that is $\frac{\partial^2 W}{\partial z^2}(0,0) = \bar{W} > 0$, and for each $z = (0, z_2)$, the following holds:

$$\frac{\partial\psi}{\partial\xi}(z,0) := M, \; g(z,0) := g_0, \text{and } \frac{\partial W}{\partial z}(z) = z_2^T\bar{W}_2$$

where M, g_0 and $\bar{W}_2$ are constant matrices, *Assumption 5.28.* □

As discussed in Chapter 5, assumptions (i) and (ii) are fundamental, while (iii) and (iv) are made for convenience to avoid separate tests of GAS and differentiability of V.

The following theorem presents the basic recursive step of forwarding.

Theorem 6.9 *(Forwarding as a recursive output feedback passivation)* Let $U(x)$ be a positive definite, radially unbounded, locally quadratic, storage function such that the system

$$\dot{\xi} = a(\xi) + b(\xi)u, \quad y_0 = (L_b U)^T(\xi) \tag{6.2.22}$$

is OFP($-\frac{1}{2}$) and ZSD. Furthermore, let the pair $(\frac{\partial a}{\partial \xi}(0), b(0))$ be stabilizable. Then, under Assumption 6.8, the cascade

$$\begin{array}{rcl} \dot{z} &=& f(z) + \psi(z,\xi) + g(z,\xi)u \\ \dot{\xi} &=& a(\xi) + b(\xi)u, \quad y = (L_g V)^T(z,\xi) \end{array} \tag{6.2.23}$$

is OFP($-\frac{1}{2}$) with a positive definite, radially unbounded storage function $V(z,\xi) = W(z) + \Psi(z,\xi) + U(\xi)$. Its cross-term is

$$\Psi(z,\xi) = \int_0^\infty \frac{\partial W}{\partial z}(\tilde{z}(s))(\psi(\tilde{z}(s),\tilde{\xi}(s)) + g(\tilde{z}(s),\tilde{\xi}(s))y_0(\tilde{\xi}(s))\, ds \tag{6.2.24}$$

evaluated along the solution $(\tilde{z}(s),\tilde{\xi}(s)) = (\tilde{z}(s;(z,\xi)),\tilde{\xi}(s;\xi))$ of the system (6.2.23) with the feedback control $u = -y_0(\xi)$.

Moreover, if the Jacobian linearization of (6.2.23) is stabilizable, the control law $u = -y = -(L_g V)^T$ achieves GAS and LES of $(z,\xi) = (0,0)$.

Proof: The system $\dot{\xi} = a(\xi) - b(\xi)L_b U(\xi)$ is GAS and LES. This follows from the OFP(-$\frac{1}{2}$) and ZSD properties of (6.2.22) and the stabilizability of its Jacobian linearization (Corollary 5.30).

The construction of $V(z,\xi)$ is an application of Theorem 5.8 to the cascade

$$\begin{array}{rcl} \dot{z} &=& f(z) + \psi(z,\xi) - g(z,\xi)(L_b U)^T(\xi) \\ \dot{\xi} &=& a(\xi) - b(\xi)(L_b U)^T(\xi) \end{array} \tag{6.2.25}$$

By this construction, the derivative of $V(z,\xi)$ along the solutions of (6.2.25) is

$$\dot{V} = L_f W(z) + \dot{U}\Big|_{u=-(L_b U)^T} \leq -\frac{1}{2}\|L_b U\|^2 \tag{6.2.26}$$

where we have used the OFP($-\frac{1}{2}$) property of the system (6.2.22).

With $u = -\frac{1}{2}y + v = -\frac{1}{2}(L_gV)^T + v$, we obtain

$$
\begin{aligned}
\dot{V}\Big|_{u=-\frac{1}{2}y+v} &= \dot{V}\Big|_{u=-(L_bU)^T} + (L_gV)(L_bU - \frac{1}{2}L_gV)^T + L_gVv \\
&\leq -\frac{1}{2}\|L_bU\|^2 + L_gV(L_bU)^T - \frac{1}{2}\|L_gV\|^2 + L_gVv \\
&\leq L_gVv = y^Tv
\end{aligned}
\tag{6.2.27}
$$

which proves that (6.2.23) is OFP($-\frac{1}{2}$).

When the Jacobian linearization of (6.2.23) is stabilizable, Theorem 5.29 and Corollary 5.30 guarantee that GAS/LES of the equilibrium $(z, \xi) = (0, 0)$ is achieved with the control law $\tilde{u} = -(L_bU)^T - (L_gV)^T$, which adds the damping control $-(L_gV)^T$ to the stable system (6.2.25). This result implies that $(z, \xi) = (0, 0)$ attracts all the solutions which start in the largest invariant set of $\dot{z} = f(z)$, $\dot{\xi} = a(\xi)$ where $\|L_gV\| \equiv \|L_bU\| \equiv 0$. Because $\dot{V}\Big|_{u=-(L_gV)^T} \leq -\frac{1}{4}\|L_bU\|^2 - \frac{1}{4}\|L_gV\|^2$, Theorem 2.21 implies that the control law $u = -(L_gV)^T$ also achieves GAS of $(z, \xi) = (0, 0)$. LES follows from the stabilizability of the Jacobian linearization of (6.2.23).

□

Feedforward systems. With a recursive application of the basic forwarding step we now construct a design procedure for systems in the form

$$
\begin{aligned}
\dot{z}_1 &= f_1(z_1) + \psi_1(z_1, z_2, \ldots, z_n, \xi) + g_1(z_1, z_2, \ldots, z_n, \xi)u \\
&\vdots \\
\dot{z}_{n-1} &= f_{n-1}(z_{n-1}) + \psi_{n-1}(z_{n-1}, z_n, \xi) + g_{n-1}(z_{n-1}, z_n, \xi)u \\
\dot{z}_n &= f_n(z_n) + \psi_n(z_n, \xi) + g_n(z_n, \xi)u \\
\dot{\xi} &= a(\xi) + b(\xi)u, \quad z_i \in I\!R^{q_i},\ i = 1, \ldots, n
\end{aligned}
\tag{6.2.28}
$$

where each z_i-block satisfies Assumption 6.8, with the required modification of notation. We point out that Assumption 6.8 imposes two fundamental restrictions on the system (6.2.28). They are the linear growth in z_i of the interconnection terms ψ_i and g_i, and the polynomial growth of the Lyapunov functions $W_i(z_i)$. Taken together they prevent the possibility of the solutions of (6.2.28) from escaping to infinity in finite time.

Proposition 6.10 *(Absence of escape in finite time)*
Consider (6.2.28) under Assumption 6.8 and let $u(t)$ be such that the solution $\xi(t)$ of the last equation exists for all $t \geq 0$. Then, the solution $(z(t), \xi(t))$ of (6.2.28) also exists for all $t \geq 0$.

Proof: We first prove that $z_n(t)$ exists for all $t \geq 0$. Proceeding as in the proof of Theorem 4.7, we use

$$\dot{W}_n = L_{f_n}W_n + L_{\psi_n}W_n + L_{g_n}W_n u(t) \leq L_{\psi_n}W_n + L_{g_n}W_n u(t) \tag{6.2.29}$$

and the fact that, for z_n large,

$$|L_{\psi_n}W_n + L_{g_n}W_n u(t)| \leq \gamma(\|(\xi(t), u(t))\|)W_n(z_n(t)) \tag{6.2.30}$$

These inequalities yield the estimate

$$\dot{W}_n \leq \gamma(\|(\xi, u)\|)W_n \tag{6.2.31}$$

which can be integrated as

$$W_n(z_n(t)) \leq e^{\int_0^t \gamma(\|(\xi(s),u(s))\|)ds} W_n(z_n(0)) \tag{6.2.32}$$

Because $\xi(t)$ and $u(t)$ exist for all $t \geq 0$, so does $W_n(z_n(t))$. Because W_n is radially unbounded, this also implies that $z_n(t)$ exists for all $t \geq 0$. The proof is analogous for each z_i, $i \leq n-1$. □

Forwarding procedure. If the Jacobian linearization of (6.2.28) is stabilizable, we can achieve GAS/LES of $(z, \xi) = (0, 0)$ in n recursive forwarding steps. The design is a bottom-up procedure in which a passivating output y_1 and the Lyapunov function V_1 for the entire system are constructed at the final step. Using the notation

$$G(z_1, \ldots, z_{n-1}, z_n, \xi) = (g_1(z_1, \ldots, z_n, \xi), \ldots, g_{n-1}(z_{n-1}, z_n, \xi), g_n(z_n, \xi), b(\xi))^T$$

we start with $y_0 = L_bU(\xi)$. The first step of forwarding yields

$$\begin{aligned}
V_n(z_n, \xi) &= W_n(z_n) + \Psi_n(z_n, \xi) + U(\xi) \\
\Psi_n(z_n, \xi) &= \int_0^\infty \frac{\partial W_n}{\partial z_n}(\tilde{z}_n)(\psi_n(\tilde{z}_n, \tilde{\xi}) - g_n(\tilde{z}_n, \tilde{\xi})L_bU(\tilde{\xi})\, ds \\
y_n &= L_G V_n(z_n, \xi)
\end{aligned}$$

where the integral is evaluated along the solutions of

$$\begin{aligned}
\dot{z}_n &= f_n(z_n) + \psi_n(z_n, \xi) + g_n(z_n, \xi)u \\
\dot{\xi} &= a(\xi) + b(\xi)u
\end{aligned}$$

with the feedback $u = -L_bU(\xi)$. For $i = n-1, \ldots, 1$, the recursive expressions are

$$\begin{aligned}
V_i(z_i, \ldots, z_n, \xi) &= W_i(z_i) + \Psi_i(z_i, \ldots, z_n, \xi) + V_{i+1}(z_{i+1}, \ldots, z_n, \xi) \\
\Psi_i(z_i, \ldots, z_1, \xi) &= \int_0^\infty \frac{\partial W_i}{\partial z_i}(\psi_i - g_i y_{i+1})\, ds \\
y_i &= L_G V_i(z_i, \ldots, z_n, \xi), \qquad i = n-1, \ldots, 1
\end{aligned}$$

where the integral is evaluated along the solutions of (6.2.28) with the control law $u = -y_{i+1}(z_{i+1}, \dots, z_n, \xi)$.

The final Lyapunov function is thus

$$V(z_1, \dots, z_n, \xi) = U + \sum_{i=n}^{1}(W_i + \Psi_i)$$

and GAS/LES of the entire system (6.2.28) is achieved with the feedback control law

$$u = -L_G V = -L_G U - \sum_{i=n}^{1} L_G(W_i + \Psi_i)$$

Stability margins. The stability margins of the forwarding design follow from its optimality. Proposition 6.9 shows that, if one starts with an OFP($-\frac{1}{2}$) system, this property is propagated through each step of forwarding. By Theorem 3.23, this means that, at each step, the control law $u_i = -L_G V_i$ minimizes a cost functional of the form

$$J = \int_0^\infty (l(z, \xi) + u^T u)dt, \quad l(z, \xi) \geq 0$$

and hence, achieves a disk margin $D(\frac{1}{2})$.

We stress that the stability margins of forwarding are achieved despite the fact that, in general, the constructed Lyapunov function V is not necessarily a CLF. The reason is that we have not imposed any restriction on the dimension of the vectors z_i and ξ so that, in general, $\dot{V}$ is rendered only negative semidefinite, rather than negative definite. As an illustration, let the last equation of (6.2.28) be

$$\dot{\xi} = A\xi + b(\xi)u, \quad A + A^T = 0 \tag{6.2.33}$$

The time-derivative of the Lyapunov function $U = \frac{1}{2}\xi^T\xi$ is

$$\dot{U} = (L_b U)u = \xi^T b(\xi)u$$

which means that V_n is a CLF for (6.2.33) only if the dimension of u is greater than or equal to that of ξ. When this is not the case, the task of finding a CLF may not be straightforward even for (6.2.33).

6.2.3 Removing the weak minimum phase obstacle

The above forwarding procedure started with the output $y_0 = L_b U$, which satisfied only the relative degree requirement. The recursive steps consisted of passivation designs for the subsystems of increasing dimensions. Only the output $y_1 = L_G V$ constructed in the final step satisfied both the relative degree one and the weak minimum phase requirements. In all the intermediate steps, the zero-dynamics subsystems for the constructed outputs can be unstable. Forwarding has thus removed the weak minimum phase obstacle to feedback passivation. In this sense, forwarding complements backstepping which has removed the relative degree obstacle.

It should be stressed, however, that the forwarding assumptions (Assumption 6.8) restrict the type of zero-dynamics instability. Instability in the Jacobian linearization can be caused only by repeated eigenvalues on the imaginary axis and, as shown in Proposition 6.10, no solution can escape to infinity in finite time. In Chapter 4 (Theorem 4.41), we have shown that with this non-minimum phase property the semiglobal stabilization of nonpeaking cascades is still possible.

Partially linear cascades. We now return to the cascade in Chapter 4:

$$\begin{aligned} \dot{z} &= f(z) + \tilde{\psi}(z, \xi)y \\ \dot{\xi} &= A\xi + Bu, \qquad y = C\xi \end{aligned} \tag{6.2.34}$$

where $\dot{z} = f(z)$ is GAS and the pair (A, B) is stabilizable. In Theorem 4.41, we have achieved *semiglobal* stabilization of (6.2.34) using partial-state feedback, under the assumption that (6.2.34) is a *nonpeaking cascade*, that is, the system (A, B, C) is nonpeaking and $\tilde{\psi}$ depends only on its nonpeaking states. We have also shown that, if either one of these conditions is not satisfied, then there exist vector fields $f(z)$ for which (6.2.34) is not semiglobally stabilizable, even by full-state feedback.

We now prove that under the same nonpeaking assumption, the cascade (6.2.34) can be *globally* stabilized. Our proof does not require an extra LES assumption of the z-subsystem $\dot{z} = f(z)$, although it involves steps of forwarding in which the state z is part of the lower subsystem. This difficulty is overcome by modifying the Lyapunov function $W(z)$ in such a way that, near the origin, the designed control laws do not depend on z. This modification ensures, at each step of forwarding, an exponential convergence of all the states involved in the construction of the cross-term.

Theorem 6.11 *(Nonpeaking cascades: global stabilization)*
Assume that (6.2.34) is a nonpeaking cascade, that is, (A, B, C) is a nonpeaking system, and ξ enters the interconnection $\tilde{\psi}(z, \xi)y$ only with its nonpeaking components: $\tilde{\psi} = \tilde{\psi}(z, y, \xi_s)$. Then the cascade (6.2.34) is globally stabilizable by full-state feedback.

Proof: As in the proof of Proposition 6.2, we assume, without loss of generality, that the ξ-subsystem has a uniform relative degree r and is in the normal form

$$\begin{aligned} \begin{bmatrix} \dot{\xi}_u \\ \dot{\xi}_s \end{bmatrix} &= \begin{bmatrix} A_u & A_J \\ 0 & A_s \end{bmatrix} \begin{bmatrix} \xi_u \\ \xi_s \end{bmatrix} + \begin{bmatrix} B_s \\ B_u \end{bmatrix} y \\ y^{(r)} &= u \end{aligned} \tag{6.2.35}$$

where A_s is Lyapunov stable, A_u has all its eigenvalues on the imaginary axis, and A_J is a part of the unstable Jordan blocks corresponding to the repeated eigenvalues on the imaginary axis.

The proof is in three parts: feedback passivation, forwarding, and backstepping.

Feedback passivation: Treating $\dot{y} = v_1$ as our virtual control, we extract the feedback passive part of the cascade (6.2.34):

$$\begin{aligned} \dot{z} &= f(z) + \tilde{\psi}(z, \xi_s, y)y \\ \dot{\xi}_s &= A_s\xi_s + B_s y \\ \dot{y} &= v_1 \end{aligned} \tag{6.2.36}$$

Proposition 4.21 yields the stabilizing feedback

$$v_1 = \alpha_1(z, y, \xi_s) = -y - (L_{\tilde{\psi}}W)^T - B_s^T P_s \xi_s \tag{6.2.37}$$

and the storage function $V_1(z, \xi_s, y) = W(z) + \xi_s^T P\xi_s + \frac{1}{2}y^T y$, where $W(z)$ is a Lyapunov function for $\dot{z} = f(z)$ and $P_s > 0$ satisfies $P_s A_s + A_s^T P_s \leq 0$. Note, however, that $(z, \xi_s, y) = (0, 0. 0)$ need not be LES.

Forwarding: If A_u is stable, we apply forwarding to the augmented subsystem

$$\begin{aligned} \dot{\xi}_u &= A_u\xi_u + A_{us}\xi_s + B_u y \\ \dot{z} &= f(z) + \tilde{\psi}(z, \xi_s, y)y \\ \dot{\xi}_s &= A_s\xi_s + B_s y \\ \dot{y} &= -y - (L_{\tilde{\psi}}W)^T - B_s^T P_s\xi_s + v_2 \end{aligned} \tag{6.2.38}$$

If A_u is not stable, we partition it and ξ_u in such a way that the ξ_u-subsystem takes the form

$$\begin{bmatrix} \dot{\xi}_{uu} \\ \dot{\xi}_{us} \end{bmatrix} = \begin{bmatrix} A_{uu} & A_{uJ} \\ 0 & A_{us} \end{bmatrix} \begin{bmatrix} \xi_{uu} \\ \xi_{us} \end{bmatrix} + \begin{bmatrix} A_{Ju} & B_{uu} \\ A_{Js} & B_{us} \end{bmatrix} \begin{bmatrix} \xi_s \\ y \end{bmatrix} \tag{6.2.39}$$

where A_{us} is stable. Now we apply forwarding to

$$\begin{aligned} \dot{\xi}_{us} &= A_{us}\xi_{us} + A_{Js}\xi_s + B_{us} y \\ \dot{z} &= f(z) + \tilde{\psi}(z, \xi_s, y)y \\ \dot{\xi}_s &= A_s\xi_s + B_s y \\ \dot{y} &= -y - (L_{\tilde{\psi}}W)^T - B_s^T P_s\xi_s + v_2 \end{aligned} \tag{6.2.40}$$

and leave the ξ_{uu}-block to be stabilized in the recursive application of forwarding.

Forwarding requires that, when $v_2 \equiv 0$, the states ξ_s and y, entering the ξ_{us}-subsystem of (6.2.40) converge exponentially to zero. However, the z-subsystem is not assumed to be exponentially stabilizable and the term $(L_{\tilde{\psi}}W)^T$ in (6.2.38) may destroy the exponential convergence of y. To eliminate the effect of this term near $z=0$, we "flatten" the Lyapunov function $W(z)$ around $z=0$ by replacing $W(z)$ with

$$\tilde{W}(z) = \int_0^{W(z)} \gamma(s)ds$$

where γ is a smooth positive function satisfying the following requirements:

$$\begin{array}{rcll}\gamma(s) & = & 0, & \text{for } s \in [0,1] \\ \gamma(s) & \geq & \delta > 0, & \text{for } \ s \geq 2\end{array} \tag{6.2.41}$$

One such function is $\gamma(s) = e^{-\frac{1}{(s-1)^2}}$ for $s \geq 1$ and $\gamma(s) = 0$ otherwise. The term $L_{\tilde{\psi}}\tilde{W}$ vanishes near the origin. The modified storage function $\tilde{V}_1 = \tilde{W} + \xi_s^T P \xi_s + \frac{1}{2}y^T y$ is radially unbounded, but it is not positive definite. It is only positive *semidefinite*. To prove that $(z, \xi_s, y) = (0,0,0)$ is GAS with the modified control

$$v_1 = \tilde{\alpha}_1(z, y, \xi_s) = -y - (L_{\tilde{\psi}}\tilde{W})^T - B_s^T P_s \xi_s, \tag{6.2.42}$$

we use Theorem 2.24. The closed-loop system (6.2.36), (6.2.42) is asymptotically stable conditionally to the set $\{(z,\xi_s,y)|\tilde{V}_1 = 0\}$ and GAS follows because $\dot{\tilde{V}}_1 \equiv 0 \Rightarrow y \equiv 0 \Rightarrow \xi_s \to 0$. In the manifold $\xi_s = 0, y = 0$ the z-dynamics reduces to $\dot{z} = f(z)$ which implies that $z \to 0$. Hence, for a given initial condition $(z(0), \xi_s(0), y(0))$ there exists T such that $\|z(t)\| \leq 1, \forall\, t \geq T$. Furthermore, for $t \geq T$, the control v_1 is independent of z and becomes $v_1 = -y - B_s P_s \xi_s$, which guarantees the exponential convergence of $\xi_s(t)$ and $y(t)$.

Now the conditions for the construction of the cross-term Ψ are satisfied and we can proceed with the forwarding design for (6.2.40). Assumption 6.8 is satisfied because the added ξ_{us}-subsystem is linear and the pair (A,B) is stabilizable. A Lyapunov function V_2 and a stabilizing feedback $v_2 = \alpha_2(z, \xi_s, \xi_{us}, y)$ are thus constructed to achieve GAS of $(z, \xi_{us}, \xi_s, y) = (0,0,0,0)$. For the next step of forwarding we extract the stable part of ξ_{uu} to augment the cascade (6.2.40). This procedure is repeated until $\xi_{u\ldots u}$ is void. The number of steps of forwarding needed is equal to the maximal multiplicity q of the eigenvalues of A_u. In the last step we obtain $v_q = \alpha_q(z,\xi,y)$, the stabilizing control law for (6.2.38), and the accompanying Lyapunov function $V_q(z,\xi,y)$.

Backstepping. If $r=1$ in (6.2.35), the control law

$$u = \alpha(z, \xi_s, \xi_u, y) = \alpha_1(z, \xi_s, y) + \alpha_q(z, \xi_s, \xi_u, y) \tag{6.2.43}$$

achieves GAS of $(z, y, \xi_u, \xi_s) = (0,0,0,0)$ and $\dot{V}_q|_{u=\alpha} \leq 0$. If $r > 1$, (6.2.43) is a virtual control law for $\dot{y}$. This control law must be backstepped through $r-1$ integrators to stabilize the entire system.

□

Although more complicated than the *partial-state* linear feedback in Theorem 4.41 which achieves semiglobal stabilization, the *full-state* feedback design in Theorem 6.11 achieves global stabilization and leads to an improvement in performance.

Example 6.12 *(Forwarding design for a nonpeaking cascade)*
We have achieved semiglobal stabilization of the nonpeaking cascade

$$\begin{aligned} \dot{z} &= -\delta z + \xi_3 z^2, \quad \delta > 0 \\ \dot{\xi}_1 &= \xi_2 \\ \dot{\xi}_2 &= \xi_3 \\ \dot{\xi}_3 &= u \end{aligned} \tag{6.2.44}$$

in Example 4.42 with the linear high-low gain feedback

$$u = -a\xi_3 - \xi_2 - \frac{1}{a}\xi_1. \tag{6.2.45}$$

The design (6.2.45) is appealing for its simplicity but we have seen that it causes the fast peaking of the control u and the slow peaking of the state ξ_1. In addition, because it does not use z for feedback, it does not improve the slow convergence of $\dot{z} = -\delta z$.

We will now show that these undesirable features can be eliminated by the full-state feedback forwarding design of Theorem 6.11. In this design we first disregard the ξ_1-equation and achieve feedback passivation of the subsystem

$$\begin{aligned} \dot{z} &= -\delta z + z^2 y \\ \dot{\xi}_2 &= \xi_3 \\ \dot{\xi}_3 &= u, \quad y = \xi_3 \end{aligned} \tag{6.2.46}$$

This subsystem meets the relative degree one and weak minimum phase requirements of Proposition 4.21 which yields the stabilizing control law

$$u = -\xi_2 - \xi_3 - z^3 \tag{6.2.47}$$

and the Lyapunov function $U = \frac{1}{2}(z^2 + \xi_2^2 + \xi_3^2)$.

To stabilize the entire system (6.2.44), we apply one step of forwarding by constructing a Lyapunov function for the augmented system

$$\begin{aligned} \dot{\xi}_1 &= \xi_2 \\ \dot{z} &= -\delta z + z^2 \xi_3 \\ \dot{\xi}_2 &= \xi_3 \\ \dot{\xi}_3 &= -\xi_2 - \xi_3 - z^3 \end{aligned} \tag{6.2.48}$$

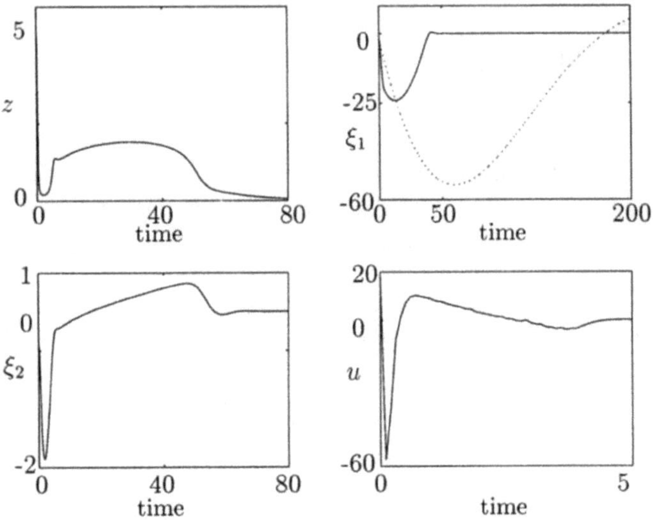

Figure 6.5: Typical response of the forwarding design for (6.2.46). Compared with the low-high gain design (dotted in ξ_1), the peaking of u and ξ_1 is significantly reduced.

Defining the new state

$$\zeta = \xi_1 + \int_0^\infty \tilde{\xi}_2(s)ds = \xi_1 + \phi_1(\xi_2, \xi_3, z) \tag{6.2.49}$$

where $\tilde{\xi}_2(s) = \tilde{\xi}_2(s; (z, \xi_2, \xi_3))$ is a solution of (6.2.48), we obtain the Lyapunov function

$$V = U + \frac{1}{2}\zeta^2 = \frac{1}{2}(z^2 + \xi_2^2 + \xi_3^2 + \zeta^2) \tag{6.2.50}$$

With the additional feedback

$$v = -\frac{\partial \phi_1}{\partial \xi_3}\zeta, \tag{6.2.51}$$

the final control law is

$$u = \xi_2 - \xi_3 - z^3 - \frac{\partial \phi_1}{\partial \xi_3}\zeta \tag{6.2.52}$$

With this control law, the time-derivative of V is

$$\dot{V} = -\delta z^2 - \xi_3^2 - \frac{\partial \phi_1}{\partial \xi_3}\zeta\xi_3 - (\frac{\partial \phi_1}{\partial \xi_3}\zeta)^2 \leq 0 \tag{6.2.53}$$

and the closed-loop system has a gain margin $(\frac{1}{2}, \infty)$.

The control law (6.2.52) contains the integrals ϕ_1 and $\frac{\partial \phi_1}{\partial \xi_3}$. As in Section 5.4.3, these integrals were numerically evaluated over the interval $T = 60$.

Figure 6.5 shows the significant improvement in performance with respect to the partial-state feedback design in Figure 4.8. The overshoot of ξ_1 is reduced in half and the settling time is reduced by an order of magnitude. By comparing the control effort we see that the partial-state feedback design is active only during a very short transient with a peak about two times larger than the full-state design. On the other hand, the full-state feedback remains active steering ξ_3 to achieve fast convergence of ξ_1 and z. Because the design does not force ξ_3 to stay small after its fast convergence, it alleviates the slow peaking of the state ξ_1. All the states converge in the same time scale. □

6.2.4 Geometric properties of forwarding

To exhibit the underlying geometry of forwarding, we consider a special class of feedforward systems

$$\begin{aligned} \dot{z}_1 &= F_1 z_1 + \psi_1(z_2, \ldots, z_n) + g_1(z_2, \ldots, z_n)u \\ &\vdots \\ \dot{z}_{n-1} &= F_{n-1} z_{n-1} + \psi_{n-1}(z_n) + g_{n-1}(z_n)u \\ \dot{z}_n &= F_n z_n + g_n u \end{aligned} \tag{6.2.54}$$

where $F_i + F_i^T = 0$, $i = 1, \ldots, n$. We call such systems *strict-feedforward systems* because they exclude any feedback connection except in $\dot{z}_i = F z_i$. It is easily verified that each subsystem of (6.2.54) satisfies Assumption 6.8 with a quadratic Lyapunov function $W_i(z_i) = \frac{1}{2} z_i^T z_i$. Hence, (6.2.54) is globally stabilizable if its Jacobian linearization is stabilizable.

At each step of forwarding, a Lyapunov function is constructed for the corresponding cascade

$$\begin{aligned} \dot{z}_i &= F_i z_i + \psi(\xi) \\ \dot{\xi} &= a(\xi), \end{aligned} \tag{6.2.55}$$

where $\dot{\xi} = a(\xi)$ is GAS and LES. As shown in Section 5.3.1, the construction of the cross-term Ψ for such a cascade is equivalent to the use of the decoupling change of coordinates

$$\zeta_i = z_i + \int_t^{\infty} e^{-F_i(\tau - t)} \psi(\xi(\tau + t; t; \xi)) d\tau, \tag{6.2.56}$$

which transforms (6.2.55) into

$$\begin{aligned} \dot{\zeta}_i &= F_i \zeta_i \\ \dot{\xi} &= a(\xi) \end{aligned}$$

In the new coordinates (ζ_i, ξ), the Lyapunov function $V_i(z,\xi) = \frac{1}{2}z_i^T z_i + \Psi(z_i,\xi) + U(\xi)$ reduces to the sum $\frac{1}{2}\zeta_i^T\zeta_i + U(\xi)$. This change of coordinates will help us to display significant geometric properties of forwarding.

Proposition 6.13 (*Geometry of forwarding*)
If the Jacobian linearization of (6.2.54) is stabilizable, then there exists a global change of coordinates $\zeta = T(z)$, such that, in the new coordinates, a storage function for (6.2.54) is quadratic

$$V(T^{-1}(\zeta)) = \frac{1}{2}\sum_{i=n}^{1}\zeta_i^T\zeta_i$$

For the output $y = L_G V$, the system (6.2.54) is OFP($-\frac{1}{2}$) and the feedback control

$$u = -y = -L_G V(T^{-1}(\zeta)) = -\sum_{i=1}^{n}\kappa_i(\zeta_{i+1},\ldots,\zeta_n)^T\zeta_i \tag{6.2.57}$$

where $\kappa_i(\zeta_{i+1},\ldots,\zeta_n) = L_G\zeta_i$, achieves GAS and LES of $z = 0$.

The resulting closed-loop system (6.2.54), (6.2.57) has the form

$$\begin{aligned}
\dot{\zeta}_1 &= (F_1 - \kappa_1\kappa_1^T)\zeta_1 \\
\dot{\zeta}_2 &= (F_2 - \kappa_2\kappa_2^T)\zeta_2 - \kappa_2\kappa_1^T\zeta_1 \\
&\vdots \\
\dot{\zeta}_{n-1} &= (F_{n-1} - \kappa_{n-1}\kappa_{n-1}^T)\zeta_{n-1} - \kappa_{n-1}(\textstyle\sum_{i=1}^{n-2}\kappa_i^T\zeta_i) \\
\dot{\zeta}_n &= (F_n - \kappa_n\kappa_n^T)\zeta_n - \kappa_n(\textstyle\sum_{i=1}^{n-1}\kappa_i^T\zeta_i)
\end{aligned} \tag{6.2.58}$$

□

A geometric interpretation of the change of coordinates $\zeta = T(z)$ is that at each step of the design, it transforms the added equation into

$$\dot{\zeta}_i = F_i\zeta_i + \kappa_i v_i \tag{6.2.59}$$

so that, for $v_i = 0$, the hyperplane $\zeta_i = 0$ is the *global stable manifold* of the augmented cascade: the solutions starting in the manifold converge to the origin. This manifold remains invariant under feedback $v_i = -\kappa_i(\xi_{i+1},\ldots,\xi_n)\zeta_i$, which renders it attractive, that is, achieves GAS of the augmented system.

Example 6.14 (*Phase portrait of a forwarding design*)
Let us reexamine the second step of our introductory example

$$\begin{aligned}
\dot{x}_2 &= x_3 + x_3^3 - x_3^2 v_2 \\
\dot{x}_3 &= -x_3 + v_2
\end{aligned} \tag{6.2.60}$$

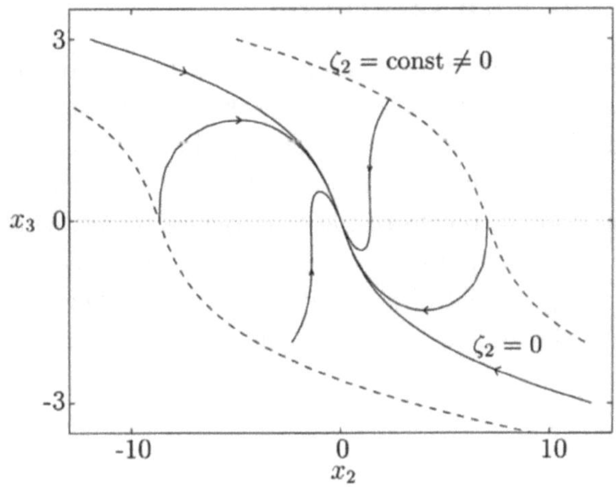

Figure 6.6: The stable manifold $\zeta_2 = 0$ of the system (6.2.60) with $v_2 = 0$ is rendered attractive by the forwarding design.

When $v_2 = 0$, then $\zeta_2 = x_2 + x_3 + x_3^3$ is constant along each solution of (6.2.60). The solutions converge to the axis $x_3 = 0$, dashed curves in Figure 6.6. With the additional feedback

$$v_2 = -\kappa_2(x_3)\zeta_2 = -x_2 - x_3 - \frac{x_3^3}{3} \tag{6.2.61}$$

the solutions converge to the globally stable manifold $\zeta_2 = 0$, solid curves in Figure 6.6.

□

The geometry of the second order system (6.2.60) is propagated through the steps of forwarding to form a sequence of nested invariant submanifolds. After n steps, the solutions are attracted first, to the manifold $\zeta_1 = 0$, which is invariant; then to the submanifold $\zeta_1 = \zeta_2 = 0$, which is also invariant. Eventually, the solutions are attracted to the submanifold $\zeta_1 = \zeta_2 = \ldots = \zeta_{n-1} = 0$, in which the feedback system is described by $\dot{\zeta}_n = (F_n - \kappa_n(\zeta)\kappa_n^T(\zeta))\zeta_n$. Each of the invariant submanifolds is the stable manifold of the cascade (6.2.55). At each step of the design, the stable manifold of the corresponding augmented system is rendered attractive by the new term $-\kappa_i(\zeta)^T\zeta_i$ added in the control law.

6.2.5 Designs with saturation

Because of the complexity of forwarding, which is due to the integrations required for the construction of cross-terms Ψ_i, simplified designs are even more desirable than in the case of backstepping. In our simplification of backstepping, the exact implementation of derivatives was avoided by employing high-gain feedback loops to enforce the convergence to desired invariant submanifolds. To avoid computation of the integrals required for forwarding, we will employ low-gain control laws with *saturation.* They let the solutions approach nested submanifolds which are different from the submanifolds of forwarding.

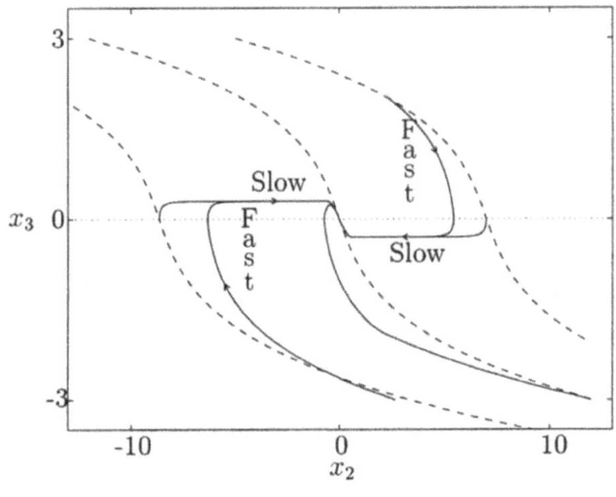

Figure 6.7: The saturation design lets the solutions of the system (6.2.60) approach the manifold $x_3 = 0$.

Example 6.15 *(Phase portrait of saturation design)*
To illustrate such a simplification of forwarding, we again consider the second order system (6.2.60). We have just seen that, with $v_2 = 0$, its solutions converge to the axis $x_3 = 0$. If v_2 is saturated at a small ϵ, then the solutions will converge to an ϵ-neighborhood of $x_3 = 0$, where $|x_3|$ is small, and the following approximations hold:

$$V_2 = \frac{x_3^2}{2} + \frac{1}{2}(x_2 + x_3 + x_3^3)^2 \approx \frac{x_3^2}{2} + \frac{1}{2}(x_2 + x_3)^2 \tag{6.2.62}$$

and

$$L_g V_2 = x_2 + x_3 + \frac{x_3^3}{3} \approx x_2 + x_3 \tag{6.2.63}$$

The ϵ-saturated control law

$$v_2 = -\sigma_\epsilon(x_2 + x_3) \tag{6.2.64}$$

has the following two properties. First, while in saturation, it lets the solutions of (6.2.60) converge to a neighborhood of $x_3 = 0$. Second, in this neighborhood, the damping control $v = -L_g V_2$ can be replaced by its linear approximation (6.2.63).

Figure 6.7 shows the phase portrait of (6.2.60) with the saturated control law (6.2.63) for $\epsilon = 0.3$. A comparison with Figure 6.6 shows the difference in geometric properties of the two designs. Instead of converging to the stable manifold $\zeta_2 = 0$, as in Figure 6.6, the solutions in Figure 6.7 mimic the uncontrolled behavior ($v_2 = 0$) until they approach the axis $x_3 = 0$. In addition, the smallness of ϵ creates a time-scale separation between the convergence rate to the manifold $x_3 = 0$, which is *fast*, and the convergence rate to the origin *along* the manifold $x_3 = 0$, which is *slow* because it is governed by the equation

$$\dot{\zeta} = -\sigma_\epsilon(\zeta) + \mathcal{O}(x_3^2), \quad \zeta = x_2 + x_3$$

The separation between the off-manifold behavior and in-manifold behavior illustrated in Figure 6.7 depends on the smallness of the parameter ϵ. Intermediate phase portraits between the extremes shown in Figures 6.6 and 6.7 can be obtained with larger values of ϵ.

The above example is important because it shows that a linear design combined with a saturation suffices to achieve global stabilization of the feedforward system

$$\begin{aligned} \dot{x}_2 &= x_3 + x_3^3 - x_3^2 v \\ \dot{x}_3 &= -x_3 + v \end{aligned} \tag{6.2.65}$$

The linear part of the saturation design is a forwarding design for the Jacobian linearization of the system (6.2.60), that is for the double integrator $\dot{x}_2 = x_3$, $\dot{x}_3 = u$. For this system, forwarding yields the Lyapunov function $V_2 = \frac{1}{2}x_3^2 + \frac{1}{2}(x_2+x_3)^2$ and the linear control $v = -x_2 - x_3$, which correspond, respectively, to the approximations (6.2.62) and (6.2.63) of the forwarding design for the original system (6.2.65). □

Nested saturation design. The design illustrated on the system (6.2.65) is now extended to the class of strict-feedforward systems (6.2.54). For the linear part of the design, we consider only the Jacobian linearization

$$\begin{aligned}\dot{z}_1 &= F_1 z_1 + M_1[z_2^T, \ldots, z_n^T]^T + G_1 u \\ &\vdots \\ \dot{z}_{n-1} &= F_{n-1} z_{n-1} + M_{n-1} z_n + G_{n-1} u \\ \dot{z}_n &= F_n z_n + g_n u\end{aligned} \tag{6.2.66}$$

and assume that it is stabilizable. A forwarding design for this linear system yields the quadratic Lyapunov function

$$V = \sum_{i=1}^{n} (z_i^T W_i z_i + z_i \Psi_i [z_{i+1}^T \ldots z_n^T]^T) =: \sum_{i=1}^{n} [z_i^T \ldots z_n^T] P_i [z_i^T \ldots z_n^T]^T$$

and the linear control law

$$u = -L_G V = -\sum_{i=1}^{n} \left([G_i^T \ldots G_n^T] P_i [z_i^T \ldots z_n^T]^T \right) := -\sum_{i=1}^{n} K_i z$$

By combining this linear design with saturations, we recover the nested saturation design of Teel [109], which was the first constructive result for the stabilization of feedforward systems.

Proposition 6.16 *(Nested saturation design)*
Consider the strict-feedforward system (6.2.54) and assume that its Jacobian linearization is stabilizable. Then, for any $\epsilon_n > 0$ there exists a sequence of saturation levels $\epsilon_n > \epsilon_{n-1} > \ldots > \epsilon_1 > 0$ of the saturation functions $\sigma_n, \ldots, \sigma_1$, such that control law

$$u = -\sigma_n(K_n z + \sigma_{n-1}(K_{n-1} z + \ldots + \sigma_1(K_1 z)) \ldots) \tag{6.2.67}$$

achieves global asymptotic stability (GAS) and local exponential stability (LES) of $z = 0$. □

It is of interest to compare the nested saturation design with the nested high-gain design of Proposition 6.5 for strict-feedback systems. Because saturations are used instead of linear low gains, the result of Proposition 6.16 is global as opposed to the semiglobal result of Proposition 6.5. As stated here, these two results are asymptotic in the sense that they are guaranteed to hold for sufficiently small values of the parameters ϵ_i. Additional effort is required to quantify these values.

Example 6.17 *(Saturation design for a nonpeaking cascade)*
The basic idea of the nested saturation design can be used in the stabilization of the cascade in Example 6.12:

$$\begin{array}{rcl}\dot{\xi}_1 & = & \xi_2 \\ \dot{z} & = & -\delta z + \xi_3 z^2 \\ \dot{\xi}_2 & = & \xi_3 \\ \dot{\xi}_3 & = & u\end{array} \tag{6.2.68}$$

Repeating the feedback passivation part of the forwarding design, and ignoring first the state ξ_1, we obtain

$$u = -\xi_2 - \xi_3 - z^3 + v \tag{6.2.69}$$

For the Lyapunov function $U = \frac{1}{2}(z^2 + \xi_2^2 + \xi_2\xi_3 + \xi_3^2)$, this feedback transformation achieves

$$\dot{U} = -\delta z^2 - \xi_2^2 - \xi_2\xi_3 - \xi_3^2 + \xi_3 v \le -\delta z^2 - \frac{1}{2}\xi_2^2 - \frac{1}{2}\xi_3^2 + \xi_3 v \tag{6.2.70}$$

Here we depart from the forwarding design and avoid the computation of the integrals in Example 6.12. We augment the control law (6.2.69) with the saturated feedback

$$v = -\sigma_\epsilon(\zeta_l), \quad \zeta_l = \xi_1 + \xi_2 + \xi_3 \tag{6.2.71}$$

where ζ_l is the linearization of ζ , obtained in the forwarding design (6.2.49). The bound $|v| \le \epsilon$ and (6.2.70) imply that $\dot{U} < 0$ provided that

$$|\xi_3| > 2\epsilon, \ |z| > \frac{\epsilon}{\sqrt{2\delta}}, \text{ or } |\xi_2| > \frac{\epsilon}{\sqrt{2}}$$

Hence, for any solution of the closed-loop system, there exists $t = t_1$ after which the states $z(t)$, ξ_2, and $\xi_3(t)$ are bounded by $\mathcal{O}(\epsilon)$. For $t \ge t_1$, we have

$$\dot{\zeta}_l = -\sigma_\epsilon(\zeta_l) + z^3 = -\sigma_\epsilon(\zeta_l) + \mathcal{O}(\epsilon^3) \tag{6.2.72}$$

which implies that ζ_l will also be bounded by ϵ after some time $t = t_2$. For $t \ge t_2$, the control law is not saturated and the system (6.2.68) is an exponentially stable linear system perturbed by higher-order terms in ξ_3 and z. For ϵ small enough, the solution is in the region of attraction of $(z, \xi) = (0, 0)$, which proves that the saturation design achieves GAS/LES of $(z, \xi) = (0, 0)$.

The above analysis does not quantify the saturation level ϵ which achieves GAS. If this level has to be chosen too small, the performance and robustness of the saturation design may be compromised. With the control law (6.2.71), simulations have shown that saturation levels higher than $\epsilon = 1$ do not improve

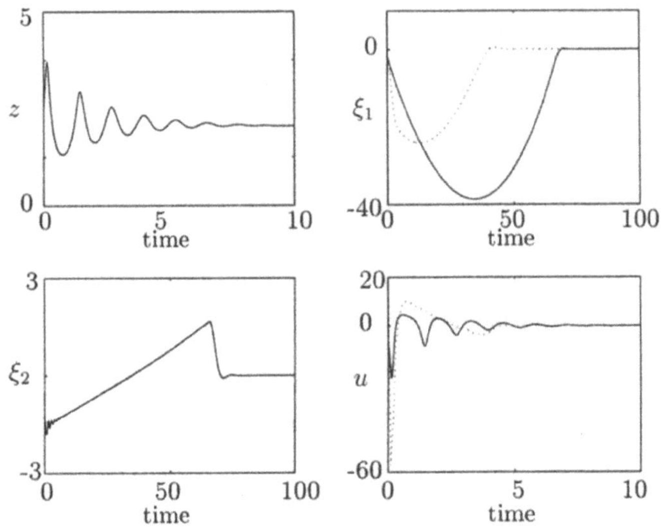

Figure 6.8: Saturation design for (6.2.68), solid curves, compared with forwarding, dashed curves.

the performance because the response becomes more oscillatory. However, $\epsilon = 1$ gives a satisfactory performance shown in Figure 6.8 for the same initial condition as in Example 6.12.

□

Relaxed forwarding. Using the relaxed construction of Section 5.3, we now provide a simplification of forwarding for a larger class than the strict-feedforward systems. The main building block in this simplification of forwarding is the cascade

$$\begin{aligned} \dot{z} &= Fz + \psi(z,\xi) + g(z,\xi)u \\ \dot{\xi} &= a(\xi) + b(\xi)u, \end{aligned} \tag{6.2.73}$$

where $\dot{z} = Fz$ is stable, and $\psi(z,\xi) = M\xi + r(z,\xi)$, with $r(z,\xi)$ second or higher order in ξ. For this system Corollary 5.26 guarantees the existence of a Lyapunov function which is obtained from the quadratic approximation of the cross term Ψ.

Proposition 6.18 *(Relaxed forwarding)*
Suppose that $U(x)$ is a positive definite, radially unbounded storage function such that the system

$$\dot{\xi} = a(\xi) + b(\xi)u, \quad y_0 = L_bU(\xi) \tag{6.2.74}$$

is OFP($-\frac{1}{2}$) and ZSD, and, moreover, $\dot{U}|_{u=-y_0}$ is locally quadratic, that is, $\dot{U}|_{u=-y_0} \leq -c\|\xi\|^2$, in some neighborhood of the origin. Let $\bar{V}(z,\xi)$ be the quadratic approximation of the Lyapunov function $V_0 = W(z)+\Psi(z,\xi)+U(\xi)$.

Then the cascade (6.2.73) with the output $y = L_G V(z,\xi)$ and the storage function

$$V(z,\xi) = U(\xi) + \ln(\bar{V}(z,\xi)+1) + \int_0^{U(\xi)} \gamma(s)ds$$

is OFP($-\frac{1}{2}$).

If the Jacobian linearization of (6.2.23) is stabilizable, the control law $u = -y = -L_G V$ achieves GAS and LES of the cascade. □

The relaxed forwarding procedure employs Proposition 6.18 as its basic step. The development of this procedure follows that of Section 6.2.2 and is not given here. We see that the main simplification in the relaxed procedure is that the Lyapunov function $V(z,\xi)$ can be computed by solving the set of algebraic equations (5.3.24) rather than evaluating the integrals needed for the cross-term $\Psi(z,\xi)$.

Relaxed forwarding applies to a larger class of systems than the nested saturation design because ψ and g in (6.2.73) are allowed to depend on z. Another important difference is that it provides a Lyapunov function for the closed-loop system. However, the control laws designed with relaxed forwarding and nested saturations have similar geometric properties because the control law $u = -L_G V$ in the above proposition is of the form

$$u = -(1+\gamma(U))L_b U - \frac{L_G \bar{V}}{1+\bar{V}} \tag{6.2.75}$$

where the function γ has to be sufficiently large to achieve domination in $\dot{V}$. As in the saturation design, the second term of the control law (6.2.75) is saturated and its gain lower than the "gain" $(1+\gamma(U))$ of the first term. The similarity with a saturation design is illustrated in the following example.

Example 6.19 *(Relaxed forwarding for a nonpeaking cascade)*
We return to the a nonpeaking cascade considered in Examples 6.12 and 6.17. The first step, feedback passivation, is the same as in the forwarding and saturation designs and we arrive at the cascade

$$\begin{aligned}
\dot{\xi}_1 &= \xi_2 \\
\dot{z} &= -\delta z + z^2 \xi_3 \\
\dot{\xi}_2 &= \xi_3 \\
\dot{\xi}_3 &= -z^3 - \xi_2 - \xi_3 + v
\end{aligned} \tag{6.2.76}$$

To construct a composite Lyapunov function, we employ the relaxed change of coordinates given in Proposition 5.22 because the nonresonance condition is satisfied. Since $M_l = 0$, we solve $NA - FN = -M$, where

$$F = 0, \quad M = \begin{bmatrix} 0 & 1 & 0 \end{bmatrix}, \quad A = \begin{bmatrix} -\delta & 0 & 0 \\ 0 & 0 & 1 \\ 0 & -1 & -1 \end{bmatrix}$$

The solution $N = [0\ 1\ 1]$ and $\zeta_1 = \xi_1 + N \begin{bmatrix} z & \xi_2 & \xi_3 \end{bmatrix}^T = \xi_1 + \xi_2 + \xi_3$ transform the system (6.2.76) into

$$\begin{aligned} \dot{\zeta}_1 &= -z^3 + v \\ \dot{z} &= -\delta z + z^2 \xi_3 \\ \dot{\xi}_2 &= \xi_3 \\ \dot{\xi}_3 &= -z^3 - \xi_2 - \xi_3 + v \end{aligned} \tag{6.2.77}$$

Since the cross-term z^3 in the ζ_1-subsystem is independent of ζ_1, we can use the Lyapunov function

$$V = \sqrt{\zeta_1^2 + 1} - 1 + \int_0^U \gamma(s)\, ds \tag{6.2.78}$$

where $U(z, \xi_2, \xi_3) = \frac{1}{2}(z^2 + \xi_2^2 + \xi_3^2)$ and γ has to be chosen to guarantee

$$\dot{V} = \frac{-z^3 \zeta_1}{\sqrt{\zeta_1^2 + 1}} + \gamma(U)(-\delta z^2 - \xi_3^2) \leq 0$$

One such γ is $\gamma(U) = \frac{1}{\delta}(1 + U)$.

Returning to the cascade (6.2.77), we employ the damping control law

$$v = -L_G V = -\frac{1}{\delta}(1 + U)\xi_3 - \frac{\zeta_1}{\sqrt{\zeta_1^2 + 1}} \tag{6.2.79}$$

Let us now compare this control law with the one obtained by the saturation design. Clearly, the second term of (6.2.79) is a saturated function of ζ_1 with saturation level one. Instead of employing a small saturation level ϵ, relaxed design increases the gain in the first term of the control law (6.2.79). Because the control law (6.2.79) is a rescaled version of the saturated control law, the responses of the two designs are similar.

□

6.2.6 Trade-offs in saturation designs

Simplification versus performance. By avoiding the computation of the cross-terms Ψ_i, saturation and relaxed forwarding designs considerably simplify forwarding but they also change its geometric properties. It has already been shown in Figures 6.6 and 6.7 that the saturation design is less active in the regions of the state space where the control law saturates and a similar conclusion applies to the relaxed forwarding design. In particular, these designs do not react to large excursions of the state z_i during the stabilization of the lower states $(z_{i+1}, \ldots, z_n)$. In fact, the stabilization of the state z_i is delayed until the solution has approached the manifold $z_{i+1} = \ldots = z_n = 0$. Along this manifold, the convergence of z_i is slow because of saturation.

The *saturation* in forwarding is dual to the *domination* in backstepping. A benefit from these simplified designs is that they tolerate more uncertainty in the form of the nonlinearities: a growth estimate is sufficient to determine the control law gains. However, this is also a limitation, because the system nonlinearities are not actively employed for stabilization.

Flexibility in the choice of the saturation levels. In forwarding, an additional feedback is designed at each step to achieve GAS of a system which is already GS. Because damping controls $v_i = -L_g V_i$ have a disk margin $D(0)$, the designer is free at each step to replace the control law $v_i = -L_g V_i$ by the control law $v_i = -\varphi(LgV_i)$ where $\varphi(.)$ is any static nonlinearity in the sector $(0, \infty)$, with $\varphi'(0) > 0$ to ensure local exponential stability. The added control law can thus be saturated at each step at a level chosen by the designer.

The situation is different in the saturation designs where the smallness of the saturation levels is dictated by system nonlinearities. This situation is dual to high-gain designs where the gains must dominate system nonlinearities and cannot be freely chosen by the designer. High gains and low saturation levels are both harmful for the robustness of the feedback system: high gains increase the sensitivity to fast unmodeled dynamics, while low saturation levels increase the sensitivity to external disturbances. Because of the saturation, the control law does not react to an instability caused by such disturbances.

Example 6.20 *(Saturation levels as design parameters)*
We consider the stabilization of the strict-feedforward system

$$\begin{aligned} \dot{x}_1 &= x_2 + 3x_2^3 - 3x_3^3 \\ \dot{x}_2 &= x_3 \\ \dot{x}_3 &= u \end{aligned} \tag{6.2.80}$$

Forwarding design. We first stabilize the (x_2, x_3)-subsystem of (6.2.80) with

$$u = -x_2 - x_3 + v_1$$

and the Lyapunov function $V_2 = \frac{1}{2}x_2^2 + \frac{1}{2}x_3^2$. To complete the design with forwarding, we evaluate

$$\zeta_1 := x_1 + \int_0^\infty (\tilde{x}_2(s) + 3\tilde{x}_2^3(s) - 3\tilde{x}_3^3(s))ds \tag{6.2.81}$$

along the solutions of the subsystem

$$\begin{aligned} \dot{\tilde{x}}_2 &= \tilde{x}_3 \\ \dot{\tilde{x}}_3 &= -\tilde{x}_2 - \tilde{x}_3 \end{aligned}$$

We obtain

$$\zeta_1 = x_1 + x_2 + x_3 + 3x_2^3 + 3x_2^2 x_3 + 3x_2 x_3^2, \tag{6.2.82}$$

and use it to construct the Lyapunov function $V = \frac{1}{2}(\zeta_1^2 + x_2^2 + x_3^2)$. The resulting damping control law

$$v_1 = -L_g V = -x_3 - \kappa_1(x_2, x_3)\zeta_1, \quad \kappa_1 = \frac{\partial \zeta_1}{\partial x_3} = 1 + 3x_2^2 + 6x_2 x_3 \tag{6.2.83}$$

achieves GAS. This follows from $\dot{V} = -x_3^2 - (L_g V)^2 \leq 0$ and the fact that $x_3 \equiv 0$ and $L_g V \equiv 0$ imply $x_2 \equiv x_1 \equiv 0$. The control law $u = -x_2 - x_3 - L_g V$ also achieves LES of $x = 0$ and $D(\frac{1}{2})$ disk margin.

If it is desirable to limit the control effort, the flexibility of forwarding allows us to saturate the nonlinear part (6.2.83) of the control law and use instead $v_1 = -\sigma_M(L_g V)$, that is

$$u = -x_2 - x_3 - \sigma_M(L_g V) \tag{6.2.84}$$

where M is the saturation level. The GAS and LES properties are preserved with (6.2.84) since $\dot{V} = -x_3^2 - L_g V \sigma_M(L_g V) \leq 0$. In contrast to the saturation design, the saturation level M introduced in forwarding is a free design parameter, not dictated by system's nonlinearities. We will see how this freedom can be used to enhance robustness and performance.

Saturation design. In this design we saturate the linearization of $\kappa_1\zeta_1$ to obtain the control law

$$u = -x_2 - x_3 - \sigma_\epsilon(x_1 + x_2 + x_3) \tag{6.2.85}$$

Proposition 6.16 guarantees GAS and LES of the closed-loop system if the saturation level ϵ is sufficiently small.

To determine ϵ required for stability, we introduce $\zeta_l = x_1 + x_2 + x_3$ and rewrite the closed-loop system as

$$\begin{aligned} \dot{\zeta}_l &= -\sigma_\epsilon(\zeta_l) + 3x_2^3 - 3x_3^3 \\ \dot{x}_2 &= x_3 \\ \dot{x}_3 &= -x_2 - x_3 - \sigma_\epsilon(\zeta_l) \end{aligned} \tag{6.2.86}$$

The linear (x_2, x_3)-subsystem is exponentially stable and its input $\sigma_\epsilon(\zeta_l)$ is bounded by ϵ, so that $|3x_2^3 - 3x_3^3| \leq 6\epsilon^3$ after some finite time. Substituting this bound into the first equation in (6.2.86) we obtain that the closed-loop system is asymptotically stable if $\epsilon < 0.408$.

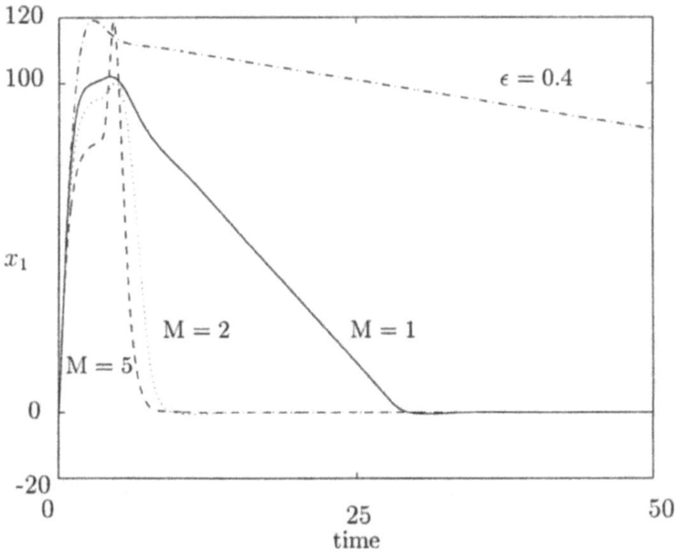

Figure 6.9: Transient of x_1 due to the initial condition $(x_1, x_2, x_3) = (-2, 3, 1)$ for forwarding and saturation designs

Performance comparison. For the saturation design, we select $\epsilon = 0.4$. This value is not conservative because our simulations show instability already at $\epsilon = 0.6$. For forwarding design (6.2.84), we let the saturation level M be our only design parameter. For $M = 1, 2, 5$, Figure 6.9 shows that the x_1-transients are superior to the transient obtained with the saturation design. Large swings in x_1 are caused by $3x_2^3 - 3x_3^3$. However, the same nonlinearity can be used to rapidly bring x_1 back, which is accomplished by forwarding. On the other hand, saturation design is incapable of exploiting this opportunity because its only information about the nonlinearity is an upper bound. Indeed, the saturation design (6.2.85) would remain the same even if the sign of the nonlinearity is reversed.

Another drawback of low saturation levels is that an external disturbance of magnitude ϵ is sufficient to destabilize the system. In the system (6.2.86), a constant disturbance $w = -0.41$ added at the input causes the state x_1 to grow unbounded. □

6.3 Interlaced Systems

6.3.1 Introductory example

With backstepping and forwarding, we are able to recursively design feedback control laws for global stabilization of strict-feedback and strict-feedforward nonlinear systems. A combination of backstepping and forwarding is now employed to achieve global stabilization of a larger class of *interlaced systems.*

To begin with, we consider the third-order interlaced system

$$\begin{aligned} \dot{x}_1 &= x_2 + x_2 x_3 \\ \dot{x}_2 &= x_3 + x_2^2 \\ \dot{x}_3 &= u + x_1 x_2 x_3 \end{aligned} \tag{6.3.1}$$

As in the other two introductory examples (6.1.1) and (6.2.1), the Jacobian linearization of (6.3.1) is a chain of integrators. However, because of the nonlinear term $x_2 x_3$, the system (6.3.1) is not in *feedback* form, nor is it in *feedforward* form, because of the terms $x_1 x_2 x_3$ and x_2^2. Nevertheless, the structure of (6.3.1) is conducive for a systematic design, starting with a scalar subsystem and proceeding with two successive augmentations.

Instead of starting from the top equation, as in Section 6.1.1, or from the bottom equation, as in Section 6.2.1, we start with the middle equation

$$\dot{x}_2 = x_3 + x_2^2 \tag{6.3.2}$$

and treat x_3 as our virtual control. For this scalar system, a Lyapunov function is $V_1 = \frac{1}{2}x_2^2$ and a stabilizing feedback is $x_3 = \alpha_1(x_2) = -x_2 - x_2^2$. We then employ one step of forwarding to stabilize the subsystem (6.3.2) augmented by the top equation of (6.3.1)

$$\begin{aligned} \dot{x}_1 &= x_2 - x_2^2 - x_2^3 + x_2 v \\ \dot{x}_2 &= -x_2 + v \end{aligned} \tag{6.3.3}$$

where the "control" x_3 has been augmented to $x_3 = \alpha_1(x_2) + v$. With $v = 0$, the equilibrium $(x_1, x_2) = (0, 0)$ of (6.3.3) is globally stable and forwarding yields the Lyapunov function

$$
\begin{aligned}
V_2 &= V_1 + \lim_{s\to\infty} \tilde{x}_1^2(s) && (6.3.4)\\
&= \frac{1}{2}x_2^2 + \frac{1}{2}\zeta_1^2, \quad \zeta_1 = x_1 + x_2 + \frac{1}{2}x_2^2 + \frac{1}{3}x_2^3 && (6.3.5)
\end{aligned}
$$

The additional feedback $v = -L_g(V_2 - V_1) = -(1 - x_2^2)\zeta_1$ achieves GAS of (6.3.3) and the augmented control law is

$$
x_3 = \alpha_1(x_2) + v = -x_2 - x_2^2 - (1 - x_2^2)\zeta_1 := \alpha_2(\zeta_1, x_2) \qquad (6.3.6)
$$

To stabilize the entire system (6.3.1), we employ one step of backstepping. With the passivating output $y = x_3 - \alpha_2(\zeta_1, x_2)$ we rewrite the system (6.3.1) as

$$
\begin{aligned}
\dot{\zeta}_1 &= (1 - x_2^2)(-\zeta_1 + y)\\
\dot{x}_2 &= \alpha_2(\zeta_1, x_2) + x_2^2 + y \qquad (6.3.7)\\
\dot{y} &= u + x_2 x_3 x_1 - \dot{\alpha}_2(\zeta_1, x_2)
\end{aligned}
$$

Augmenting V_2 by $\frac{y^2}{2}$ we obtain the CLF $V_3 = V_2 + \frac{y^2}{2} = \frac{1}{2}(x_2^2 + \zeta_1^2 + y^2)$ and employ it to design a control law $u = \alpha_3(\zeta_1, x_2, y)$ which achieves GAS of (6.3.7), and hence, of (6.3.1).

We have solved the stabilization problem for the interlaced system (6.3.1) by using first one step of forwarding and then one step of backstepping. For an interlaced system of the form

$$
\begin{aligned}
\dot{x}_1 &= x_2 + a_1(x_2, x_3) + g_1(x_2, x_3)u\\
\dot{x}_2 &= x_3 + x_2^2 \qquad (6.3.8)\\
\dot{x}_3 &= u + a_3(x_2, x_3)
\end{aligned}
$$

we proceed in the reverse order: first one step of backstepping for the subsystem

$$
\begin{aligned}
\dot{x}_2 &= x_3 + x_2^2\\
\dot{x}_3 &= u + a_3(x_2, x_3)
\end{aligned} \qquad (6.3.9)
$$

and then one step of forwarding for the entire system (6.3.8).

With the examples (6.1.1), (6.2.1), (6.3.1), and (6.3.8) we have illustrated four different decompositions of the stabilization problem for a third-order system. In each of these examples, the sequence of design steps was determined by system interconnections, that is, by the states which enter the nonlinearities. The growth of the nonlinearities is unrestricted and uncertainties, such as the unknown sign of the parameter θ in the system (6.1.1), can be accommodated.

When a system configuration does not permit a decomposition into a sequence of backstepping/forwarding steps, then additional properties, like the *growth* or the *sign* of the nonlinearities, become important, as illustrated by the system

$$\begin{aligned} \dot{x}_1 &= x_2 + a_1(x_1, x_3) \\ \dot{x}_2 &= x_3 \\ \dot{x}_3 &= u \end{aligned} \tag{6.3.10}$$

This system has the same Jacobian linearization as the four previous examples but differs in the structure of its nonlinear term $a_1(x_1, x_3)$. Because this nonlinearity depends on x_1 and x_3, the design can not be decomposed as before. For global stabilizability we need a further characterization of the nonlinearity $a_1(x_1, x_3)$. In forwarding we impose a *linear growth* assumption of a_1 as a function of x_1 and a *stability* condition $a_1(x_1, 0)x_1 \leq 0$. Without such restrictions, the global stabilization may be impossible. For example, in the case $a_1(x_1, x_3) = x_1^2 + x_3^2$, using $\eta = x_1 + x_2$, we obtain

$$\dot{\eta} = x_3 + x_3^2 + \eta + (x_1^2 - x_1) \geq -1 + \eta \tag{6.3.11}$$

This shows that, irrespective of the choice of the control, initial conditions which satisfy $\eta(0) > 1$ cannot be driven to the origin.

6.3.2 Non-affine systems

Thus far, our presentation of backstepping and forwarding has been restricted to nonlinear systems *affine* in the control, that is, $\dot{x} = f(x) + g(x)u$. This restriction is not essential and we now briefly discuss non-affine situations. Even if the entire system is affine in the control variable u, non-affine situations are likely to occur at intermediate steps of interlaced designs, as in the following example:

$$\begin{aligned} \dot{x}_1 &= x_2 + x_3^2 \\ \dot{x}_2 &= x_3 + x_2^2 \\ \dot{x}_3 &= u \end{aligned} \tag{6.3.12}$$

This system can be stabilized by one step of forwarding followed by one step of backstepping. However, the first step of forwarding is for the subsystem

$$\begin{aligned} \dot{x}_1 &= x_2 + x_3^2 \\ \dot{x}_2 &= x_3 + x_2^2 \end{aligned} \tag{6.3.13}$$

which is not affine in the "control" x_3.

For backstepping, we will only be interested in the non-affine case

$$\begin{aligned} \dot{z} &= f(z, \xi) \\ \dot{\xi} &= a(z, \xi) + b(z, \xi)u \end{aligned} \tag{6.3.14}$$

where we assume that a virtual control law $\xi = \alpha(z)$ is designed to achieve GAS of the non-affine z-subsystem. As before, the new variable $y = \xi - \alpha(z)$ is a passivating output for the system (6.3.14) which can be rewritten in the form

$$\begin{aligned} \dot{z} &= f(z, \alpha(z)) + \tilde{g}(x, y)y \\ \dot{y} &= a(z, y + \alpha) + b(z, y + \alpha)u - \dot{\alpha}(x, y) \end{aligned} \tag{6.3.15}$$

Backstepping is then pursued as in the affine case.

To apply forwarding to the non-affine system

$$\begin{aligned} \dot{z} &= \gamma(z, \xi, u) \\ \dot{\xi} &= a(\xi, u) \end{aligned} \tag{6.3.16}$$

we assume that the control $u = \alpha(\xi)$ transforms it into the cascade of a GS subsystem $\dot{z} = \gamma(z, 0, 0)$ with a GAS/LES subsystem $\dot{\xi} = a(\xi, \alpha(\xi))$, and the interconnection term $\psi(z, \xi) = \gamma(z, \xi, \alpha(\xi)) - \gamma(z, 0, 0)$. If this cascade satisfies Assumptions 5.1 and 5.2 of Chapter 5, a Lyapunov function with cross-term can be constructed as before. What differs from the affine case is the design of the additional control v for the system

$$\begin{aligned} \dot{z} &= \gamma(z, \xi, \alpha(\xi) + v) \\ \dot{\xi} &= a(\xi, \alpha(\xi) + v) \end{aligned} \tag{6.3.17}$$

Instead of the damping control $v = -L_g V$ used in the affine case, a control law v must be designed to enhance the negativity of $\dot{V}|_{v=0} \leq 0$. This is achieved with the help of the following proposition by Lin [66].

Proposition 6.21 (*Damping control for non-affine systems*)
Consider the system

$$\dot{x} = f(x, u) = f(x, 0) + g_0(x)u + \mathcal{O}(x, u) \tag{6.3.18}$$

where $\mathcal{O}(x, u)$ contains only quadratic and higher-order terms in u. Assume that $V(x)$ is a C^1 positive definite radially unbounded function such that $L_{f(x,0)}V(x) \leq 0$. If $\dot{x} = f(x, 0) + g_0(x)u$ with output $y = L_{g_0}V$ is ZSD, then a nonlinear gain $\sigma(x)$ can be constructed such that the damping control

$$u = -\sigma(x)(L_{g_0}V)^T(x)$$

achieves GAS of (6.3.18). □

6.3.3 Structural conditions for global stabilization

We now characterize interlaced systems by certain properties of the configuration matrix

$$P(x,u) = \frac{\partial f}{\partial(x,u)} = \begin{pmatrix} \frac{\partial f}{\partial x} & \frac{\partial f}{\partial u} \end{pmatrix} \tag{6.3.19}$$

of the general nonlinear system

$$\dot{x} = f(x,u), \quad u \in I\!R, \quad x \in I\!R^n \tag{6.3.20}$$

The zero entries of $P(x,u)$ determine the feedback and feedforward connections which are absent from a block-diagram representation of the system. In our introductory examples, this structural characterization of the nonlinearities was sufficient to determine the sequence of backstepping and forwarding steps needed for the stabilization task. The same sequence applies to other systems if their configuration matrices have the zero entries.

With one step of backstepping or forwarding, the configuration matrix is augmented by one additional row and one additional column. Thus, when one step of backstepping augments the system (6.3.20) to

$$\begin{array}{rcl} \dot{x} & = & f(x,\zeta) \\ \dot{\zeta} & = & a(x,\zeta) + b(x,\zeta)u, \quad \zeta \in I\!R \end{array} \tag{6.3.21}$$

its configuration matrix P undergoes a *top-down augmentation* to

$$P_{bst} = \begin{pmatrix} & & & 0 \\ & P & & \vdots \\ & & & 0 \\ \tilde{p}_1 & \dots & \tilde{p}_{n+1} & b \end{pmatrix} \tag{6.3.22}$$

The zeros in the last column are necessary to apply one step of backstepping. In a dual manner, when in forwarding the system (6.3.20) is augmented to the form

$$\begin{array}{rcl} \dot{\zeta} & = & \gamma(x,u) \\ \dot{x} & = & f(x,u), \quad \zeta \in I\!R \end{array} \tag{6.3.23}$$

its configuration matrix P undergoes a *bottom-up augmentation* to

$$P_{fwd} = \begin{pmatrix} 0 & \tilde{p}_1 & \dots & \tilde{p}_n \\ 0 & & & \\ \vdots & & P & \\ 0 & & & \end{pmatrix} \tag{6.3.24}$$

where the zero entries in the first column are necessary to apply one step of forwarding.

Backstepping imposes restrictions on some nonzero entries of the configuration matrix. For the system (6.3.21), we must have $b(x,\zeta) \neq 0$ for all $(x,\zeta) \in I\!R^{n+1}$. This implies that, if the Jacobian linearization of (6.3.20) is stabilizable, so is the Jacobian linearization of the augmented system (6.3.21). Forwarding requires stabilizability of the Jacobian linearization.

Examining all the configuration matrices which can be generated by repeated top-down augmentations of the type (6.3.22) or bottom-up augmentations of the type (6.3.24), we arrive at the following characterization of interlaced systems.

Definition 6.22 *(Interlaced systems)*
A system (6.3.20) is called *interlaced* if its Jacobian linearization is stabilizable and its configuration matrix $P(x,u)$ satisfies the following requirements:

(i) If $j > i+1$ and $p_{ij} \not\equiv 0$, then $p_{kl}(x) \equiv 0$ for all $k \geq l$, $k \leq j-1$, and $l \leq i$.

(ii) If $p_{ij} \not\equiv 0$ for some $j \leq i$, then $p_{i\,i+1}$ is independent of $x_{i\,i+1}$ and $p_{i\,i+1}(x) \neq 0$ for all x.

□

Definition 6.22 characterizes interlacing by (i) and excludes degenerate situations in which the lack of stabilizability occurs in the Jacobian linearization or in which backstepping cannot be applied because of a nonglobal relative degree (condition (ii)).

Example 6.23 *(Three-dimensional interlaced systems)*
For third-order systems, the four different types of configuration matrix which satisfy the requirement (i) of Definition 6.22 are listed below with the two-step sequences of backstepping (bst) and forwarding (fwd):

$$\text{bst + bst}: \begin{pmatrix} * & * & 0 & 0 \\ * & * & * & 0 \\ * & * & * & * \end{pmatrix} \qquad \text{fwd + fwd}: \begin{pmatrix} 0 & * & * & * \\ 0 & 0 & * & * \\ 0 & 0 & * & * \end{pmatrix}$$

$$\text{bst + fwd}: \begin{pmatrix} 0 & * & * & * \\ 0 & * & * & 0 \\ 0 & * & * & * \end{pmatrix} \qquad \text{fwd + bst}: \begin{pmatrix} 0 & * & * & 0 \\ 0 & * & * & 0 \\ * & * & * & * \end{pmatrix}$$

Only these four configuration matrices can be generated with two top-down and/or bottom-up augmentations. In each case, a sequence of backstepping and forwarding steps to be followed can be determined from the matrix configuration. □

Theorem 6.24 (*Stabilization of interlaced systems*)
Every interlaced system is globally stabilizable by a sequence of scalar steps of backstepping and/or forwarding. The design simultaneously yields the construction of a globally stabilizing feedback and of a CLF. □

Definition 6.22 guarantees that global stabilization can be achieved without any restriction on the sign or the growth of the nonlinearities. Global stabilizability of systems which are not interlaced cannot be guaranteed.

Theorem 6.25 (*Loss of stabilizability in noninterlaced systems*)
Let $P(x,u)$ be a configuration matrix for a system whose Jacobian linearization is a chain of integrators. If $P(x,u)$ does not satisfy the condition (i) of Definition 6.22, then there exists a system, with the configuration matrix which contains all the zeros of $P(x,u)$, and satisfies the other requirements of Definition 6.22, which is not globally stabilizable.

Proof: Let $p_{ij}(x)$ and $p_{kl}(x)$ be two nonzero entries of $P(x,u)$ such that $j > i+1$, $k \geq l$, $k \leq j-1$, and $l \leq i$. We consider a system consisting of a chain of integrators $\dot{x}_1 = x_2, \ldots, \dot{x}_n = u$, except for its i-th and k-th equations, which are

$$\begin{array}{rcl} \dot{x}_i & = & x_{i+1} + x_j^2 \\ \dot{x}_k & = & x_{k+1} + x_l^2 \end{array} \tag{6.3.25}$$

In the configuration matrix $P(x,u)$ of this system the only nonzero entries are the off-diagonal entries $p_{m\,m+1} \equiv 1$, $m \in \{1,\ldots,n\}$, and the two entries $p_{ij}(x) = 2x_j$, $p_{kl}(x) = 2x_l$. Therefore, it satisfies the conditions (ii) and (iii) of Definition 6.22 and contains all the zeros of $P(x,u)$.

We will now show that this system is not globally stabilizable. Using the fact that $l \leq k, i \leq j-1$, we define the new state $\eta = x_l + \ldots + x_{j-1}$ which satisfies

$$\dot{\eta} = (x_{l+1} + \ldots + x_j) + x_l^2 + x_j^2 \tag{6.3.26}$$

$$= \eta - x_l + x_l^2 + x_j + x_j^2 \geq \eta - 1 \tag{6.3.27}$$

This proves that initial conditions $\eta(0) > 1$ cannot be driven to zero, irrespective of the choice of the control. □

For the cascade systems of Chapter 4, we have obtained stabilization results with the help of a structural characterization of nonpeaking cascades, which excludes the peaking states from the interconnection term. It sets a structural limit to global stabilization with cascade designs because, if this structural characterization is missing, interconnection growth must be restricted.

The characterization of interlaced systems plays a similar role in the recursive designs of this chapter. If its structural restrictions are relaxed, global stabilizability cannot be guaranteed without additional conditions. An illustration is given by the *nonstrict*-feedforward systems of this chapter: they do not satisfy the interlacing condition (i) of Definition 6.22 because they allow certain feedback loops. However, a forwarding design is still possible for these systems because the relaxation in the structural requirements is compensated for by additional growth and stability restrictions.

6.4 Summary and Perspectives

Backstepping and forwarding are the two building blocks for recursive construction of Lyapunov functions and globally stabilizing control laws. By successive augmentations of smaller systems, recursive designs achieve global stabilization of larger systems. They overcome the structural limitations of feedback passivation: the relative degree one and the weak minimum phase requirements. Backstepping provides the construction of a CLF, which can be employed to ensure desired stability margins. Forwarding has an optimality property which guarantees a desired disk margin.

Various simplifications of backstepping and forwarding reduce their complexity by forcing the solutions to converge towards nested invariant manifolds in different time scales. This geometric property stems from high-gain feedback in simplifications of backstepping and low-gain saturation in simplifications of forwarding. Excessive gain separation may be harmful for both performance and robustness.

With the characterization of interlaced systems, which combine feedback and feedforward connections, we have reached the limit of systematic nonlinear designs which exploit the structural properties of interconnections but do not restrict the growth of the nonlinearities. However, our characterization of interlaced systems is coordinate dependent, and hence, not complete from a geometrical point of view.

Stabilizability and controller design of noninterlaced nonlinear systems are largely open. For the cascade systems, growth restrictions and stronger stability assumptions are alternatives to the structural nonpeaking conditions to guarantee global stabilization. Possibilities for such relaxations of the structure of interlaced systems are yet to be explored.

With their different emphasis on analysis and geometry, the design procedures presented in this book reveal structural limitations of nonlinear designs and stress the need for trade-offs between performance, robustness, and complexity. A systematic treatment of these issues is yet to be undertaken.

6.5 Notes and References

The developement of nonlinear *recursive* designs is recent. The first backstepping design of Saberi, Kokotović, and Sussmann [92] removed the relative degree obstacle in the global stabilization of partially linear cascades. The backstepping methodology has since become popular and is presented in several recent textbooks [61, 73, 43]. The recursive semiglobal high-gain design for strict-feedback systems was developed by Teel and Praly [112].

For strict-feedforward systems, a recursive design with nested saturations introduced by Teel [109] has led to further advances in this direction. Mazenc and Praly [75] extended it with a Lyapunov design for feedforward systems. The forwarding design presented in this chapter was developed by the authors in [46, 95].

Our new characterization of interlaced systems was inspired by the work by Wei [118] dealing with robust stabilization of linear systems which contain uncertain entries in the matrices A and b. Initial steps toward interlaced designs of nonlinear systems were made by Qu [90].

Appendix A

Basic geometric concepts

A.1 Relative Degree

For SISO linear systems, the *relative degree* r is the difference between the number of poles and zeros in the transfer function

$$H(s) = k\frac{q_0 + q_1 s + \ldots + s^{n-r}}{p_0 + p_1 s + \ldots + s^n} \tag{A.1.1}$$

The systems with $r \geq 0$ are called *proper*, and with $r > 0$, *strictly proper*. In this book we do not consider systems with $r < 0$. To interpret r for a state-space representation

$$\begin{aligned} \dot{x} &= Ax + bu \\ y &= cx + du, \quad x \in I\!R^n, \quad u, y \in I\!R \end{aligned} \tag{A.1.2}$$

we expand $H(s)$ as

$$\begin{aligned} H(s) &= d + c(sI - A)^{-1}b \\ &= d + cb\frac{1}{s} + cAb\frac{1}{s^2} + \ldots + cA^{r-1}b\frac{1}{s^r} + \ldots \end{aligned} \tag{A.1.3}$$

When $r = 0$ we see from (A.1.3) that $H(\infty) = d \neq 0$, that is, the system has a nonzero infinite frequency throughput. For strictly proper systems ($r > 0$), the throughput is zero, $d = 0$, and r is determined by the two conditions

$$cA^k b = 0, \quad \text{for } 0 \leq k \leq r-2, \quad \text{and } cA^{r-1}b \neq 0 \tag{A.1.4}$$

The meaning of these two conditions in the time domain becomes clear from the r-th derivative $y^{(r)}$ of the output:

$$\begin{array}{rclcl} \dot{y} & = & c\dot{x} & = & cAx + cb\,u & = & cAx \\ \vdots & & & \vdots & & & \vdots \\ y^{(r-1)} & = & cx^{(r-1)} & = & cA^{r-1}x + cA^{r-2}b\,u & = & cA^{r-1}x \\ y^{(r)} & = & cx^{(r)} & = & cA^{r}x + \underbrace{cA^{r-1}b}_{\neq 0}\,u \end{array} \tag{A.1.5}$$

The statement that "the system has relative degree r" means that the input appears explicitely for the first time in the r-th derivative of the output.

This definition of the relative degree admits a direct extension to nonlinear systems. The nonlinear system

$$\begin{array}{rcl} \dot{x} & = & f(x) + g(x)u \\ y & = & h(x) + j(x)u, \quad x \in I\!R^n; \quad u, y \in I\!R \end{array} \tag{A.1.6}$$

has a relative degree zero at $x = x_0$ if $j(x_0) \neq 0$. If $j(x) \equiv 0$ in a neighborhood of x_0, we differentiate the output

$$\dot{y} = \frac{\partial h}{\partial x}\dot{x} = L_f h(x) + L_g h(x) u$$

If $L_g h(x_0) \neq 0$, then $\dot{y}$ explicitely depends on u near $x = x_0$, and, hence, $r = 1$. If $L_g h(x) \equiv 0$ near $x = x_0$, one more differentiation of y yields

$$\ddot{y} = \frac{\partial}{\partial x}(L_g h)\dot{x} = L_f^2 h(x) + L_g L_f h(x) u$$

Now, if $L_g L_f h(x_0) \neq 0$ then $r = 2$. We see, therefore, that $L_g h(x_0)$, $L_g L_f h(x_0)$, are the nonlinear analogs of cb, cAb. Likewise, $L_g L_f^k h(x)$ is the nonlinear analog of $cA^k b$.

Definition A.1 *(Relative degree of SISO systems)*
The relative degree of the nonlinear sysem (A.1.6) at $x = x_0$ is the integer r such that

(i) $L_g L_f^k h(x) \equiv 0$, for $k = 0, \ldots, r-2$, and x in a neighborhood of $x = x_0$;

(ii) $L_g L_f^{(r-1)} h(x_0) \neq 0$. □

For nonlinear systems, the relative degree is a *local* concept, defined in some neighborhood of $x = x_0$. If conditions (i) and (ii) hold globally, we say that the system (A.1.6) has a *global* relative degree r. In contrast to the linear case, the relative degree of a nonlinear system may not be defined at some point $x = x_0$. Thus, for the system $\dot{x} = u$, $y = \sin x$, a relative degree is not defined at $x_0 = \frac{\pi}{2}$.

A MIMO system with m inputs and m outputs

$$\begin{aligned} \dot{x} &= f(x) + g(x)u \\ y &= h(x) + j(x)u, \quad x \in I\!R^n, \quad u, y \in I\!R^m \end{aligned} \tag{A.1.7}$$

has relative degre zero at $x = x_0$ if $j(x_0)$ is nonsingular. If $j(x) \equiv 0$ near $x = x_0$, we associate to each output y_i an integer r_i which is the number of differentiations of the output y_i needed for one of the inputs to appear explicitely.

Definition A.2 *(Relative degree of MIMO systems)*
The MIMO system (A.1.7) has a relative degree $\{r_1, \ldots, r_m\}$ at $x = x_0$ if

(i) $L_{g_j} L_f^k h_i(x) = 0$ for all $1 \leq i, j \leq m$, for all $k < r_i - 1$, and for all x in a neighborhood of $x = x_0$,

(ii) the $m \times m$ matrix

$$R(x) = \left[\frac{\partial y_i^{(r_i)}}{\partial u_j}\right]_{1 \leq i,j \leq m} = \begin{pmatrix} L_{g_1} L_f^{r_1-1} h_1(x) & \ldots & L_{g_m} L_f^{r_1-1} h_1(x) \\ \vdots & & \vdots \\ L_{g_1} L_f^{r_m-1} h_1(x) & \ldots & L_{g_m} L_f^{r_m-1} h_m(x) \end{pmatrix} \tag{A.1.8}$$

is nonsigular at $x = x_0$. □

Condition (ii) is the MIMO generalization of the condition $L_g L_f^{r-1} h(x_0) \neq 0$ in the SISO case. If $r_1 = r_2 = \ldots = r_m$, we say that the system (A.1.7) has a *uniform* relative degree r_1.

A.2 Normal Form

When the relative degree r of the SISO system (A.1.6) is defined at $x = x_0$, then a change of coordinates

$$(\xi, z) = T(x), \quad \xi \in I\!R^r, \quad z \in I\!R^{n-r} \tag{A.2.1}$$

which transforms the nonlinear system $\dot{x} = f(x) + g(x)u$ in a *normal form* exists near $x = x_0$. We assume $f(x_0) = 0$, set $T(x_0) = (0, 0)$, and define the first r components $T_i(x)$ of $T(x)$ as

$$\begin{aligned} \xi_1 &= T_1(x) &= \; & y &= \; & h(x) \\ \xi_2 &= T_2(x) &= \; & \dot{y} &= \; & L_f h(x) \\ &\vdots & & & & \vdots \\ \xi_r &= T_r(x) &= \; & y^{(r-1)} &= \; & L_f^{r-1} h(x) \end{aligned} \tag{A.2.2}$$

Because, by assumption, $L_gL_f^{(r-1)}h(x_0) \neq 0$, the new coordinates ξ_i satisfy

$$\begin{aligned} \dot{\xi}_1 &= \xi_2 \\ &\vdots \\ \dot{\xi}_{r-1} &= \xi_r \\ \dot{\xi}_r &= L_f^r h(x) + L_gL_f^{(r-1)}h(x)\, u \end{aligned} \tag{A.2.3}$$

Proposition A.3 *(Linear independence of output derivatives)*
If a SISO system has relative degree r at $x = x_0$, then the row vectors

$$\left\{ \frac{\partial T_1}{\partial x}(x_0), \ldots, \frac{\partial T_r}{\partial x}(x_0) \right\} \tag{A.2.4}$$

are linearly independent.

Proof: By contradiction, suppose that there exists constants c_k such that

$$\frac{\partial T_r}{\partial x}(x_0) = \sum_{k=1}^{r-1} c_k \frac{\partial T_i}{\partial x}(x_0) \tag{A.2.5}$$

Then we have

$$\begin{aligned} L_gL_f^{(r-1)}h(x_0) &= \frac{\partial T_r}{\partial x}(x_0)g(x_0) \\ &= \sum_{k=1}^{r-1} c_k \frac{\partial T_k}{\partial x}(x_0)g(x_0) \\ &= \sum_{k=1}^{r-1} c_k L_gL_f^{(k-1)}h(x_0) = 0 \end{aligned}$$

which contradicts the relative degree assumption that $L_gL_f^{(r-1)}h(x_0) \neq 0$. □

In the MIMO case, we associate in a similar way r_i components of $T(x)$ to the output y_i and its first $(r_i - 1)$ derivatives, that is,

$$\xi_1^i = T_1^i(x) = y_i,\ \xi_2^i = T_2^i(x) = L_f h_i(x),\ \ldots,\ \xi_{r_i}^i = L_f^{r_i-1}h_i(x)$$

The proof of Proposition A.3 is easily extended to show that the so defined $r = \sum_{i=1}^m r_i$ components of the change of coordinates are linearly independent at $x = x_0$ if the relative degree is $\{r_1, \ldots, r_m\}$.

In general, the change of coordinates needs to be completed by $n - r$ functions $T_{r+1}(x), \ldots, T_n(x)$ such that the matrix

$$\left(\frac{\partial T}{\partial x}\right)(x_0) \tag{A.2.6}$$

is nonsingular. This is necessary for $T(x)$ to qualify as a local change of coordinates. Using the notation

$$
\begin{aligned}
\xi^i &= (\xi_1^i, \ldots, \xi_{r_i}^i)^T,\ \xi = (\xi^1, \ldots, \xi^m), \\
z &= (T_{r+1}(x), \ldots, T_n(x))^T \\
b_{ij}(z,\xi) &= L_{g_j} L_f^{(r_i-1)} h_i(T^{-1}(z,\xi)) \quad \text{for } 1 \le i,j \le m && \text{(A.2.7)} \\
a_i(z,\xi) &= L_f^{r_i} h_i(T^{-1}(z,\xi)) \quad \text{for } 1 \le i \le m && \text{(A.2.8)}
\end{aligned}
$$

we rewrite the system $\dot{x} = f(x) + g(x)u$, $y = h(x)$, in the *normal form*

$$
\begin{aligned}
\dot{z} &= q(z,\xi) + \gamma(z,\xi)u \\
\dot{\xi}_1^i &= \xi_2^i \\
&\ \vdots \\
\dot{\xi}_{r_i-1}^i &= \xi_{r_i}^i \\
\dot{\xi}_{r_i}^i &= a_i(z,\xi) + \textstyle\sum_{j=1}^m b_{ij}(z,\xi)u_j, \\
y_i &= \xi_1^i, \qquad\qquad 1 \le i \le m
\end{aligned}
\tag{A.2.9}
$$

In special situations, including the SISO case, it is possible to select the coordinates z such that $\gamma(z,\xi) \equiv 0$, and, hence, $\dot{z} = q(z,\xi)$.

The coefficients $b_{ij}(z,\xi)$ in (A.2.7) are the elements of the matrix

$$
R(x) = \left[\frac{\partial y_i^{(r_i)}}{\partial u_j}\right]_{1 \le i,j \le m} = \begin{pmatrix} L_{g_1} L_f^{r_1-1} h_1(x) & \ldots & L_{g_m} L_f^{r_1-1} h_1(x) \\ \vdots & & \vdots \\ L_{g_1} L_f^{r_m-1} h_1(x) & \ldots & L_{g_m} L_f^{r_m-1} h_m(x) \end{pmatrix}
$$

which, by the relative degree definition, is invertible near $x = x_0$. Thus, $b^{-1}(z,\xi)$ exists and the feedback transformation

$$
u = b^{-1}(z,\xi)(-a(z,\xi) + v) \tag{A.2.10}
$$

which is well-defined in the neighborhood of $(z,\xi) = (0,0)$, transforms the ξ-subsystem of (A.2.9) into m decoupled integrator chains

$$
\dot{\xi}_1^i = \xi_2^i, \ \ldots, \ \dot{\xi}_{r_i-1}^i = \xi_{r_i}^i, \ \dot{\xi}_{r_i}^i = v_i, \ 1 \le i \le m \tag{A.2.11}
$$

Each output $y_i = \xi_1^i$ is controlled by the new input v_i through a chain of r_i integrators.

Thus, when $\dot{x} = f(x) + g(x)u$, $y = h(x)$, has a well-defined relative degree near $x = 0$, then a change of coordinates $(z,\xi) = T(x)$ and a feedback transformation

$$
u = \alpha(x) + \beta(x)v, \ \ \beta(x) \text{ invertible,} \tag{A.2.12}
$$

can make its input-ouput behavior near $x = x_0$ be the same as that of the m chains of integrators (A.2.11). In that sense, the relative degree is a structural invariant of the nonlinear system: it cannot be altered by changes of coordinates or feedback transformations.

The relative degree can be modified by *dynamic* feedback transformations. Tthe addition of integrators at the input v increases the relative degree. In particular, we can make the m chains of integrators in (A.2.11) to be of equal length by defining $r^* = \max\{r_1, \ldots, r_m\}$ and by adding $r^* - r_i$ integrators to each chain:

$$\dot{v}_i = \zeta_1^i, \; \dot{\zeta}_1^i = \zeta_2^i, \; \ldots, \; \dot{\zeta}_{r^*-r_i-1}^i = w_i \tag{A.2.13}$$

Then the augmented system with the new input w and the old output y has a uniform relative degree r^*.

A.3 The Zero Dynamics

The relative degree property is useful for input-ouput linearization, decoupling, output tracking, and similar control tasks. However, the feasibility of these tasks depends critically on the subsystem

$$\dot{z} = q(z, \xi) + \gamma(z, \xi)u \tag{A.3.1}$$

The state z of this subsystem is rendered unobservable by the control law (A.2.10) which cancels all the z-dependent terms in the ξ-subsystem of (A.2.9).

To see the importance of the subsystem (A.3.1), we analyze it when the output y of (A.2.9) is maintained at zero, that is, when $\xi(0) = 0$ and the control (A.2.10) is chosen to satisfy $y(t) \equiv 0$, that is,

$$u = -b^{-1}(z, 0)a(z, 0) \tag{A.3.2}$$

The subsystem (A.3.1) then becomes an autonomous system

$$\dot{z} = q(z, 0) - \gamma(z, 0)b^{-1}(z, 0)a(z, 0) =: f_{zd}(z) \tag{A.3.3}$$

with an equilibrium at $z = 0$. Its solutions are the dynamics of the system (A.2.9) which remain upon "zeroing the output" $y(t) \equiv 0$, hence the term *zero dynamics*.

The zero dynamics of a SISO linear system (A.1.2) are determined by the zeros of its transfer function $H(s)$, as we now show using the state-space representation

$$A = \begin{pmatrix} 0 & 1 & 0 & \dots & 0 \\ 0 & 0 & 1 & \dots & 0 \\ \vdots & \vdots & \vdots & \ddots & \vdots \\ 0 & 0 & 0 & \dots & 1 \\ -p_0 & -p_1 & -p_2 & \dots & -p_{n-1} \end{pmatrix}, \quad b = \begin{pmatrix} 0 \\ 0 \\ \vdots \\ 0 \\ k \end{pmatrix}$$

$$c = \begin{pmatrix} q_0 & q_1 & \dots & q_{n-r-1} & 1 & 0 \dots & 0 \end{pmatrix}$$

With the first r coordinates $\xi_1 = cx$, $\xi_2 = cAx$, $\dots$, $\xi_r = cA^{r-1}x$ and the remaining $n-r$ coordinates $z_1 = x_1$, $z_2 = x_2$, $\dots$, $z_{n-r} = x_{n-r}$, the normal form (A.2.9) becomes

$$\begin{aligned} \dot{z} &= Qz + ey \\ \dot{\xi}_1 &= \xi_2 \\ &\vdots \\ \dot{\xi}_r &= \rho_1^T \xi + \rho_2^T z + k\,u \end{aligned} \tag{A.3.4}$$

The constraint $y(t) \equiv 0$ is enforced with $\xi_1(0) = \xi_2(0) = \dots = \xi_r(0) = 0$ and the feedback

$$u = -\frac{1}{k}(\rho_1^T \xi + \rho_2^T z) \tag{A.3.5}$$

The zero-dynamics subsystem is

$$\dot{z} = Qz \tag{A.3.6}$$

and the above calculation shows that

$$Q = \begin{pmatrix} 0 & 1 & 0 & \dots & 0 \\ 0 & 0 & 1 & \dots & 0 \\ \vdots & \vdots & \vdots & \ddots & \vdots \\ 0 & 0 & 0 & \dots & 1 \\ -q_0 & -q_1 & -q_2 & \dots & -q_{n-r-1} \end{pmatrix} \quad b = \begin{pmatrix} 0 \\ 0 \\ \vdots \\ 0 \\ 1 \end{pmatrix}$$

This means that the eigenvalues of Q are the zeros of $H(s)$. With the feedback (A.3.5), the eigenvalues of Q become $n-r$ eigenvalues of the system (A.3.4), the other r eigenvalues being zero. With this pole-zero cancellation, the zero dynamics $z(t)$ are rendered unobservable. When the zero-dynamics subsystem is unstable, that is, $H(s)$ is a nonminimum phase transfer function, the pole-zero cancellation destabilizes the whole system and must be avoided. When the zero-dynamics subsystem is asymptotically stable (stable), the system is minimum phase (weakly minimum phase), and pole-zero cancellations do not cause instability.

For a nonlinear system, the zero-dynamics subsystem (A.3.3) is also determined by the constraint $y(t) \equiv 0$, and its properties are not altered by a change of coordinates or a feedback transformation. The zero-dynamics subsystem (A.3.3) is thus another structural invariant of the nonlinear system (A.1.7).

A.4 Right-Invertibility

With the normal form (A.2.9) we can solve the *tracking problem* in which we force $y(t)$ to track a reference signal $y_R(t)$. The requirement $y(t) = y_R(t)$ constraints the state ξ in (A.2.9):

$$\xi(t) \equiv \xi_R(t) = (\xi_R^1(t), \xi_R^2(t), \ldots, \xi_R^m(t))^T \tag{A.4.1}$$

where $\xi_R^i(t) = \left(y_{iR}(t), \dot{y}_{iR}(t), \ldots, y_{iR}^{(r_i-1)}(t)\right)^T$

The constraint (A.4.1) is enforced with an initial condition

$$\xi(0) = (\xi_R^1(0), \xi_R^2(0), \ldots, \xi_R^m(0))^T$$

and the input

$$u = u_R(t) = b^{-1}(\xi_R(t), z(t))(-a(z(t), \xi_R(t)) + \begin{bmatrix} y_{1R}^{(r_1)}(t) \\ \vdots \\ y_{mR}^{(r_m)}(t) \end{bmatrix}) \tag{A.4.2}$$

where $z(t)$ is the solution of

$$\dot{z} = q(z, \xi_R(t)) + p(z, \xi_R(t))b^{-1}(z, \xi_R(t))(-a(z, \xi_R(t)) + \begin{bmatrix} y_{1R}^{(r_1)}(t) \\ \vdots \\ y_{mR}^{(r_m)}(t) \end{bmatrix}) \tag{A.4.3}$$

with any initial condition $z(0)$.

The expressions (A.4.2) and (A.4.3) define an *inverse system* which for a given $y(t) = y_R(t)$ at its input generates $u_R(t)$ at its output. The required number of derivatives of $y_R(t)$ is determined by the relative degree $\{r_1, \ldots, r_m\}$. They drive the "inverse-dynamics" subsystem (A.4.3), which, for $y_R(t) \equiv 0$, reduces to the zero-dynamics subsystem (A.3.3).

A system for which the tracking problem can be solved is called *right-invertible*, and the system (A.4.3) with input $y_R(t)$, output $u(t)$, and state

$z(t)$, is a *right-inverse* of the original system. Right-invertibility is thus a property implied by the existence of a relative degree.

The concepts of relative degree, zero dynamics, and right-invertibility, are extended in a straightforward manner to "non-square" MIMO systems with m inputs and p outputs, provided that $m \geq p$.

A.5 Geometric properties

Here, and elsewhere in the book, we call certain dynamic system properties *geometric*, if they cannot be altered by the choice of coordinates. We have seen that relative degree, zero dynamics, and right-invertibility are input-output geometric properties which remain invariant under feedback.

Appendix B

Proofs of Theorems 3.18 and 4.35

B.1 Proof of Theorem 3.18

We first prove a converse stability result.

Lemma B.1 (*Converse stability with parameters*)
Consider the system

$$\dot{x} = f(x, \theta), \qquad x \in \mathbb{R}^n \tag{B.1.1}$$

where $\theta \in R^p$ is constant, f is a C^r function, and $f(0, \theta) = 0$ for all $\theta \in \Theta$ where $\Theta \subset R^p$ may be unbounded. If the equilibrium $x = 0$ is GAS and LES for all $\theta \in \Theta$, then there exists a C^r function $V(x, \theta)$ such that, for all $\theta \in \Theta$,

$$\begin{aligned} c(\theta)\|x\|^2 \leq V(x, \theta) \leq \gamma_\theta(\|x\|) \\ \frac{\partial V}{\partial x} f(x, \theta) \leq -\|x\|^2 \end{aligned} \tag{B.1.2}$$

where $c(\theta) > 0$ is a continuous function and $\gamma_\theta(\cdot) \in \mathcal{K}_\infty$.

Proof: In the system

$$\dot{\bar{x}} = \frac{1}{1 + \|f(\bar{x}, \theta)\|^2} f(\bar{x}, \theta) =: \bar{f}(\bar{x}, \theta) \tag{B.1.3}$$

the globally Lipschitz vector field $\bar{f}$ has the same direction as f in (B.1.1) at each x. In rescaled time, the solutions of (B.1.1) and (B.1.3) coincide. Therefore, the equilibrium $\bar{x} = 0$ of (B.1.3) is GAS and LES.

We let $\bar{x}(s; x_0, 0)$ denote the solution of (B.1.3) with the initial condition $\bar{x}(0; x_0, 0) = x_0$. For all $\theta \in \Theta$

$$V(x_0, \theta) = \int_0^\infty \|\bar{x}(s; x_0, 0)\|^2 \, ds \tag{B.1.4}$$

is a C^r function. Thanks to the global Lipschitz property of $\bar{f}$

$$\|\bar{f}(x,\theta)\| \le L(\theta)\|x\|, \quad L(\theta) = \max\left\{1, \sup_{\|x\|\le 1} \left\|\frac{\partial f}{\partial x}(x,\theta)\right\|\right\}$$

we can use

$$\frac{d}{ds}\|\bar{x}\|^2 = 2\bar{x}^T \bar{f}(\bar{x},\theta) \ge -2L(\theta)\|\bar{x}\|^2$$

to obtain $\|\bar{x}(s)\|^2 \ge e^{-2L(\theta)s}\|x_0\|^2$ and prove

$$V(x_0,\theta) \ge \int_0^\infty e^{-2L(\theta)s}\|x_0\|^2\, ds = c(\theta)\|x_0\|^2$$

On the other hand, the GAS/LES properties of (B.1.3) guarantee the existence of $\kappa_\theta(\cdot) \in \mathcal{K}$ and $\lambda(\theta) > 0$ such that

$$\|\bar{x}(s)\| \le \kappa_\theta(\|x_0\|)e^{-\lambda(\theta)s}\|x_0\|, \quad \forall \theta \in \Theta$$

Substituting in (B.1.4), we obtain

$$V(x_0,\theta) \le \int_0^\infty \kappa_\theta^2(\|x_0\|)e^{-2\lambda(\theta)s}\|x_0\|^2\, ds = \frac{\kappa_\theta^2(\|x_0\|)}{2\lambda(\theta)}\|x_0\|^2 =: \gamma_\theta(\|x_0\|)$$

The time-derivative of V along the solutions of (B.1.3) is, by construction,

$$\dot{V}(x_0,\theta)\Big|_{(B.1.3)} = -\|x_0\|^2.$$

Finally, the time-derivative of V along the solutions of the original system (B.1.1) is

$$\dot{V}(x_0,\theta)\Big|_{(B.1.1)} = \frac{\partial V}{\partial x} f(x_0,\theta) = -(1 + \|f(x_0,\theta)\|^2)\|x_0\|^2 \le -\|x_0\|^2$$

□

We now proceed with the proof of Theorem 3.18. Introducing $\zeta = z - \bar{z}(x)$ we rewrite the singularly perturbed system (3.2.13),(3.2.14) as

$$\begin{aligned} \dot{x} &= f(x,\bar{z}(x)) + p(x,\zeta) \\ \mu\dot{\zeta} &= q(x,\zeta) + \mu\tfrac{\partial \bar{z}}{\partial x}(f(x,\bar{z}(x)) + p(x,\zeta)) \end{aligned} \tag{B.1.5}$$

By Lemma B.1, for the subsystem $\dot{x} = f(x,\bar{z}(x))$ there exists a C^2 function $W_1(x)$ such that

$$\begin{aligned} & c_1\|x\|^2 \le W_1(x) \le \gamma_1(\|x\|) \\ & \frac{\partial W}{\partial x}[f(x,\bar{z}(x))] \le -\|x\|^2 \end{aligned} \tag{B.1.6}$$

Likewise, for the subsystem $\dot{\zeta} = q(x, \zeta)$, with x as the parameter, there exists a C^2 function $W_2(\zeta, x)$ such that

$$\begin{aligned} & c_2(x)\|\zeta\|^2 \leq W_2(\zeta, x) \leq \gamma_{2,x}(\|\zeta\|) \\ & \frac{\partial W}{\partial \zeta} q(\zeta, x) \leq -\|\zeta\|^2 \end{aligned} \tag{B.1.7}$$

As a Lyapunov function for the system (B.1.5) we use

$$V(x, \zeta) = W_1(x) + W_2(\zeta, x) \tag{B.1.8}$$

which is positive definite and radially unbounded in both x and ξ. For the compact sets $\mathcal{C}_x$ and $\mathcal{C}_z$ defined in the theorem, there exists a compact set $\mathcal{C}_\zeta$ such that, whenever $x \in \mathcal{C}_x$, $z \in \mathcal{C}_z$, then $\zeta \in \mathcal{C}_\zeta$. Thus, there exists a real number $N > 0$ such that the set $\mathcal{N} := \{(x, \zeta) : V(x, \zeta) \leq N\} \supset \mathcal{C}_x \times \mathcal{C}_\zeta$ is compact.

The differentiability properties of W_1, W_2, p, f, $\bar{z}$ and $W_2(0, x) = 0$, $p(x, 0) = 0$, $\forall x$, imply that there exists $M > 0$ independent of μ, such that, for any $(x, \zeta) \in \mathcal{N}$,

$$\begin{array}{ll} \left\|\frac{\partial W_1}{\partial x}\right\| \leq M\|x\| & \left\|\frac{\partial W_2}{\partial x}\right\| \leq M\|\zeta\| \\ \left\|\frac{\partial W_2}{\partial \zeta}\right\| \leq M\|\zeta\| & \left\|\frac{\partial \bar{z}}{\partial x}\right\| \leq M \\ \|p\| \leq M\|\zeta\| & \|f\| \leq M\|x\| \end{array}$$

Using these bounds we obtain

$$\begin{aligned} \dot{V} &= \frac{\partial W_1}{\partial x} f + \frac{\partial W_1}{\partial x} p + \frac{1}{\mu}\frac{\partial W_2}{\partial \zeta} q + \left(\frac{\partial W_2}{\partial x} + \mu \frac{\partial W_2}{\partial \zeta}\frac{\partial \bar{z}}{\partial x}\right)(f + p) \\ &\leq [\|x\| \ \|\zeta\|] \begin{bmatrix} -1 & M^2 + \frac{\mu}{2}M^3 \\ M^2 + \frac{\mu}{2}M^3 & -\frac{1}{\mu} + M^2 + \mu M^3 \end{bmatrix} \begin{bmatrix} \|x\| \\ \|\zeta\| \end{bmatrix} \end{aligned}$$

This proves that, for μ sufficiently small, $\dot{V}$ is negative definite for all $(x, \zeta) \in \mathcal{N}$. Therefore, the equilibrium $(x, \zeta) = (0, 0)$ of the system (B.1.5) is asymptotically stable and the set $\mathcal{C}_x \times \mathcal{C}_\zeta$ is in its region of attraction.

□

B.2 Proof of Theorem 4.35

We first consider the case when the matrix A_0 is Lyapunov stable, that is when the system (A, B, C) is weakly minimum phase. Let $A_s = \text{diag}\{A_h, A_c\}$ where A_h is Hurwitz, while all the eigenvalues of A_c are on the imaginary axis, so that $A_c = -A_c^T$. Let $(\xi_c \ \xi_h)^T$ and $(B_c \ B_h)^T$ be the corresponding decompositions of the state ξ_0 and of the matrix B, respectively. The Lyapunov function $W_c = \frac{1}{2}\xi_c^T\xi_c$ is constant along the trajectories of $\dot{\xi}_c = A_c\xi_c$. This suggests that the subsystem

$$\dot{\xi}_c = A_c\xi_c + B_c u_0,$$

be stabilized with the damping control of Section 3.5.2, that is with $u_0 = K_0(a)\xi_c = -\frac{1}{a}B_c^{\mathrm{T}}\frac{\partial W}{\partial \xi_c} = -\frac{1}{a}B_c^{\mathrm{T}}\xi_c$. With this feedback, the derivative of $W_c = \frac{1}{2}\xi_c^T\xi_c$ is nonpositive

$$\dot{W}_c = -\frac{1}{a}\xi_c^T B_c B_c^T \xi_c \leq 0 \tag{B.2.1}$$

and asymptotic stability is guaranteed because (B_c^T, A_c) is observable. The observability follows from our assumption that (A_c, B_c) is stabilizable and the fact that $-A_c^T = A_c$. Hence, if A_0 is Lyapunov stable, the low-gain feedback matrix $K_0(a)$ is simply $\mathrm{diag}\{0, -\frac{1}{a}B_c^T\}$. For a fixed $T > 0$, (B.2.1) and the observability of (B_c^T, A_c) imply that, along the trajectories of the closed-loop system, $W_c(\xi_c(t+T)) - W_c(\xi_c(t)) \leq -\frac{\beta}{a}W_c(\xi_c(t))$, where $\beta > 0$ is independent of a. We conclude that, for a large enough,

$$\|e^{(A_s+B_sK_0(a))t}\| \leq \gamma_1 e^{-\frac{\beta}{a}t} \tag{B.2.2}$$

To prove that γ_1 is independent of a, let $P > 0$ satisfy $PA_h + A_h^TP = -I$ and note that the derivative of the Lyapunov function $W = kW_c + \xi_h^TP\xi_h$ is

$$\dot{W} = -\frac{k}{a}\xi_c^T B_c B_c^T \xi_c - \xi_h^T\xi_h - 2\xi_h^T P B_h(\frac{1}{a}B_c^T\xi_c)$$

Completing the squares, we show that $\dot{W}$ is negative semidefinite if $k > \|PB_h\|^2$. The observability of the pair (B_c^T, A_c) and the fact that W is independent of a, yield an estimate $\|\xi_s(t)\| \leq \gamma_1\|\xi_s(0)\|$ for some constant γ_1 independent of a. This proves (4.5.19) and (4.5.20).

In the case when A_0 is unstable due to repeated eigenvalues on the imaginary axis, we apply a preliminary feedback $u = \frac{1}{a}K_s + v$ to stabilize the ξ_s-subsystem, which yields

$$\begin{pmatrix} \dot{\xi}_u \\ \dot{\xi}_s \end{pmatrix} = \begin{pmatrix} A_u & \frac{1}{a}B_uK_s + A_J \\ 0 & A_s + \frac{1}{a}B_sK_s \end{pmatrix}\begin{pmatrix} \xi_u \\ \xi_s \end{pmatrix} + \begin{pmatrix} B_u \\ B_s \end{pmatrix} v \tag{B.2.3}$$

The matrix $A_s + \frac{1}{a}B_sK_s$ is Hurwitz for all $a > 0$ and a change of coordinates of the form $\tilde{\xi}_u = \xi_u + T\xi_s$ exists that diagonalizes (B.2.3) as

$$\begin{pmatrix} \dot{\tilde{\xi}}_u \\ \dot{\xi}_s \end{pmatrix} = \begin{pmatrix} A_u & 0 \\ 0 & A_s + \frac{1}{a}B_sK_s \end{pmatrix}\begin{pmatrix} \tilde{\xi}_u \\ \xi_s \end{pmatrix} + \begin{pmatrix} \tilde{B}_u \\ B_s \end{pmatrix} v \tag{B.2.4}$$

A construction of a low-gain $K_u(a)$ such that $A_u + \tilde{B}_u K_u(a)$ is Hurwitz for all $a > 0$ and

$$\|K_u(a)\tilde{\xi}_u(t)\| = \|K_u(a)e^{(A_u+\tilde{B}_u K_u(a))t}\tilde{\xi}_u(0)\| \leq \frac{\gamma_1}{a}e^{-\sigma(a)t}\|\tilde{\xi}_u(0)\| \qquad \text{(B.2.5)}$$

is available from [68]. With the feedback $v = K_u(a)\tilde{\xi}_u$, the solution $\xi_s(t)$ of (B.2.4) satisfies

$$\xi_s(t) = e^{(A_s+\frac{1}{a}B_s K_s)t}\xi_0 + \int_0^t e^{(A_s+\frac{1}{a}B_s K_s)(t-\tau)}B_s K_u(a)\tilde{\xi}_u(\tau)d\tau$$

Using (B.2.2) and (B.2.5), we have for all $t \geq 0$

$$\begin{aligned} \|\xi_s(t)\| &\leq \gamma e^{-\frac{\beta}{a}s}\|\xi_s(0)\| + \frac{\gamma_1}{a}\|\tilde{\xi}_u(0)\| \textstyle\int_0^\infty e^{-\frac{\beta}{a}\tau}d\tau \\ &\leq \gamma\|\xi_s(0)\| + \gamma'\|\tilde{\xi}_u(0)\| \leq \gamma_2\|\xi_0(0)\| \end{aligned}$$

for some constants γ, γ', and γ_2 independent of a. We conclude that the state ξ_s does not peak and that the low-gain feedback $K_0(a)\xi_0 = \frac{1}{a}K_s\xi_s + K_u(a)\tilde{\xi}_u$ satisfies (4.5.19). □

Bibliography

[1] B.D.O. Anderson, J.B. Moore, *Optimal Control: Linear Quadratic Methods.* Prentice-Hall, Englewood Cliffs, NJ, 1990.

[2] B.D.O. Anderson, S. Vongpanitlerd, *Network Analysis and Synthesis.* Prentice-Hall, Englewood Cliffs, NJ, 1973.

[3] M.A. Aizerman and F.R. Gantmacher, *Absolute stability of regulator systems*, Holden-Day, San Francisco, 1964.

[4] Z. Artstein, "Stabilization with relaxed controls," *Nonlinear Analysis*, vol. 7, pp. 1163-1173, 1983.

[5] A. Bacciotti, "Further remarks on potentially global stabilizability," *IEEE Trans. on Automatic Control*, 34, pp. 637-638, 1989.

[6] E.A. Barbashin, N.N. Krasovski, "On the stability of motion in the large," *Dokl. Akad. Nauk. SSSR*, vol. 86, pp. 453-456, 1952 (in Russian).

[7] S. Boyd, L. El Ghaoui, E. Feron, V. Balakrishnan, *Linear Matrix Inequalities in System and Control Theory.* SIAM Studies in Appl. Math., vol. 15, Philadelphia, 1994.

[8] J.H. Braslavsky, R.H. Middleton, "Global and semi-global stabilizability in certain cascade nonlinear systems," *IEEE Trans. on Automatic Control*, vol. 41, pp. 876-880, 1996.

[9] R.W. Brockett, *Finite dimensional linear systems.* Wiley, New-York, 1969.

[10] R.W. Brockett, "Asymptotic stability and feedback stabilization," *Differential Geometric Control Theory*, R.W. Brockett, R.S. Millman, H.J. Sussmann eds., in Progress in Mathematics, vol 27, pp. 181-191, 1983.

[11] Y.N. Bibikov, *Local theory of Nonlinear Analytic Ordinary Differential Equations.* Lectures Notes in Mathematics 702, Springer-Verlag, Berlin Heidelberg New-York, 1979.

[12] R.T. Bupp, C-J Wan, V.T. Coppola, D.S. Bernstein, "Design of rotational actuator for global stabilization of translational motion," *Proc. Active Control of Vibration Noise*, ASME Winter Mtg., 1994.

[13] C.I. Byrnes, A. Isidori, "New results and examples in nonlinear feedback stabilization," *Systems & Control Letters*, vol. 12, pp. 437-442, 1989.

[14] C.I. Byrnes, A. Isidori, "Asymptotic stabilization of minimum phase nonlinear systems," *IEEE Trans. on Automatic Control*, vol. 36, pp. 1122-1137, 1991.

[15] C.I. Byrnes, A. Isidori, J.C. Willems, "Passivity, feedback equivalence, and global stabilization of minimum phase systems," *IEEE Trans. on Automatic Control*, vol. 36, pp. 1228-1240, 1991.

[16] J. Carr, *Applications of Center Manifold Theory.* Springer-Verlag, New-York, 1981.

[17] M.J. Corless, G. Leitmann, "Continuous state feedback guaranteeing uniform ultimate boundedness," *IEEE Trans. on Automatic Control*, vol. 26, pp. 1139-1144, 1981.

[18] C. A. Desoer, M. Vidyasagar, *Feedback Systems: Input-Output Properties*, Academic Press, New York, 1975.

[19] F. Esfandiari, H.K. Khalil, "Output feedback stabilization of fully linearizable systems," *Internation Journal of Control*, vol. 56, pp. 1007-1037, 1992.

[20] B.A. Francis, K. Glover, "Bounded peaking in the optimal linear regulator with cheap control," *IEEE Trans. on Automatic Control*, vol.23, pp. 608-617, 1978.

[21] J.C. Doyle, B.A. Francis, A.R. Tannenbaum. *Feedback Control Theory.* MacMillan, New York, NY, 1992.

[22] R.A. Freeman, P.V. Kokotović, "Backstepping design of robust controllers for a class of nonlinear systems," *Proceedings of IFAC Nonlinear Control Systems Design Symposium*, Bordeaux France, pp. 307-312, 1992.

[23] R.A. Freeman, P.V. Kokotović, "Design of 'softer' robust nonlinear control laws," *Automatica*, vol. 29, pp. 1425-1437, 1993.

[24] R.A. Freeman, P.V. Kokotović, "Optimal nonlinear controllers for feedback linearizable systems," *Proceedings of Workshop on Robust Control via Var. Struct. & Lyapunov Techniques*, Italy, 1994.

[25] R.A. Freeman, P.V. Kokotović, "Optimality of robust nonlinear feedback controls," *SIAM J. Control & Optimization*, to appear.

[26] R.A. Freeman, P.V. Kokotović, *Robust Control of Nonlinear Systems*, Birkhauser, Boston, 1996.

[27] R. Ghanadan, "Nonlinear control system design via dynamic order reduction," *Proceedings of 1994 CDC*, Lake Buena Vista, FL, pp. 3752-3757, 1994.

[28] S.T. Glad, "On the gain margin of nonlinear and optimal regulators," *IEEE Trans. on Automatic Control*, vol. 29, pp. 615-620, 1984.

[29] S.T. Glad, "Robustness of nonlinear state feedback – a survey," *Automatica*, vol. 23, pp. 425-435, 1987.

[30] M. Green, D. Limebeer, *Linear Robust Control*. Prentice-Hall, Englewood Cliffs, NJ, 1995.

[31] M.J. Grimble, T.J. Owens, "On improving the robustness of LQ regulators," *IEEE Trans. on Automatic Control*, vol. 31, pp. 54-55, 1986.

[32] J. Guckenheimer, P. Holmes, *Nonlinear Oscillations, Dynamical Systems, and Bifurcations of Vector Fields*. Applied Mathematics Sciences, vol. 42, Springer-Verlag, New-York, 1983.

[33] W. Hahn, *Stability of Motion*. Springer Verlag, Berlin, Heidelberg, 1967.

[34] J. Hauser, S. Sastry, G. Meyer, "Nonlinear control design for slightly nonminimum phase systems," *Automatica*, vol. 28, pp. 665-679, 1992.

[35] J. Hauser, S. Sastry, P.V. Kokotović, "Nonlinear control via approximate input-output linearization," *IEEE Trans. on Automatic Control*, vol. 37, pp. 392-398, 1992.

[36] K. Hedrick, Private communication regarding the Ph.D. thesis by N.H. Getz, U.C. Berkeley, 1996.

[37] D.J. Hill, P.J. Moylan, "The stability of nonlinear dissipative systems," *IEEE Trans. on Automatic Control*, vol. 21, pp 708-711, 1976.

[38] D.J. Hill, P.J. Moylan, "Stability results for nonlinear feedback systems," *Automatica*, vol. 13, pp. 377-382, 1977.

[39] M.W. Hirsch, S. Smale, *Differential Equations, Dynamical Systems, and Linear Algebra.* Academic Press, 1974.

[40] P.A. Ioannou, J. Sun, *Stable and Robust Adaptive Control.* Prentice-Hall, Englewood Cliffs, NJ, 1995.

[41] P.A. Ioannou, G. Tao, "Frequency domain conditions for strictly positive real functions," *IEEE Trans. on Automatic Control*, vol. 32, pp. 53-54, 1987.

[42] A. Iggidr, B. Kalitine, R. Outbib, "Semi-sefinite Lyapunov functions, stability and stabilization," *Mathematics of Control, Signals, and Systems*, to appear, 1996.

[43] A. Isidori, *Nonlinear Control Systems.* 2nd. ed., Springer-Verlag, Berlin, 1989.

[44] D.H. Jacobson, *Extensions of Linear-Quadratic Control, Optimization and Matrix Theory.* Academic Press, New York, 1977.

[45] M. Jankovic, D. Fontaine, P.V. Kokotović, "TORA example: cascade and passivity based control designs," *IEEE Trans. on Control Systems Technology*, vol. 4, pp. 292-297, 1996.

[46] M. Jankovic, R. Sepulchre, P.V. Kokotović, "Constructive Lyapunov stabilization of nonlinear cascade systems," *IEEE Trans. on Automatic Control*, to appear, 1996.

[47] M. Jankovic, R. Sepulchre, P.V. Kokotović, "Global adaptive stabilization of cascade nonlinear systems," *Proceedings of 13-th IFAC World Congress*, San Francisco, pp. 311-316, 1996.

[48] Z.P. Jiang, A.R. Teel, L. Praly, "Small-gain theorem for ISS systems and applications," *Mathematics of Control, Signals, and Systems*, vol. 7, pp. 95-120, 1994.

[49] V. Jurdjevic, J.P. Quinn. "Controllability and stability," *J. Differential Equations*, vol. 28, pp. 381-389, 1978.

[50] R.E. Kalman, "Contributions to the theory of optimal control," *Bol. Soc. Mat. Mex.*, vol. 5, pp. 102-119, 1960.

[51] R.E. Kalman, "Lyapunov functions for the problem of Lur'e in automatic control," *Proc. Math. Acad. Science USA*, vol. 49, pp. 201-205, 1963.

[52] R.E. Kalman, "When is a linear control system optimal?," *Trans. ASME Ser. D: J. Basic Eng.*, vol. 86, pp. 1-10, 1964.

[53] I. Kanellakopoulos, P.V. Kokotović, A. S. Morse, "Systematic design of adaptive controllers for feedback linearizable systems," *IEEE Trans. Automatic Control*, vol. 36, pp. 1241-1253, 1991.

[54] I. Kanellakopoulos, P.V. Kokotović, A. S. Morse, "Adaptive feedback linearization of nonlinear systems," 311-346, *Foundations of Adaptive Control*, P.V. Kokotović ed., Springer-Verlag, 1991.

[55] I. Kanellakopoulos, "Block backstepping for adaptive nonlinear control," *Preprints of the 1995 IFAC Nonlinear Control Systems Design Symposium*, Tahoe City, CA, June 1995, pp. 113-118, also submitted to *Automatica*.

[56] H.K. Khalil, *Nonlinear Systems*. MacMillan, New York, NY, 1992.

[57] P.V. Kokotović, H.K. Khalil, J. O'Reilly, *Singular Perturbations Methods in Control: Analysis and Design*. Academic Press, New York, 1986.

[58] P.V. Kokotović, R. Marino, "On vanishing stability regions in nonlinear systems with high-gain feedback," *IEEE Trans. Automatic Control*, vol. 31, pp. 967-970, 1986.

[59] P.V. Kokotović, H.J. Sussmann, "A positive real condition for global stabilization of nonlinear systems," *Systems & Control Letters*, vol. 13, pp. 125-133, 1989.

[60] A.A. Krasovski, "A new solution to the problem of a control system analytical design," *Automatica*, vol. 7, pp. 45-50, 1971.

[61] M. Krstić, I. Kanellakopoulos, P.V. Kokotović, *Nonlinear and Adaptive Control Design*. Wiley, New York, 1995.

[62] M. Krstić, J. Sun, P.V. Kokotović, "Robust control of nonlinear systems with input unmodeled dynamics," *IEEE Trans. on Automatic Control*, vol. 41, pp. 913-920, 1996.

[63] J.P. LaSalle, "Stability theory for ordinary differential equations," *Journal of Differential Equations*, vol. 4, pp. 57-65, 1968.

[64] N.A. Lehtomaki, N.R. Sandell, M. Athans, "Robustness results in linear quadratic gaussian based multivariable control," *IEEE Trans. on Automatic Control*, vol. 26, pp. 1-10, 1981.

[65] W. Lin, "Input saturation and global stabilization by output feedback for affine systems," *Proceedings of 33-rd CDC*, Lake Buena Vista, FL, pp. 1323-1328, 1994.

[66] W. Lin, "Global asymptotic stabilization of general nonlinear systems with stable free dynamics by using passivity and smooth saturation," *Automatica*, vol. 32, pp. 915-924, 1996.

[67] Z. Lin, A. Saberi, "Semi-global stabilization of partially linear composite systems via feedback of the state of the linear part," *Systems & Control Letters*, vol. 20, pp. 199-207, 1993.

[68] Z. Lin, A. Saberi, "Semi-global exponential stabilization of linear systems subject to "input saturation" via linear feedback," *Systems & Control Letters*, vol. 21, pp. 225-240, 1993.

[69] R. Lozano, B. Brogliato, I.D. Landau, "Passivity and global stabilization of cascaded nonlinear systems," *IEEE Trans. on Automatic Control*, vol. 37, pp. 1386-1388, 1992.

[70] A.I. Lurie. *Some Nonlinear Problems in the Theory of Automatic Control.* Moscow, Gostekhizdat., 1951.

[71] I.M. Mareels, D.J. Hill, "Monotone stability of nonlinear feedback systems," *Journal of Mathematical Systems, Estimation, and Control*, vol. 2, pp. 275-291, 1992.

[72] R. Marino, P. Tomei, "Robust stabilization of feedback linearizable time-varying uncertain systems," *Automatica*, vol. 29, pp. 181-189, 1993.

[73] R. Marino, P. Tomei, *Nonlinear Control Design.* Prentice Hall International, (UK), 1995.

[74] P. Martin, B. D'Andrea-Novel, R. Sepulchre, "Full linearization of a class of mechanical systems via dynamic state feedback," *Proceedings of MTNS*, Kobe, 1991.

[75] F. Mazenc, L. Praly, "Adding an integration and global asymptotic stabilization of feedforward systems," *Proceedings of 1994 CDC*, Lake Buena Vista, FL, pp. 121-126, (1994); also to appear in *IEEE Trans. on Automatic Control.*

[76] A. Megretski, A. Rantzer, "System analysis via integral quadratic constraints," preprint, 1994.

[77] A. Michel, R. Miller, *Qualitative Analysis of Large Scale Dynamical Systems.* Academic, New York, 1977.

[78] R. Middleton, "Trade-offs in linear control system design," *Automatica*, vol 27, pp. 281-292, 1991.

[79] T. Mita, "On zeros and responses of linear regulators and linear observers," *IEEE Trans. on Automatic Control*, vol. 22, pp. 423-428, 1977.

[80] P.J. Moylan, "Implications of passivity in a class of nonlinear systems," *IEEE Trans. on Automatic Control*, vol. 19, pp. 373-381, 1974.

[81] P.J. Moylan, B.D.O. Anderson, "Nonlinear regulator theory and an inverse optimal control problem," *IEEE Trans. Automatic Control*, vol. 18, pp. 460-465, 1973.

[82] K. S. Narendra, A. M. Annaswamy, *Stable Adaptive Systems.* Prentice Hall, Englewood Cliffs, NJ, 1989.

[83] B.N. Naumov, Y.Z. Tsypkin, "A frequency criterion for absolute process stability in nonlinear automatic control systems," *Automation and Remote Control*, vol. 25, pp. 406-408, 1964.

[84] H. Nijmeijer, A. van der Schaft, *Nonlinear Dynamical Control Systems.* Springer-Verlag, New York, 1990.

[85] R. Ortega, "Passivity properties for stabilization of cascaded nonlinear systems," *Automatica*, vol 27, pp. 423-424, 1989.

[86] V.A. Pliss, *Certain problems in the theory of stability in the large.* published by Leningrad State University, 1958.

[87] V.M. Popov, "Absolute stability of nonlinear control systems of automatic control," *Automation and Remote Control*, vol. 22, pp. 857-875, 1962.

[88] V.M. Popov, *Hyperstability of Control Systems*, Springer-Verlag, 1973.

[89] L. Praly, Y. Wang, "Stabilization in spite of matched unmodeled dynamics and an equivalent definition of input-to-state stability," *Mathematics of Control, Signal and Systems*, to appear.

[90] Z. Qu, "Robust control design for nonlinear uncertain systems without matching condition," *IEEE Trans. on Automatic Control*, vol. 40, pp. 1453-1460, 1995.

[91] M.G. Safonov, *Stability and Robustness of Multivariable Feedback Systems. MIT Press*, 1980.

[92] A. Saberi, P.V. Kokotović, H.J. Sussmann, "Global stabilization of partially linear composite systems," *SIAM J. Control and Optimization*, vol. 28, pp. 1491-1503, 1990.

[93] A. Saberi, P. Sannuti, "Squaring down by static and dynamic compensators," *IEEE Trans. on Automatic Control*, vol. 33, pp. 358-365, 1988.

[94] I.W. Sandberg, "A frequency domain condition for the stability of systems containing a single time-varying nonlinear element," *Bell Systems Tech. Journal*, vol. 43, pp. 1601-1638, 1964.

[95] R. Sepulchre, M. Jankovic, P.V. Kokotović, "Integrator forwarding: a new recursive nonlinear robust design," *Proceedings of 13-th IFAC World Congress*, San Francisco, pp. 85-90, 1996.

[96] D. Siljak, *Large-Scale Dynamic Systems. North-Holland*, New York, 1978.

[97] J. Slotine, K. Hedrick, "Robust input-output feedback linearization," *Internation Journal of Control*, vol. 57, pp. 1133-1139, 1993.

[98] E.D. Sontag, "A Lyapunov-like characterization of asymptotic controllability," *SIAM J. on Control and Optimization*, vol. 21, pp. 462-471, 1983.

[99] E.D. Sontag, "Smooth Stabilization implies coprime factorization," *IEEE Trans. on Automatic Control*, vol. 34, pp. 435-443, 1988.

[100] E.D. Sontag, "Remarks on stabilization and input-to-state stability," *Proceedings of 28-th CDC*, Tampa, FL, pp. 1376-1378, 1989.

[101] E.D. Sontag, "A universal construction of Artstein's theorem on nonlinear stabilization," *Systems & Control Letters*, vol. 13, pp. 117-123, 1989.

[102] E.D. Sontag, Y. Wang, "On characterizations of the input-to-state stability property," *Systems & Control Letters*, vol. 24, pp. 351-359, 1995.

[103] E.D. Sontag, "State-space and I/0 stability for nonlinear systems," in: *Feedback Control, Nonlinear Systems, and Complexity*, B.A. Francis and A.R. Tannenbaum eds., Lectures Notes in Control and Information Sciences 202, Springer Verlag, 1995.

[104] M. Spong, K. Khorasani, P.V. Kokotović, "An integral manifold approach to the feedback control of flexible joint robots," *IEEE Journal of Robotics and Automation*, vol. RA-3, pp. 291-300, 1987.

[105] H.J. Sussmann, P.V. Kokotović, "The peaking phenomenon and the global stabilization of nonlinear systems," *IEEE Trans. on Automatic Control*, vol. 36, pp. 424-439, 1991.

[106] H.J. Sussmann, E.D. Sontag, Y. Yang, "A general result on the stabilization of linear systems using bounded controls," *IEEE Trans. on Automatic Control*, vol. 39, pp. 2411-2425, 1994.

[107] G. Tao, P.A. Ioannou, "Strictly positive real matrices and the Lefschetz-Kalman-Yakubovich lemma," *IEEE Trans. on Automatic Control*, vol.33, pp. 1183-1185, 1988.

[108] J.H. Taylor, "Strictly positive-real functions and the Lefschetz-Kalman-Yakubovich (LKY) lemma," *IEEE Trans. on Circuits and Systems*, pp. 310-311, March 1974.

[109] A.R. Teel, *Feedback Stabilization: Nonlinear Solutions to Inherently Nonlinear Problems.* PhD Dissertation, University of California, Berkeley, 1992.

[110] A.R. Teel, "Semi-global stabilization of minimum phase nonlinear systems in special normal forms," *Systems & Control Letters*, vol. 19, pp. 187-192, 1992.

[111] A.R. Teel, "A nonlinear small gain theorem for the analysis of control systems with saturation," *IEEE Trans. on Automatic Control*, to appear in vol. 41, no. 9, 1996.

[112] A.R. Teel, L. Praly, "Tools for semiglobal stabilization by partial state feedback and output feedback," *SIAM J. Control and Optimization*, vol. 33, pp. 1443-1488, 1995.

[113] J. Tsinias, "Sufficient Lyapunov like conditions for stabilization," *Mathematics of Control, Signals, and Systems*, vol. 2, pp. 343-357, 1989.

[114] J.N. Tsitsiklis, M. Athans, "Guaranteed robustness properties of multivariable nonlinear stochastic optimal regulators," *IEEE Trans. on Automatic Control*, vol. 29, pp. 690-696, 1984.

[115] M. Vidyasagar, *Nonlinear Systems Analysis.* Marcel Dekker, New York, 1993.

[116] M. Vidyasagar, "Decomposition techniques for large-scale systems with nonadditive interactions: stability and stabilizability," *IEEE Trans. on Automatic Control*, vol. 25, pp. 773-779, 1980.

[117] C-J Wan, D.S. Bernstein, V.T. Coppola, "Global stabilization of the oscillating eccentric rotor," *Proceedings 33-rd IEEE CDC*, Lake Buena Vista, FL, pp. 4024-4029, 1994.

[118] K. Wei, "Stabilization of linear time-invariant interval systems via constant state feedback control," *IEEE Trans. on Automatic Control*, vol. 39, pp. 22-32, 1994.

[119] J. L. Willems, *Stability theory of dynamical systems.* Wiley, New-York, 1970.

[120] J. C. Willems, "Dissipative dynamical systems, parts I and II " *Arch. Rational Mechanics and Analysis*, vol. 45, pp. 321-393, 1972.

[121] V.A. Yakubovich, "Solutions to some matrix inequalities occurring in the theory of automatic control," *Dokl. Akad. Nauk SSSR*, vol. 143, 1962 (in Russian).

[122] T. Yoshizawa, *Stability Theory by Lyapunov Second Method.* The Mathematical Society of Japan, Tokyo, 1966.

[123] G. Zames, "On the input-output stability of time-varying nonlinear feedback systems - parts I and II," *IEEE Trans. on Automatic Control*, vol. 11, pp. 228-238 and 465-476, 1966.

Index

MIX
Papier aus verantwortungsvollen Quellen
Paper from responsible sources
FSC® C105338

If you have any concerns about our products,
you can contact us on
ProductSafety@springernature.com

In case Publisher is established outside the EU,
the EU authorized representative is:
Springer Nature Customer Service Center GmbH
Europaplatz 3, 69115 Heidelberg, Germany

Printed by Libri Plureos GmbH
in Hamburg, Germany